PROGRAMED REVIEWS OF CHEMICAL PRINCIPLES

SECOND EDITION

JEAN D. LASSILA • GORDON M. BARROW
MALCOLM E. KENNEY • ROBERT L. LITLE
WARREN E. THOMPSON

W. A. BENJAMIN, INC. PUBLISHERS

W. A. Benjamin, Inc.
Menlo Park, California • Reading, Massachusetts
London • Amsterdam • Don Mills, Ontario • Sydney

 Printed in the United States of America. Published simultaneously in Canada. Library of Congress Catalog Card No. 73-5376.

ISBN 0-8053-6026-3
BCDEFGHIJ-DO-7987654

PREFACE

To learn chemistry, students should master the basic building blocks: stoichiometry, nomenclature, introductory atomic and molecular structure, general properties of solutions, and equilibrium calculations. An understanding of these subjects requires the ability to apply the basic principles to solving problems. This book reviews these principles and provides sound techniques for applying to specific problems the general concepts of chemistry.

The reviews are intended to be supplementary; that is, the topics discussed are those we considered most essential, and their treatment is intended to give the student a firm basis for further study. Some knowledge of the concepts and terminology pertinent to each section is assumed and the unifying material necessary for the complete development of the subject must be provided by lectures or reading assignments.

This edition has been revised to follow the order of topics in the second edition of *Chemical Principles,* by Richard E. Dickerson, Harry B. Gray, and Gilbert P. Haight, Jr. Cross referencing between this book and *Chemical Principles* occurs when appropriate. (You should refer also to *A Study Guide to Chemical Principles,* by Wilbert Hutton, for more detailed discussion of difficult topics in the text.)

Nevertheless, an understanding of a section is not dependent on the material in *Chemical Principles:* These reviews (adapted from *Understanding Chemistry* by Barrow *et al.*) can be used with any general chemistry text or lecture course.

Programed Reviews of Chemical Principles is a programed instruction book, not a textbook or problem manual. The material is divided into chapters (called "reviews") and, when appropriate, into sections. The objective and relevance of each review and section is stated in the opening paragraphs, which also tell how a student can determine whether he has already mastered the material. Then there is a logical sequence of short, numbered items that provide information and require the student to answer a question. By participating in the development of each topic, the student learns, through independent study, topics that otherwise might cause him difficulty.

Individual items taken out of context may seem trivial, but to evaluate programed material, you should read through a complete sequence of items and note how a topic is developed. Use of programed instruction in chemistry is rather recent, but there is clear indication that, when properly used, this type of instruction is an effective teaching aid.

JEAN D. LASSILA
GORDON M. BARROW
MALCOLM E. KENNEY
ROBERT L. LITLE
WARREN E. THOMPSON

NOTE TO THE STUDENT

As you have probably noticed, the layout for this book is rather unusual; hence, the procedure for using the book may be new to you. The material presented here is programed; that is, it consists of a logical sequence of short numbered items (one or two sentences) that provide information and also require you to answer a question, carry out a calculation, or insert a word or number.

The instructions are simple. Read and complete the items in numerical order. On the third page after a given item is the correct answer (or answers). To derive the most benefit from the book, you should write your answers in the blanks and then turn the page to see whether you are on the right track. (In some items the blank is too small for the answer, but there is space beneath the item or at the right for large answers and computations.)

An understanding of each item depends on your comprehension of the preceding items; therefore, when you make an error, you should not go on to the next item until you know the reason for your mistake and have corrected it. Answers to problems can be calculated with a slide rule, and all answers have been rounded off to the correct number of significant figures. You will find that units are sometimes given with reciprocal numbers.

For example, moles liter^{-1} means moles per liter.

Each review leads to questions that are, in effect, summary problems that should indicate whether you have mastered the material. Then, if you want to try your skill, you can work the problems in the book by Ian S. Butler and Arthur E. Grosser, *Relevant Problems for Chemical Principles.* The average completion time for most sections is between one and two hours.

CONTENTS

REVIEW 1 GRAM-ATOMS, MOLES, AND EMPIRICAL FORMULAS

Early in your study of chemistry—and probably as long as you continue it—you will find it necessary to measure amounts of chemicals. You will need to know how much of one substance will react with a measured amount of another substance. For that elementary and practical information, you must know what chemical symbols mean and how they are related both to measured weights of elements and compounds and to numbers of atoms and molecules.

This review will examine in detail the concepts of the gram-atom (Section 1–1) and the mole (Section 1–2), the quantities which relate weights to number of atoms or molecules. Section 1–3 will show how gram-atoms and moles relate the percentage by weight composition of a compound to its empirical formula.

To use this book effectively, you should read Item 1 on page 4 and write the answer in the blank. (In some items there will be too little room to write the answer in the blank, but there will be space underneath or beside the item for it. This space can be used also for calculation.) When you turn the page, you will find the correct answer to Item 1 at the top, right-hand side (page 7). Item 2 is across from it, on the left-hand side (page 6). Right away you will understand how this works. It allows you to check your answer easily, yet not be able to see answers to later items. At Item 9 you should return to page 4 to work through the second row of items.

1–1 GRAM-ATOMS

In Chapter 1 of *Chemical Principles* you learned about chemical symbols and about gram-atoms and moles, the quantities which relate weights of elements and compounds to numbers of atoms and molecules. This section will review in detail the meaning and use of the gram-atom. This term is no longer used by everyone (some chemists use the word "mole" for this quantity when discussing elements as well as compounds). However, an understanding of the concept is necessary and fundamental.

The items in the section will seem trivial at first, but work through each one in the numbered order. Write down the answer before you check it. Although some atomic weights are given to four significant figures, as is usually done in atomic weight tables, values for all quantities needed in the problems can be used with three significant figures so slide rule calculation is possible.

When you have finished the review you will be able to solve problems such as "Given that the atomic weight of sulfur is 32.1, calculate the number of atoms and gram-atoms in 5.21 g of sulfur." If you can do this calculation and the items beginning with number 50 easily and logically, you have an adequate mastery of the gram-atom concept and can proceed directly to Section 1–2.

ITEM 1

Symbols are used to represent the chemical elements. For example, Cu is the chemical symbol for ________.

ITEM 10

It is more convenient, however, to have the symbols imply different masses of the different elements but always the same number of ________ of the elements.

ITEM 19

A table of atomic weights can be regarded as giving either the relative masses of the atoms of different elements or the actual masses of the atoms of various elements expressed in ________.

ITEM 28

A gram atomic weight, often shortened to *gram-atom,* is the weight of the element in grams that is numerically equal to the mass of ________ atom of the element expressed in ________.

ITEM 37

When the symbols are used to imply *pound atomic weights,* as they sometimes are in engineering calculations, the symbol C implies ________ of carbon.

ITEM 46

A 3.70-g sample of copper contains ________ g-atoms of copper. Since each gram-atom of any element contains ________ atoms, this sample contains ________ atoms of copper.

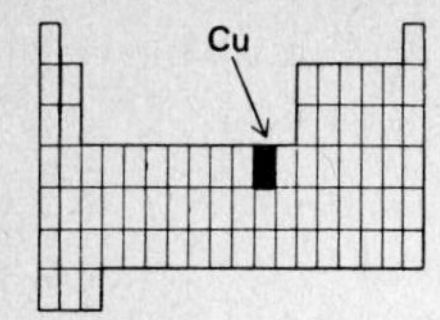
Cu

ITEM 2

The smallest unit of an element that we are concerned with in chemical studies is the ________. The chemical symbol, therefore, is often used to represent one ________.

ITEM 11

When the chemical symbols imply one atom, the mass represented by the symbol O is different from the mass represented by H. In fact, O represents a mass that is ________ than that represented by H by a factor of about ________.

ITEM 20

Since one atom of hydrogen has a mass of about 1.67×10^{-24} g, or about ________ amu, we see that an atomic mass unit is very ________ compared to weights used in the laboratory.

ITEM 29

12.01 g of carbon is one ________ of carbon.

ITEM 38

Then 1 pound of hydrogen, 12 pounds of carbon, and 16 pounds of oxygen each would correspond to about 1 ________ of the element, and there would be approximately ________ number of atoms of each element in these amounts.

ITEM 47

A 1.34-g sample of hydrogen contains ________ g-atoms, or ________ atoms of hydrogen.

ANSWER 1 copper

ANSWER 10 atoms

ANSWER 19 amu (atomic mass units)

ANSWER 28 one
amu

ANSWER 37 12.01 pounds (1 pound atomic weight)

ANSWER 46

$$\frac{3.70 \text{ g}}{63.5 \text{ g g-atom}^{-1}} = 0.0583 \text{ g-atom}$$

6.02×10^{23}

0.0583 g-atom $(6.02 \times 10^{23}$ atoms g-atom$^{-1})$
$= 3.51 \times 10^{22}$ atoms

ITEM 3

For example, when symbols are used in a structural formula, as in the one at the right for the water molecule, each individual symbol represents one __________ of the element.

ITEM 12

The *relative* masses of atoms of different elements are listed in tables, or are presented in periodic charts (see inside front cover), and are generally referred to as __________ weights.

ITEM 21

Tables of atomic weights used by chemists can be interpreted as giving the mass, in __________, of one *average* atom of the element as it occurs in nature.

ITEM 30

Samples containing 12 g of carbon or 1.0 g of hydrogen or 63.5 g of copper will all have about the same __________.

ITEM 39

A *pound atomic weight,* then, is defined as the weight of an element, expressed in __________, that is numerically equal to the __________.

ITEM 48

In general, the number of gram-atoms in a sample is calculated by dividing the __________ of the sample by the __________ of the element.

$$\begin{array}{ccc} H & & H \\ & \diagdown\ \diagup & \\ & O & \end{array}$$

ANSWER 2 atom
atom

ANSWER 11 greater
16

ANSWER 20 1
small

ANSWER 29 gram-atom
(or gram atomic weight)

ANSWER 38 pound atomic weight
the same

ANSWER 47

$$\frac{1.34 \text{ g}}{1.01 \text{ g g-atom}^{-1}} = 1.33 \text{ g-atoms}$$

$$1.33 \text{ g-atoms } (6.02 \times 10^{23} \text{ atoms g-atom}^{-1}) = 8.01 \times 10^{23} \text{ atoms}$$

ITEM 4

Since one hydrogen atom has a mass of about 1.67×10^{-24} g, the chemical symbol H, if it implies one atom, would correspond to a mass of ________ g.

ITEM 13

Atomic weights do not give us the actual masses of the atoms of the elements in grams; what they do give are the ________ masses of the atoms of different elements.

ITEM 22

The value given for copper in a table of atomic weights is 63.54. This means that the mass of ________.

ITEM 31

When dealing with laboratory amounts of chemicals, it is convenient to have the chemical symbol for an element imply one ________ of the element.

ITEM 40

A *gram atomic weight* of an element is defined as ________.

ITEM 49

To calculate the weight of an element that would be required to give x g-atoms of the element, we would ________ x by the ________ of the element.

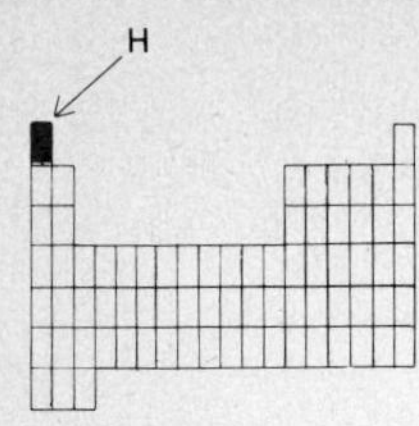

ANSWER 3 atom

ANSWER 12 atomic

ANSWER 21 amu

ANSWER 30 number of atoms

ANSWER 39 pounds
atomic weight

ANSWER 48 weight (or mass) in grams
atomic weight

ITEM 5

The oxygen atom is about 16 times as heavy as the hydrogen atom, and if the chemical symbol O implies one atom of oxygen, it would correspond to a mass of __________ g.

ITEM 14

Although, historically, these relative atomic weights were based on the mass of an atom of the lightest element __________ being assigned the value 1, and later an atom of oxygen being assigned the value __________, they are now based on the carbon isotope of mass number 12 being assigned the value __________.

ITEM 23

The atomic weight given for carbon is 12.01115 (we shall use 12.01). This means that, in addition to the atoms that have a mass of exactly 12 amu, there must be at least one naturally occurring isotope with a mass __________ than 12 amu.

ITEM 32

With this understanding, we can say that C not only designates the element as __________, but also implies __________ g of carbon.

ITEM 41

Since the number of atoms in 1 gram atomic weight is 6.02×10^{23} (using three significant figures), the number of atoms implied by the chemical symbol could be one or, more conveniently for laboratory purposes, could be __________.

ITEM 50

What is the mass in grams of one atom of xenon (at. wt: 131)?

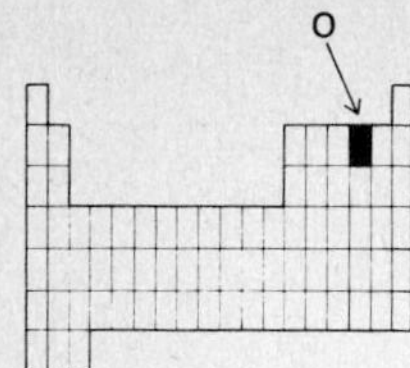

Avogadro's number: 6.022×10^{23}

ANSWER 4 1.67×10^{-24}

ANSWER 13 relative

ANSWER 22 an average atom of naturally occurring copper is 63.54 amu

ANSWER 31 gram atomic weight (or gram-atom)

ANSWER 40 the weight of the element in grams that is numerically equal to its atomic weight

ANSWER 49 multiply atomic weight

ITEM 6

When chemical symbols represent one atom of the element, each symbol represents a different mass because the average mass per atom of every element is ________ from that of every other element.

ITEM 15

A table of atomic weights gives the value for copper as 63.54. This implies that, if one atom of carbon, ^{12}C, weighs exactly 12 units, one atom of copper weighs ________ units.

ITEM 24

If it were possible to measure samples of carbon and copper weighing 1201 and 6354 amu, respectively, there would be ________ atoms in each of these samples.

ITEM 33

In the same manner, Cu implies ________ g of ________.

ITEM 42

In general, the symbol C, when not used in a structural formula to indicate one atom, implies ________ g and ________ atoms of carbon.

ITEM 51

Calculate the number of atoms contained in a sample of nickel (at. wt: 58.7) weighing 7.30×10^{-3} g.

ANSWER 5 $16(1.67 \times 10^{-24}) = 27 \times 10^{-24}$
or 2.7×10^{-23}

ANSWER 14 hydrogen
16
12 (exactly)

ANSWER 23 greater

ANSWER 32 carbon
12.01

ANSWER 41 6.02×10^{23}

ANSWER 50

$$\frac{131 \text{ g g-atom}^{-1}}{6.02 \times 10^{23} \text{ atoms g-atom}^{-1}} = 2.18 \times 10^{-22} \text{ g atom}^{-1}$$

ITEM 7

The masses of individual atoms are of the order of (i.e., in terms of powers of 10) __________ g.

ITEM 16

Since most naturally occurring elements are made up of atoms with different masses, that is, most elements are composed of two or more __________, the atomic mass used by chemists is an average value.

ITEM 25

Since individual atoms of carbon and copper have average masses that are in the ratio C/Cu = __________/__________, larger samples of carbon and copper will have equal numbers of atoms so long as the weights of the samples are in the ratio __________/__________.

ITEM 34

Since each atom of carbon has a mass that is 12 times greater than that of each atom of hydrogen, the number of atoms of carbon in 12 g of carbon will be the same as the number of atoms of hydrogen in __________ of hydrogen.

ITEM 43

Similarly, the symbol O implies __________ g and __________ atoms of oxygen.

ITEM 52

Calculate the number of gram-atoms and atoms in a sample of cadmium (at. wt: 112) weighing 72.2 g.

Isotope	Natural abundance, %
^{64}Zn	48.9
^{66}Zn	27.8
^{67}Zn	4.1
^{68}Zn	18.6
^{70}Zn	0.6

ANSWER 6 different

ANSWER 15 63.54

ANSWER 24 100

ANSWER 33 63.54
copper

ANSWER 42 12.01
6.02×10^{23}

ANSWER 51

$$\frac{7.30 \times 10^{-3}\text{ g}}{58.7\text{ g g-atom}^{-1}} = 1.24 \times 10^{-4}\text{ g-atoms}$$

number of atoms
$= 6.02 \times 10^{23}$ atoms g-atom^{-1} (1.24×10^{-4} g-atoms)
$= 7.46 \times 10^{19}$ atoms

ITEM 8

If the chemical symbol is used to refer to an amount of material that is convenient for laboratory work, it must imply a larger weight of the element and a __________ number of atoms of the element than is implied when the symbol represents one atom.

ITEM 17

A unit of mass, called the *atomic mass unit,* is defined so one atom of ^{12}C has a mass of 12 of these units. Therefore, the table of atomic weights expresses the actual mass of __________ atom of each of the elements as measured in __________ units.

ITEM 26

If 12.01 g of carbon contain x atoms, 63.54 g of copper will contain __________ atoms.

ITEM 35

A gram-atom of one element contains the same number of __________ as a gram-atom of any other element.

ITEM 44

The mass of one average atom of oxygen can be calculated from the values of the previous item to be __________ g (to three significant figures).

ITEM 53

How many grams of sulfur (at. wt: 32.1) must be weighed to make up a sample containing 0.0250 g-atom?

ANSWER 7 10^{-23} (10^{-24} to 10^{-22})

ANSWER 16 isotopes

ANSWER 25 12.01/63.54
12.01/63.54

ANSWER 34 1 g

ANSWER 43 15.9994 (or about 16.00)
6.02×10^{23}

ANSWER 52

$$\frac{72.2\ \text{g}}{112\ \text{g g-atom}^{-1}} = 0.645\ \text{g-atom}$$

$$6.02 \times 10^{-23}\ \text{atoms g-atom}^{-1}\ (0.645\ \text{g-atom}) = 3.88 \times 10^{23}\ \text{atoms}$$

ITEM 9

One could require the chemical symbol to signify a convenient weight (1 g or 1 pound, for instance) of each element. Then, since the atoms of different elements have different masses, the chemical symbols for the different elements all would refer to __________ numbers of atoms.

ITEM 18

If the abbreviation amu is used for this unit that measures masses of atoms, we would say that one atom of ^{12}C weighs 12 __________, one atom of hydrogen weighs about __________, and one atom of copper weighs 63.54 __________.

ITEM 27

Weights of carbon of 12.01 g and of copper of 63.54 g are amounts known as *gram atomic weights.* This term is used because they correspond to weights in grams that are numerically equal to the masses of individual atoms expressed in __________.

ITEM 36

The chemical symbols for various elements imply the same __________ but different __________ of the elements.

ITEM 45

One gram-atom (g-atom) of copper contains __________ atoms.

ITEM 54

What weight of iron (at. wt: 55.9) must be measured in order to have a sample that contains 0.384×10^{23} atoms?

ANSWER 8 larger

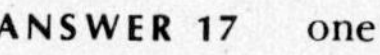

ANSWER 17 one
atomic mass

ANSWER 26 x

ANSWER 35 atoms

ANSWER 44

$$\frac{16.00 \text{ g g-atom}^{-1}}{6.02 \times 10^{23} \text{ atoms g-atom}^{-1}}$$

$$= 2.66 \times 10^{-23} \text{ g atom}^{-1}$$

ANSWER 53

$0.0250 \text{ g-atom } (32.1 \text{ g g-atom}^{-1})$
$= 0.803 \text{ g}$

ANSWER 9 different

(*Item 10 is on page 4.*)

ANSWER 18 amu
1 amu
amu

(*Item 19 is on page 4.*)

ANSWER 27 amu

(*Item 28 is on page 4.*)

ANSWER 36 number of atoms
weights (or masses)

(*Item 37 is on page 4.*)

ANSWER 45 6.02×10^{23}

(*Item 46 is on page 4.*)

ANSWER 54

number of gram-atoms

$$= \frac{0.384 \times 10^{23} \text{ atoms}}{6.02 \times 10^{23} \text{ atoms g-atom}^{-1}} = 0.0638 \text{ g-atom}$$

Therefore, the weight of sample should be 0.0638 g-atom (55.9 g g-atom^{-1}) $= 3.57$ g.

You have learned how symbols of elements and atomic weights are related to measured weights and numbers of atoms. You know what a gram-atom is and that the number of atoms in a gram-atom is Avogadro's number, 6.02×10^{23}. Now you are ready to broaden this concept and to learn about molecules and the ever-necessary mole.

1-2 MOLES

In calculations involving elements and atoms, as in Section 1–1, you need not be concerned about whether an element is made up of single atoms, pairs of atoms, or large crystals of atoms. However, some calculations involving polyatomic elements and most calculations involving compounds must be done on the basis of molecules, rather than atoms. These calculations, as you will see, are analogous to those that you have already done for weights, atoms, and gram-atoms in a sample of an element. Chemical formulas will be used to represent amounts of compounds in exactly the same way that chemical symbols are used to represent amounts (one atom or 1 g-atom) of elements.

Again, we often use four significant figures for atomic weights, as they are given in tables. Nevertheless, to facilitate computation, all problems can be done by using three significant figures.

ITEM 1

When the water molecule is represented by the structural formula at the right, the formula indicates that two hydrogen ________ and one oxygen ________ are arranged as shown.

ITEM 7

A gram molecular weight, or mole, of a chemical compound was defined, then, as the weight of the compound, expressed in grams, that is numerically equal to ________ expressed in ________.

ITEM 13

Because in one molecule of chlorine, Cl_2, there are ________ atoms of chlorine, there are ________ g-atoms of chlorine in 1 mole of chlorine.

ITEM 19

When you hear of moles of electrons, moles of photons, moles of positive charges, and so forth, you should realize that the word "mole" is being used to imply ________ electrons, photons, positive charges, and so on.

ITEM 25

A mole of oxygen atoms contains ________ atoms.

ITEM 31

A cylinder holding 0.13 g of ethane, C_2H_6, contains ________ mole(s) of C_2H_6, ________ molecules of C_2H_6, ________ atoms of C, and ________ atoms of H.

O
H
H

ITEM 2

More often we would simply write H_2O and could view this formula as indicating that one molecule of water contains __________ of hydrogen and __________ of oxygen.

ITEM 8

One molecule of H_2O contains __________ of hydrogen and __________.

ITEM 14

Since one molecule of sulfur in some forms of the element contains eight atoms, 1 mole of sulfur will be equivalent to __________ g-atom(s).

ITEM 20

An 18-g sample of water contains __________ mole(s) of water, or __________ g-atom(s) of hydrogen and __________ g-atom(s) of oxygen.

ITEM 26

When not specified otherwise, a mole of oxygen refers to a mole of oxygen molecules, so the number of atoms in a mole of oxygen is __________. But the terms "1 mole O" or "a mole of O atoms" indicate that the number of atoms referred to is __________.

ITEM 32

When the chemical formula of a compound is thought of as representing one molecule, it tells us the number of __________.

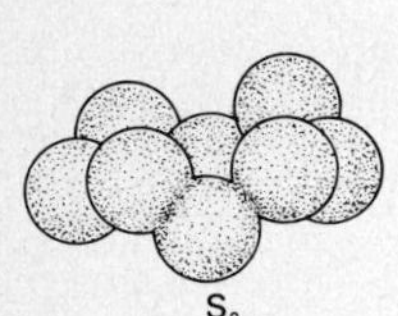

ANSWER 1 atoms
atom

ANSWER 7 the mass of one molecule
amu

ANSWER 13 two
2

ANSWER 19 6.02×10^{23}

ANSWER 25 6.02×10^{23}

ANSWER 31

$$\frac{0.13 \text{ g}}{30 \text{ g mole}^{-1}} = 0.0043 \text{ mole}$$

0.0043 mole (6.02×10^{23} molecules mole^{-1})
$= 2.6 \times 10^{21}$ molecules

2.6×10^{21} molecules (2 C atoms molecule^{-1})
$= 5.2 \times 10^{21}$ C atoms

2.6×10^{21} molecules (6 H atoms molecule^{-1})
$= 1.6 \times 10^{22}$ H atoms

ITEM 3

The mass of one molecule of water, then, is the sum of the masses of __________. Expressed in amu, the mass of one molecule of H_2O is __________.

ITEM 9

One mole of water contains __________ g-atom(s) of hydrogen and __________.

ITEM 15

Just as 1 g-atom of any element contains __________ atoms, 1 mole of any compound contains __________.

ITEM 21

An 18.02-g sample of water contains __________ molecules of water, or __________ atoms of hydrogen and __________ atoms of oxygen.

ITEM 27

Using the periodic table to obtain atomic weights, calculate the weight of a mole of each of the following compounds:

CH_4 __________
$BaCl_2$ __________
P_2O_5 __________

ITEM 33

For the amounts of materials that are used in the laboratory, it is more convenient to have the chemical symbol for an element denote 1 g-atom of the element and the formula for a compound denote __________ of the compound.

ANSWER 2 two atoms
one atom

ANSWER 8 two atoms
one atom of oxygen

ANSWER 14 8

ANSWER 20 1
2
1

ANSWER 26 12.04×10^{23}
6.02×10^{23}

ANSWER 32 atoms of the various elements that compose one molecule of the compound

ITEM 4

The sum of the atomic masses, in amu, of all the atoms of a molecule gives the mass of __________ in __________ units.

ITEM 10

Since 2 g-atoms of hydrogen weigh __________ g and 1 g-atom of oxygen weighs __________ g, 1 mole of water weighs __________ g.

ITEM 16

In fact the term "mole" is synonymous with the number 6.02×10^{23} to the extent that a collection of Avogadro's number of *anything* is referred to as a __________.

ITEM 22

Oxygen gas consists of diatomic molecules that are indicated by the formula __________. One molecule of oxygen contains __________ atom(s).

ITEM 28

3.70 moles of methane, CH_4, weigh __________ and contain __________ molecules.

ANSWER 3 two atoms of hydrogen and one atom of oxygen

$16.00 + 2(1.01) = 18.02$

ANSWER 9 2

1 g-atom of oxygen

ANSWER 15 6.02×10^{23}

6.02×10^{23} molecules

ANSWER 21 6.02×10^{23}

12.04×10^{23}

6.02×10^{23}

ANSWER 27

CH_4: $12.01 + 4(1.01) = 16.05$ g

$BaCl_2$: $137.34 + 2(35.45) = 208.24$ g

P_2O_5: $2(30.97) + 5(16.00) = 141.94$ g

ANSWER 33 1 mole

(*Turn to page 40.*)

ITEM 5

Just as it was convenient to deal with a quantity designated by the term "one gram atomic weight," or "1 g-atom," rather than "one atom," it is also convenient to use one *gram molecular weight* rather than one __________.

ITEM 11

When the symbol H_2O stands for one molecule of water, it indicates that one molecule of water contains __________ and has a mass of __________ amu.

ITEM 17

The mole now is defined as the amount of substance containing as many entities as there are atoms in exactly 12 grams of carbon-12. This number of atoms is __________.

ITEM 23

One mole of O_2 weighs __________ g and contains __________ molecules.

ITEM 29

One mole of sulfuric acid, H_2SO_4, weighs __________ g.

ANSWER 4 one molecule
atomic mass

ANSWER 10 2.02
16.00
18.02

ANSWER 16 mole

ANSWER 22 O_2
two

ANSWER 28

3.70 moles (16.1 g $mole^{-1}$) = 59.6 g
3.70 moles (6.02×10^{23} molecules $mole^{-1}$)
$= 22.3 \times 10^{23}$

ITEM 6

The term "gram molecular weight" (usually abbreviated "mole") is used when we deal with molecules and compounds, just as the term ________ is used in dealing with atoms and elements.

ITEM 12

More conveniently for chemical calculations, the symbol H_2O can denote 1 mole of water. The symbol then represents a mass of ________.

ITEM 18

Thus, we speak of a mole of water (6.02×10^{23} molecules) or a mole of argon (6.02×10^{23} atoms). For a monatomic element, a mole is equivalent to a ________.

ITEM 24

One mole of oxygen molecules contains ________ g-atom(s) of oxygen, or ________ atoms of oxygen.

ITEM 30

A beaker containing 257 g of H_2SO_4 contains ________ mole(s) of H_2SO_4. The number of gram-atoms of hydrogen present is ________.

ANSWER 5 molecule

ANSWER 11 two atoms of hydrogen and one atom of oxygen
18.02

ANSWER 17 6.02×10^{23}

ANSWER 23 $16.0 \times 2 = 32.0$
6.02×10^{23}

ANSWER 29

H_2SO_4: $2(1.01) + 32.06 + 4(16.00) = 98.08$ g

ANSWER 6 gram atomic weight
(or gram-atom)

(*Item 7 is on page 26.*)

ANSWER 12 18.02 g

(*Item 13 is on page 26.*)

ANSWER 18 gram-atom
(or gram atomic weight)

(*Item 19 is on page 26.*)

ANSWER 24 2
$2(6.02 \times 10^{23}) = 12.04 \times 10^{23}$

(*Item 25 is on page 26.*)

ANSWER 30

$$\frac{257 \text{ g}}{98.1 \text{ g mole}^{-1}} = 2.62 \text{ moles}$$

$$2.62 \text{ moles } (2 \text{ g-atoms mole}^{-1}) = 5.24 \text{ g-atoms}$$

(*Item 31 is on page 26.*)

Just as you learned in Section 1–1 to relate the number of grams, the number of gram-atoms, and the number of atoms in a sample, now you have learned to do similar calculations involving grams, moles, and molecules. Furthermore, you know how to tie together these two aspects by calculating, for example, the number of atoms of the elements in a sample of a compound containing a certain number of moles or having a given weight.

1-3 EMPIRICAL FORMULAS

As you learned in *Chemical Principles* (Section 1-9), atomic weights and gram-atoms or moles are important in the determination of empirical formulas of compounds from data obtained from elemental analysis. In this section you can test your knowledge of gram-atoms and moles and prove to yourself that you can do empirical formula calculations.

Because in empirical formula problems the relative numbers of atoms must be determined, the gram-atom concept is used constantly. However, a common practice is to use the term "mole," instead of "gram-atom," and to specify that a mole of atoms is meant. The terms will be used interchangeably in this section, but always with the qualification that when a mole of atoms is meant, it will be designated as such. Remember that for some elements, such as oxygen, fluorine, and hydrogen, it is essential to indicate those instances when "mole" means "mole of atoms."

When you do empirical formula determinations in the laboratory, it will be necessary to use as many significant figures as possible. The use of four or five significant figures may help to minimize the number of possible formulas. However, for the purposes of this section, two or three significant figures and slide rule computation will be adequate.

ITEM 1

When you know the atomic weights of hydrogen (1.0) and oxygen (16.0), it is easy to calculate that 1 mole of water weighs ________ g, of which hydrogen atoms constitute ________ g and oxygen atoms constitute ________ g.

ITEM 6

When the ratio of the number of gram-atoms of hydrogen to gram-atoms of oxygen is translated from the calculated value of 11/5.6 to small whole numbers, it becomes H/O = ________.

ITEM 11

Since the number of gram-atoms of H is 5.9 and of O is 5.9, the ratio of hydrogen atoms to oxygen atoms in any size sample is ________.

ITEM 16

Since the compound contains 22.6% phosphorus, the number of grams of phosphorus in 100 g of compound is ________, and the number of moles of phosphorus atoms is ________. The number of grams of chlorine is ________, and the number of moles of chlorine atoms is ________.

ITEM 21

A sample of compound XB21 weighs 0.00530 g and contains C (0.00205 g), N (0.00239 g), and H. The number of gram-atoms of carbon is ________, of nitrogen is ________, and of hydrogen is ________.

ITEM 26

Compound YD25 was prepared by a method expected to give $CH_3C_6H_4COOH$. Calculate the percentage composition expected for $C_8H_8O_2$ and compare it with the one calculated in Item 25.

ITEM 2

The percentage by weight of hydrogen in an 18.0-g sample of water is __________, and the percentage by weight of oxygen is __________.

ITEM 7

The ratio of gram-atoms of hydrogen to gram-atoms of oxygen in any sample of pure water is __________; in a one-molecule sample, the ratio of the number of atoms of H to the number of atoms of O is

__________.

ITEM 12

The empirical formula of a compound is the simplest kind of formula, giving only the relative numbers of atoms in a molecule in the smallest possible whole numbers. The empirical formula of the compound in Items 8, 9, 10, and 11 is __________.

ITEM 17

The ratio (in small whole numbers) of the number of moles P to the number of moles Cl is __________/__________. (The ratios calculated from the numbers of gram-atoms or moles are not always easy to translate into small whole numbers, but a slide rule is invaluable in doing so. Check this one by slide rule.)

ITEM 22

The ratio of the number of gram-atoms of hydrogen to the number of gram-atoms of carbon is

__________.

ITEM 27

Topaz has the following composition: 29.3% aluminum, 15.3% silicon, 34.8% oxygen, and 20.7% fluorine. The empirical formula of this mineral is

__________.

ANSWER 1 18.0
2.0
16.0

ANSWER 6 2/1

ANSWER 11 1/1

ANSWER 16

22.6

$$\frac{22.6\text{ g}}{31.0\text{ g mole}^{-1}\text{ P}} = 0.73\text{ mole P}$$

77.4

$$\frac{77.4\text{ g}}{35.5\text{ g mole}^{-1}\text{ Cl}} = 2.18\text{ moles Cl}$$

ANSWER 21

$$\frac{0.00205\text{ g}}{12.0\text{ g g-atom}^{-1}\text{ C}} = 1.71 \times 10^{-4}\text{ g-atom C}$$

$$\frac{0.00239\text{ g}}{14.0\text{ g g-atom}^{-1}\text{ N}} = 1.71 \times 10^{-4}\text{ g-atom N}$$

$$\frac{0.00530\text{ g} - 0.00444\text{ g}}{1.0\text{ g g-atom}^{-1}\text{ H}} = 8.6 \times 10^{-4}\text{ g-atom H}$$

ANSWER 26

$$\frac{8\text{ g-atoms }(12.0\text{ g g-atom}^{-1})(100)}{8(12.0)\text{ g} + 8(1.0)\text{ g} + 2(16.0)\text{ g}} = 70.6\%\text{ C}$$

$$\frac{2\text{ g-atoms }(16.0\text{ g g-atom}^{-1})(100)}{8(12.0)\text{ g} + 8(1.0)\text{ g} + 2(16.0)\text{ g}} = 23.5\%\text{ O}$$

$$\frac{8\text{ g-atoms }(1.0\text{ g g-atom}^{-1})(100)}{8(12.0)\text{ g} + 8(1.0)\text{ g} + 2(16.0)\text{ g}} = 5.9\%\text{ H}$$

Within the accuracy of the analysis, the percentages are the same.

ITEM 3

In any pure sample of natural water, the percentage by weight of hydrogen is ________, and the percentage by weight of oxygen is ________.

ITEM 8

In another compound of hydrogen and oxygen, the percentage by weight of hydrogen is 5.9. In a 100-g sample of this compound the number of grams of hydrogen is ________. The number of gram-atoms of hydrogen is ________.

ITEM 13

The known molecular weight of this compound is 34.0. Can the empirical formula be the actual formula of the molecule?

ITEM 18

The empirical formula of the compound in Item 17 is ________.

ITEM 23

The empirical formula of compound XB21 is

________.

ANSWER 2 $\frac{2.0\ g}{18\ g} \times 100 = 11$

$\frac{16\ g}{18\ g} \times 100 = 89$

ANSWER 7 2/1
2/1

ANSWER 12 HO

ANSWER 17 1/3

ANSWER 22 $\frac{8.6 \times 10^{-4}}{1.7 \times 10^{-4}} = \frac{5}{1}$

ANSWER 27

$$\frac{29.3\ g}{27.0\ g\ mole^{-1}\ Al} = 1.09\ moles\ Al$$

$$\frac{15.3\ g}{28.1\ g\ mole^{-1}\ Si} = 0.54\ mole\ Si$$

$$\frac{34.8\ g}{16.0\ g\ mole^{-1}\ O} = 2.18\ moles\ O$$

$$\frac{20.7\ g}{19.0\ g\ mole^{-1}\ F} = 1.09\ moles\ F$$

empirical formula is $Al_2SiO_4F_2$

ITEM 4

In 100 g of H_2O the number of grams of hydrogen is __________, and the number of gram-atoms of hydrogen is __________.

ITEM 9

If the compound contains 5.9% hydrogen, the percentage by weight of oxygen is __________.

ITEM 14

Since the molecular weight of the compound is 34.0, the actual formula of the molecule must be __________.

ITEM 19

Aluminum oxide contains 47.0% oxygen. The number of moles of aluminum in 100 g of this oxide is __________, and the number of moles of oxygen atoms is __________.

ITEM 24

Every problem involving the percentage by weight composition of a compound is basically a restatement of the familiar relationship: no. of moles of atoms of an element or no. of gram-atoms of an element = no. of grams of the element divided by the __________.

ANSWER 3 11
89

ANSWER 8 5.9

$$\frac{5.9\ \text{g}}{1.0\ \text{g g-atom}^{-1}} = 5.9\ \text{g-atoms}$$

ANSWER 13 No, because the molecular weight of HO is 17.0.

ANSWER 18 PCl_3

ANSWER 23 CH_5N

ITEM 5

In 100 g of H_2O the number of grams of oxygen is ________, so the number of gram-atoms is ________.

ITEM 10

The number of grams of oxygen atoms in 100 g of the compound is ________, and the number of gram-atoms is ________.

ITEM 15

A compound contains 22.6% phosphorus and 77.4% chlorine. The ratio of P to Cl atoms is the same in any size sample, but it is most convenient to calculate the ratio in a sample weighing ________ g.

ITEM 20

The ratio of moles Al to moles O is ________, and the empirical formula is ________.

ITEM 25

A carefully purified sample of compound YD25 weighs 0.00840 g. Analysis shows that it contains 4.94×10^{-4} mole C and 1.25×10^{-4} mole O. The other element present is hydrogen. What is the percentage by weight of C, H, and O?

ANSWER 4 11

$$\frac{11\text{ g}}{1.0\text{ g g-atom}^{-1}} = 11$$

ANSWER 9 94.1

ANSWER 14 H_2O_2

ANSWER 19

$$\frac{53.0\text{ g}}{27.0\text{ g mole}^{-1}\text{ Al}} = 1.96\text{ moles Al}$$

$$\frac{47.0\text{ g}}{16.0\text{ g mole}^{-1}\text{ O}} = 2.94\text{ moles O}$$

ANSWER 24 atomic weight

ANSWER 5 89

$$\frac{89\text{ g}}{16\text{ g g-atom}^{-1}} = 5.6$$

(*Item 6 is on page 42.*)

ANSWER 10 94.1

$$\frac{94.1\text{ g}}{16.0\text{ g g-atom}^{-1}} = 5.9\text{ g-atoms}$$

(*Item 11 is on page 42.*)

ANSWER 15 100

(*Item 16 is on page 42.*)

ANSWER 20 2/3
Al_2O_3

(*Item 21 is on page 42.*)

ANSWER 25

4.94×10^{-4} mole C(12.0 g mole^{-1} C)
= 0.00593 g of C

$$\frac{0.00593\text{ g}(100)}{0.00840\text{ g}} = 70.6\%\text{ C}$$

1.25×10^{-4} mole O(16.0 g mole^{-1} O)
= 0.00200 g of O

$$\frac{0.00200\text{ g}(100)}{0.00840\text{ g}} = 23.8\%\text{ O}$$

0.00840 g − 0.00200 g − 0.00593 g
= 0.00047 g of H

$$\frac{0.00047\text{ g}(100)}{0.0084\text{ g}} = 5.6\%\text{ H}$$

(*Item 26 is on page 42.*)

REVIEW 2 THE IDEAL GAS LAWS

From Chapter 1 of *Chemical Principles* it is clear that the study of gases is of critical importance to the development of our knowledge of the number of atoms in molecules and of the concept of the mole. Of particular importance is Amedeo Avogadro's hypothesis that the volume occupied by a gas at a given temperature and pressure depends only on the number of molecules present and not on any other property of the gas.

In Chapter 2 you read about the development of laws relating the volume, temperature, pressure, and number of moles of a gas. The first part of the chapter, concerning the ideal gas laws, is essential to your further study of chemistry; for that reason it is reviewed here. You realize, of course, that the statements and the equations apply only to what are called ideal gases. Although they describe exactly the behavior of no real gas (except at zero pressure), they constitute extremely useful approximations at the temperatures and pressures encountered in many experiments.

At the conclusion of this review you will be able to make calculations of the following type: "What is the molecular weight of a gas whose density at 0°C and 0.1 atm is 0.390 g liter^{-1}?" If you can work correctly the items beginning with number 80 at the end of the review, it will not be necessary for you to do the others.

ITEM 1

When the pressure on a sample of gas is increased while the temperature is held constant, the gas is compressed; that is, the volume of the gas ________.

ITEM 16

If the temperature is raised and the pressure kept constant, the gas will expand. That is, the volume will ________.

ITEM 31

One (1.00) liter of a gas at 0°C will expand, when heated to 100°C, to a volume of ________ liter(s).

ITEM 46

At 1 atm and 0°C the volume of 1.81 g of nitrogen is 1.45 liters. The density of nitrogen under these conditions is ________.

ITEM 61

One mole of chlorine, Cl_2, contains ________ molecules and weighs ________ g.

ITEM 76

The volume of 0.376 g of CO_2 at STP is ________ liter(s).

ITEM 2

Conversely, when the pressure is decreased the gas will expand; that is, the volume will ________.

ITEM 17

If the temperature of a gas sample is increased from 20°C to 40°C, the volume is found to increase by about 7%, not 100%. Hence, the volume of a gas is not directly proportional to the temperature expressed in ________.

ITEM 32

To cause a gas sample to contract from a volume of 3.8 liters to 1.4 liters, the absolute temperature can be lowered to ________ times its initial value or the pressure can be ________ to ________ times its initial value.

ITEM 47

At the same ________ and ________, the density of nitrogen is ________ times greater than the density of hydrogen.

ITEM 62

One mole of any substance contains the ________ number of molecules as 1 mole of any other substance.

ITEM 77

A gas sample occupies a volume of 546 ml at STP. It contains, therefore, ________ mole(s) of gas.

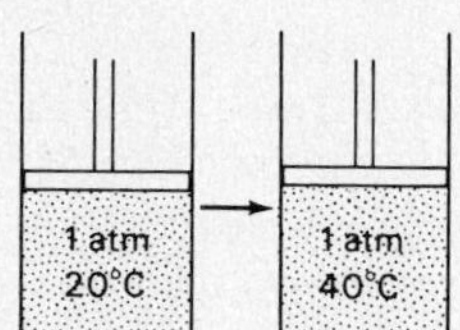

Density (g liter^{-1}) at 1 atm and 0°C	
H_2	$D = 0.090$
N_2	$D = 1.25$

ANSWER 1 decreases

ANSWER 16 increase

ANSWER 31 1.00 liter (373 deg/273 deg) = 1.37 liters

ANSWER 46 1.81 g/1.45 liters = 1.25 g liter^{-1}

ANSWER 61 6.022×10^{23}
2(35.45) = 70.90

ANSWER 76

$V = nRT/P$

$$= \frac{\dfrac{0.376\text{ g}}{44.0\text{ g mole}^{-1}} \times 0.0821 \dfrac{\text{liter atm}}{\text{deg mole}} \times 273\text{ deg}}{1.00\text{ atm}}$$

$= 0.191$ liter

(or $\dfrac{0.376\text{ g}}{44.0\text{ g mole}^{-1}} \times 22.4$ liters mole^{-1}

$= 0.191$ liter)

ITEM 3

The relation between the pressure and the volume of a constant number of moles of a gas at constant temperature is shown graphically at the right. If the pressure on a gas is doubled, the volume is reduced to half its original value. If the pressure is decreased to one fifth of its original value, the volume will __________ by a factor of __________.

ITEM 18

In fact, if the temperature is raised from 20°C to 40°C, the volume of a gas sample increases by an amount that is 20/273 of its volume at 0°C. In general, for every 1°C rise in temperature, the volume of a gas increases by __________ of its value at 0°C.

ITEM 33

Boyle's law states that: *The volume of a constant number of moles of a gas* __________. This can be written as __________ = constant (n, T constant).

ITEM 48

The equation $V = kw$, in which k is a constant, can be written for any individual gas at a fixed temperature and pressure. Since the density D is defined as w/V, the constant k is related to the density as $k =$ __________.

ITEM 63

One mole of any gas will occupy, at a given temperature and pressure, the same __________ as 1 mole of any other gas.

ITEM 78

A 0.264-g sample of H_2S gas contains __________ mole(s) and occupies, at 100°C and 743 torr pressure, a volume of __________ liter(s). (Recall that 1 atm = 760 torr.)

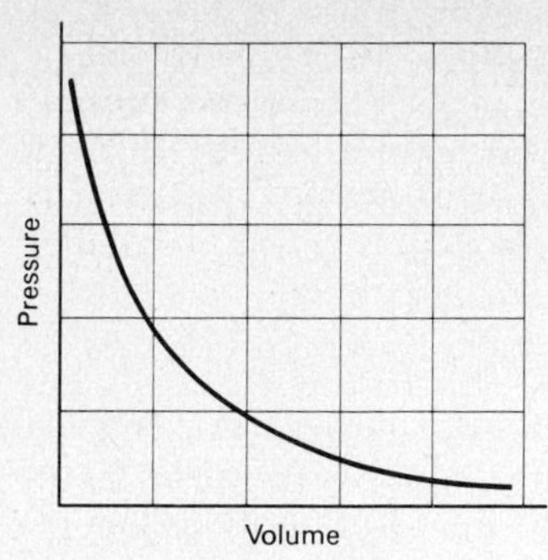

ANSWER 2 increase

ANSWER 17 °C (degrees centigrade)

ANSWER 32 1.4 liters/3.8 liters = 0.37
increased
3.8 liters/1.4 liters = 2.7

ANSWER 47 temperature
pressure

$$\frac{1.25}{0.090} = 14$$

ANSWER 62 same

ANSWER 77

$$n = PV/RT$$

$$= \frac{(1.00 \text{ atm})(0.546 \text{ liter})}{\left(0.0821 \dfrac{\text{liter atm}}{\text{deg mole}}\right)(273 \text{ deg})}$$

$$= 0.0244 \text{ mole}$$

$$\left(\text{or } \frac{0.546 \text{ liter}}{22.4 \text{ liter mole}^{-1}} = 0.0244 \text{ mole}\right)$$

ITEM 4

This relationship is described by the generalization: The volume of an ideal gas is ________ proportional to the pressure if the number of moles remains constant and the ________ remains constant. This statement is known as *Boyle's law.*

ITEM 19

Thus, for a gas sample at fixed pressure, the plot of V versus t°C has the form shown at the right. (The dashed line represents the extrapolation that is necessary because gases become liquids and solids at low temperature.) According to the extrapolation, the volume would be zero at a temperature of ________ °C.

ITEM 34

Charles' law states that: *The volume of a constant number of moles of a gas is* ________. This can be written as ________ = constant (n, P constant).

ITEM 49

The relation between the volume and mass of a given gas at a fixed pressure and temperature can then be written $V =$ ________ w.

ITEM 64

At a given T and P, the volume of any ideal gas sample will not be determined by the size, shape, mass, or chemical properties of the gas molecules but will be determined by the ________ in the sample.

ITEM 79

The volume of 22.4 liters is occupied at STP by 1 mole, or ________ molecule(s), of any gas.

Boyle's law

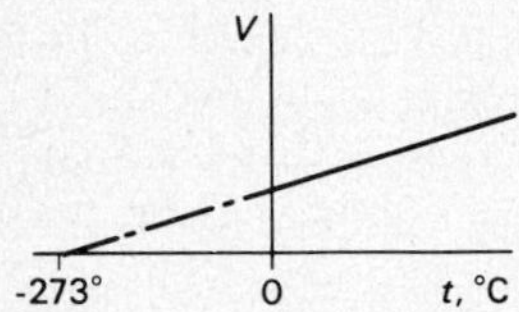

$V = kw$
$k = 1/D$

ANSWER 3 increase
five

ANSWER 18 1/273

ANSWER 33 is inversely proportional to the pressure if the temperature remains constant.
PV

ANSWER 48 $1/D$

ANSWER 63 volume

ANSWER 78

0.264 g/34.1 g $\text{mole}^{-1} = 0.00774$
$V = nRT/P$

$$= \frac{0.00774 \text{ mole}(373 \text{ deg})\left(0.0821 \dfrac{\text{liter atm}}{\text{deg mole}}\right)}{(743/760) \text{ atm}}$$

$= 0.242$ liter

ITEM 5

When the pressure on a sample of a gas (at a fixed temperature) is increased from 1.2 atm to 6.0 atm, the new volume will be what fraction of the original volume?

ITEM 20

A straight line as shown at the right represents the equation $y = m(x + b)$, in which m is the slope of the line. A comparison of this graph and the graph of the preceding item shows that, for constant pressure and constant number of moles of gas, we can write the equation $V = m($__________ + __________$)$.

ITEM 35

These laws can be combined to show the dependence of V on both P and T by the proportionality equation $V \propto T/P$ (n constant), or, with the proportionality constant α, by $V = \alpha$ __________.

ITEM 50

Since, at the same T and P, the densities of different gases are __________, the proportionality constants between volume and mass for different gases are different.

ITEM 65

In relating gas volumes to numbers of moles or molecules, we should, for convenience, have a set of standard conditions. The standard temperature and pressure (STP) generally agreed upon are $T = 0°C$, or __________ °K, and $P = 1$ atm.

ITEM 80

If the size, shape, and weight of molecules do not affect ideal gas behavior, it follows that, even if there are different kinds of molecules making up the gas sample, the relation $PV = nRT$ will apply, in which n is the total __________ of gas present.

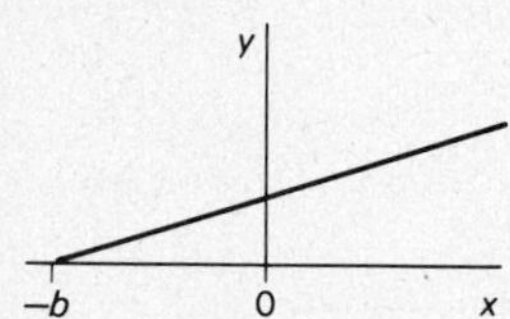

ANSWER 4 inversely
temperature

ANSWER 19 −273

ANSWER 34 directly proportional to the absolute temperature if the pressure remains constant.
V/T

ANSWER 49 $(1/D)$

ANSWER 64 number of molecules
(or number of moles)

ANSWER 79 6.022×10^{23}

ITEM 6

To increase the volume of a sample of gas, the pressure must be ________. For the volume to change from 1.56 liters to 4.28 liters, the pressure must be ________ from its original value to ________ times its original value.

ITEM 21

Thus, for a gas sample at fixed pressure, V is proportional not to t°C but to ________.

ITEM 36

This equation can be rearranged to show that, for a given sample of gas, the function of P, V, and T that remains constant is ________.

ITEM 51

Therefore, when more than one gas is considered, we cannot expect to find a general formula for the volume in terms of only the temperature, pressure, and ________ of the gases.

ITEM 66

The density of N_2 at STP is 1.25 g per liter, and therefore 1 mole of N_2 (that is, ________ g) occupies ________ liter(s) at STP.

ITEM 81

A mixture of 3.00 moles of hydrogen and 1.00 mole of argon contains a total of ________ mole(s) and occupies ________ liter(s) at STP.

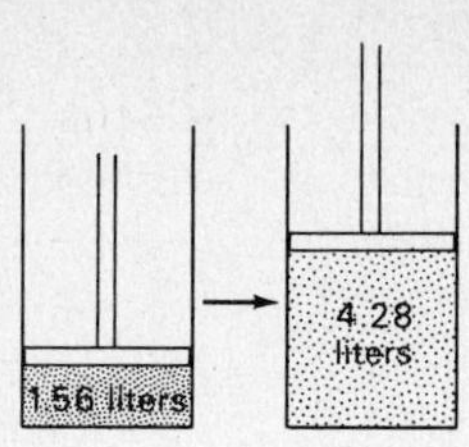

$V = \alpha(T/P)$

ANSWER 5 1.2 atm/6.0 atm = 0.20

ANSWER 20 $(t°C + 273)$

ANSWER 35 T/P (n constant)

ANSWER 50 different

ANSWER 65 273

ANSWER 80 number of moles

ITEM 7

To compress a gas sample that occupies 2.5 liters at a pressure of 1.8 atm to a volume of 1.5 liters (temperature remaining constant), the pressure must be ______ to a value of 1.8 atm × ______ = ______ atm.

ITEM 22

The absolute, or Kelvin (K), temperature scale is related to the centigrade (C) scale as follows: $T_{°K} = t_{°C} + 273$. Thus, 10°C = ______ °K.

ITEM 37

Thus, Boyle's and Charles' laws can be combined into the convenient mathematical expression ______ = constant (*n* constant).

ITEM 52

This goal can be achieved, to the extent that gases behave ideally, with the following added assumption, first proposed by Avogadro: *Equal volumes of different gases at the same temperature and pressure contain ______ number of molecules.*

ITEM 67

Similarly, once allowances are made for nonideal behavior of other gases, it is found that 1 mole of any gas at STP should occupy the volume of ______ liter(s).

ITEM 82

A mixture of 2.76 g of CH_4 and 9.34 g of NH_3 contains ______ mole(s) and, at 200°C and 3.00 atm, occupies ______ liter(s).

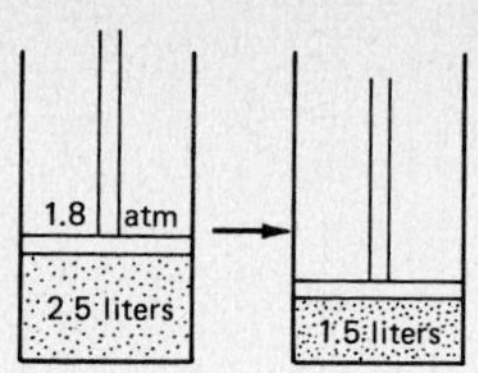

ANSWER 6 decreased
decreased
1.56 liters/4.28 liters = 0.364

ANSWER 21 (t°C + 273) or (t + 273)°C

ANSWER 36 PV/T

ANSWER 51 mass (or density)

ANSWER 66 28.0

$$\frac{28.0 \text{ g mole}^{-1}}{1.25 \text{ g liter}^{-1}} = 22.4$$

ANSWER 81 4.00
22.4 liters mole^{-1} (4.00 moles)
= 89.6 liters

ITEM 8

Boyle's law states that: *The* __________ *of a gas sample (constant number of moles) is* __________ *to the* __________ *of the gas if the temperature remains constant.*

ITEM 23

Since V is directly proportional to (t°C + 273), V is also __________ to the Kelvin temperature.

ITEM 38

For initial values V_1, P_1, and T_1, and final values V_2, P_2, and T_2, this relation can be written __________ = __________ (n constant).

ITEM 53

At the same temperature and pressure, 1 liter of nitrogen and 1 liter of hydrogen have __________ masses but __________ number of molecules.

ITEM 68

The volume of 3.2 moles of cyanogen gas (C_2N_2) at STP is __________ liter(s).

ITEM 83

Air consists very nearly of 80 mole % N_2 and 20 mole % O_2; that is, in 100 moles of air, 80 will be N_2 and 20 will be O_2.

(a) One mole of air at 25°C and 1 atm has a volume of __________ liters. It contains __________ mole(s), or __________ g, of N_2 and __________ mole(s), or __________ g, of O_2.

(b) From (a), the total mass of N_2 and O_2 in 1 mole of air is calculated to be __________ g. Thus, the density of air at 25°C and 1 atm is __________.

ANSWER 7 increased
2.5 liters/1.5 liters
3.0

ANSWER 22 283

ANSWER 37 PV/T

ANSWER 52 the same

ANSWER 67 22.4

ANSWER 82

$$\text{no. of moles} = \frac{2.76\ \text{g}}{16.0\ \text{g mole}^{-1}} + \frac{9.34\ \text{g}}{17.0\ \text{g mole}^{-1}}$$

$$= 0.722\ \text{mole}$$

$$V = nRT/P$$

$$= \frac{0.722\ \text{mole}}{3.00\ \text{atm}} \left(0.0821\ \frac{\text{liter atm}}{\text{deg mole}}\right) 473\ \text{deg}$$

$$= 9.35\ \text{liters}$$

ITEM 9

Boyle's law is written $V \propto 1/P$ (n, T constant), or, with the proportionality constant a, $V =$ __________, or $PV =$ __________.

ITEM 24

The volume of a gas sample at constant pressure varies with the temperature according to $V = m(t°\text{C} + 273) = m$ __________, in which m is a proportionality constant.

ITEM 39

This equation can be rearranged as $V_2 = V_1 \times T_2/T_1 \times$ __________ (n constant) and in this form shows explicitly its relation to Charles' and Boyle's laws.

ITEM 54

At a given temperature and pressure, the volume occupied by N molecules of nitrogen gas is __________ the volume occupied by the same number of hydrogen molecules, or any kind of gas molecules.

ITEM 69

The volume of a sample of gas is directly proportional to the __________ and to the __________ and inversely proportional to the __________.

ITEM 84

The volume of gas in a meteorological balloon is 7.00 ft^3 at ground level, where the pressure is 1 atm and the temperature is 25°C. At an altitude of 22 miles the pressure is about 4 torr and the temperature is −12°C. The volume of the gas in the balloon is then __________ ft^3.

$$\frac{P_1V_1}{T_1} = \frac{P_2V_2}{T_2}$$

ANSWER 8 volume
inversely proportional
pressure

ANSWER 23 directly proportional

ANSWER 38 P_1V_1/T_1
P_2V_2/T_2

ANSWER 53 different
the same

ANSWER 68 (22.4 liters mole^{-1})(3.2 moles)
= 72 liters

ANSWER 83

(a)

$V = 22.4$ liters (298 deg/273 deg)
= 24.5 liters

[or $V = nRT/P$

$$= 1 \text{ mole} \left(0.0821 \frac{\text{liter atm}}{\text{deg mole}}\right) \frac{298 \text{ deg}}{1 \text{ atm}}$$

= 24.5 liters]

0.80

0.80 mole (28 g mole^{-1}) = 22.4 g

0.20

0.20 mole (32 g mole^{-1}) = 6.4 g

(b)

22.4 g + 6.4 g = 28.8 g

$D = w/V$ = (28.8 g mole^{-1})/(24.5 liters mole^{-1})
= 1.2 g liter^{-1}

ITEM 10

The relation $PV = a$ shows that, for a given sample of gas, the product of the pressure and volume is constant if the __________ remains constant.

ITEM 25

If the pressure on a sample of gas remains constant, the volume is __________ proportional to the __________ temperature. This statement is known as *Charles'*, or *Gay-Lussac's, law.*

ITEM 40

When the pressure on a 7.63-liter sample of a gas is increased from 2.35 atm to 7.60 atm and the temperature is decreased from 400°K to 300°K, the volume will __________, and the new volume will be __________ liters.

ITEM 55

Avogadro's hypothesis states that: *At the same conditions of temperature and pressure equal volumes of different gases* __________.

ITEM 70

If the number of moles in a sample of gas is indicated by n, the dependence of V on n, P, and T can be written as $V \propto$ __________.

ITEM 85

One (1.00) liter of an unknown gas at 0.425 atm and 30°C has a mass of 0.548 g. The molecular weight (mol wt) of the gas is __________, and the gas might be __________.

Charles' law

$V_2 = V_1 \times \frac{T_2}{T_1} \times \frac{P_1}{P_2}$

ANSWER 9 a/P (n, T constant)
a (n, T constant)

ANSWER 24 $T_{°K}$ (or T)

ANSWER 39 P_1/P_2

ANSWER 54 equal to

ANSWER 69 absolute temperature
number of moles
pressure

ANSWER 84

$P_1V_1/T_1 = P_2V_2/T_2$

$$V_2 = \left(\frac{760 \text{ torr}}{4 \text{ torr}}\right)\left(\frac{261 \text{ deg}}{298 \text{ deg}}\right)(7.00 \text{ ft}^3)$$

$= 1160$ (approximately 1200)

ITEM 11

For a sample of gas at pressure P_1 and volume V_1, $P_1V_1 =$ __________. For the same sample of gas under different conditions, P_2 and V_2, it is also true that $P_2V_2 =$ __________. Hence, $P_1V_1 =$ __________ (n, T constant).

ITEM 26

If the absolute temperature of a gas sample is tripled and the pressure is kept constant, the volume __________ to __________ times its original value.

ITEM 41

When the temperature of a gas sample is increased from 0°C to 250°C, that is, from __________ °K to __________ °K, and the pressure on the gas is doubled, the volume will be changed by a factor of __________.

ITEM 56

The volume of a gas sample, at given T and P, is proportional to the amount of gas, which can be measured by the mass of the gas or by the number of __________ in the sample.

ITEM 71

If the proportionality constant R is introduced, the relation can be written as $V =$ __________.

ITEM 86

An ideal gas with a volume of 20.0 liters at STP is compressed to 8.50 liters and, at the same time, is heated to 200°C. What is the final pressure exerted by the gas?

$$\frac{P_1V_1}{T_1} = \frac{P_2V_2}{T_2}$$

$V \propto nT/P$

ANSWER 10 temperature
(and number of moles)

ANSWER 25 directly
absolute (Kelvin)

ANSWER 40

decrease

$$7.63 \text{ liters } \frac{(300 \text{ deg})(2.35 \text{ atm})}{(400 \text{ deg})(7.60 \text{ atm})} = 1.77 \text{ liters}$$

ANSWER 55 *contain the same number of molecules*

ANSWER 70 nT/P

ANSWER 85

$$n = \frac{PV}{RT} = \frac{(0.425 \text{ atm})(1.00 \text{ liter})}{\left(0.0821 \dfrac{\text{liter atm}}{\text{deg mole}}\right)(303 \text{ deg})}$$

$$= 0.0171 \text{ mole}$$

mol wt $= 0.548$ g/0.0171 mole
$= 32.0 \text{ g mole}^{-1}$

O_2

ITEM 12

This result can be rearranged to $V_1/V_2 =$ __________ (n, T constant), which shows that the volume varies inversely as the pressure.

ITEM 27

To cause a gas to contract from a volume of 250 milliliters (ml) to 170 ml at a constant pressure of 1 atm, the initial absolute temperature must be multiplied by a factor of __________.

ITEM 42

At the same T and P, the volume occupied by 2 g of H_2 is twice the volume occupied by 1 g of H_2. The volume is proportional to the __________ of gas.

ITEM 57

For a given gas at fixed T and P, one can write $V \propto$ (number of molecules) or, using the proportionality constant k', $V =$ __________ (P, T constant).

ITEM 72

This equation, representing the *ideal gas law,* is usually written $PV =$ __________. The constant R, known as the gas constant, must be __________ for all ideal gases.

ITEM 87

What volume is occupied by 1.34 g of gaseous C_2H_6 at 25°C and 747 torr?

$P_1V_1 = P_2V_2$
(n, T constant)

ANSWER 11 a
a
P_2V_2

ANSWER 26 increases
three

ANSWER 41 273
$273 + 250 = 523$
$V_2/V_1 = (P_1/P_2)(T_2/T_1)$

$$= \frac{1}{2}\left(\frac{523 \text{ deg}}{273 \text{ deg}}\right)$$

$$= 0.958$$

ANSWER 56 molecules or moles

ANSWER 71 RnT/P (or nRT/P)

ANSWER 86 $P_1V_1/T_1 = P_2V_2/T_2$

$$P_2 = \frac{20.0 \text{ liters}}{8.50 \text{ liters}}\left(\frac{473°}{273°}\right)1.00 \text{ atm}$$

$$= 4.08 \text{ atm}$$

ITEM 13

The pressure on a gas sample is increased from 1.0 atm to 12 atm. If the original volume was 3.8 liters, the new volume could be calculated as $V_2 = V_1(P_1/P_2)$ or $V_2 = 3.8$ liters × ________ = ________ liter(s).

ITEM 28

The mathematical formulation of Charles' law is $V \propto$ ________ (n, P constant); or, using the proportionality constant c, $V =$ ________ or $V/T =$ ________.

ITEM 43

If the amount of gas is measured by its mass w, then at constant temperature and pressure $V \propto$ ________ or, if the proportionality constant k is introduced, $V = k$ ________.

ITEM 58

Since equal volumes of different gases at the same temperature and pressure contain ________, the constant k' must be ________ for different gases.

ITEM 73

The gas constant R can have various numerical values, depending on the units in which it is expressed. If $PV = nRT$ is to be used with P in atmospheres, V in liters, T in degrees Kelvin, and n in moles, the units of R will be ________.

ITEM 88

Calculate the density of uranium hexafluoride (UF_6) vapor at 56°C and 1.00 atm.

$$\frac{V_1}{V_2} = \frac{P_2}{P_1}$$

$PV = nRT$

ANSWER 12 P_2/P_1

ANSWER 27 170 ml/250 ml = 0.680

ANSWER 42 mass
(amount or number of moles)

ANSWER 57 k' (number of molecules)

ANSWER 72 nRT
the same

ANSWER 87

mol wt of $C_2H_6 = 2(12.0) + 6(1.01) = 30.1$
$V = nRT/P$

$$= \frac{\left(\frac{1.34\text{ g}}{30.1\text{ g mole}^{-1}}\right)\left(0.0821\ \frac{\text{liter atm}}{\text{deg mole}}\right)(298\text{ deg})}{(747/760)\text{ atm}}$$

= 1.11 liters

ITEM 14

To increase the volume of a gas sample from 2.84 liters to 7.56 liters at constant temperature, the pressure must be __________ from 4.20 atm to __________ atm.

ITEM 29

For initial conditions V_1 and T_1, $V_1/T_1 =$ __________; for final conditions V_2 and T_2, $V_2/T_2 =$ __________. Hence, $V_1/T_1 =$ __________ (n, P constant).

ITEM 44

By definition, the density D of a gas is the mass per unit volume; that is, $D =$ __________.

ITEM 59

The conventional unit for measuring numbers of molecules is Avogadro's number, $N = 6.022 \times 10^{23}$. This is the number of molecules in 1 __________ of any substance.

ITEM 74

Since 1 mole of any gas occupies 22.4 liters at STP, you can calculate the value of the gas constant R. $R =$ __________ liter atm deg^{-1} mole^{-1}.

ITEM 89

Calculate the molecular weight of an unknown gas, given that 1.51 g of the gas occupy a volume of 405 ml at 100°C and 735 torr.

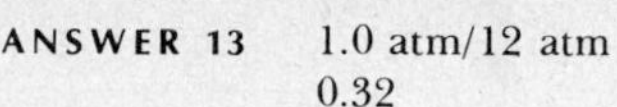

ANSWER 13 1.0 atm/12 atm
0.32

ANSWER 28 T
cT (n, P constant)
c (n, P constant)

ANSWER 43 w
w

ANSWER 58 the same number of molecules
the same

For R values in other units, see *Chemical Principles,* Section 2–5.

ANSWER 73 liter atm deg^{-1} mole^{-1}

ANSWER 88

volume of 1.00 mole = 1.00 RT/P

$$= \frac{1.00 \text{ mole}\left(0.0821 \dfrac{\text{liter atm}}{\text{deg mole}}\right)329 \text{ deg}}{1.00 \text{ atm}}$$

$$= 27.0 \text{ liters}$$

$$\left[\text{or } V_{\text{mole}} = 22.4 \text{ liters}\left(\frac{329°}{273°}\right) = 27.0 \text{ liters}\right]$$

$$\text{density} = \frac{352 \text{ g mole}^{-1}}{27.0 \text{ liters mole}^{-1}} = 13.0 \text{ g liter}^{-1}$$

ITEM 15

If the pressure on a gas sample is kept constant and no molecules of gas are added or subtracted, the volume can be changed by changing the __________ of the gas.

ITEM 30

When a gas with an initial volume of 14.7 liters is heated from 300°K to 450°K and the pressure is kept constant, the volume __________ to __________ liter(s).

ITEM 45

At 1 atm and 0°C, 0.26 g of hydrogen occupies 2.88 liters. The density under these conditions is __________.

ITEM 60

One mole of O_2 weighs 32.0 g and contains ________ molecule(s).

ITEM 75

Problems involving the *PVT* behavior of gases can be solved either by using Boyle's law, Charles' law, and the fact that 1 mole of any gas occupies ________ at STP or by using the equation __________, which summarizes these features.

ITEM 90

What volume of helium gas at STP is required to produce 1.00 liter of liquid helium? The density of liquid helium is 0.122 g ml^{-1}.

$$\frac{V_1}{T_1} = \frac{V_2}{T_2}$$

ANSWER 14

decreased
4.20 atm (2.84 liters/7.56 liters)
= 1.58 atm

ANSWER 29 c
c
V_2/T_2

ANSWER 44 w/V

ANSWER 59 mole

ANSWER 74

$R = PV/nT$

$$= \frac{(1.00 \text{ atm})(22.4 \text{ liters})}{(1.00 \text{ mole})(273 \text{ deg})}$$

$= 0.0821$

ANSWER 89

$$n = \frac{PV}{RT} = \frac{(735/760) \text{ atm } (0.405 \text{ liter})}{\left(0.0821 \dfrac{\text{liter atm}}{\text{deg mole}}\right)(373 \text{ deg})}$$

$= 0.0128$ mole
mol wt $= 1.51$ g/0.0128 mole $= 118$ g mole^{-1}

ANSWER 15 temperature

(Item 16 is on page 56.)

ANSWER 30 increases

14.7 liters (450 deg/300 deg)
= 22.1 liters

(Item 31 is on page 56.)

ANSWER 45 0.26 g/2.88 liters
= 0.090 g $liter^{-1}$

(Item 46 is on page 56.)

ANSWER 60 6.022×10^{23}

(Item 61 is on page 56.)

ANSWER 75 22.4 liters

$PV = nRT$ [or $PV = n(0.0821)T$ if P is in atm, V is in liters, and T is in degrees Kelvin]

(Item 76 is on page 56.)

ANSWER 90

wt of He = (1.00 liter)(122 g $liter^{-1}$) = 122 g
moles of He = (122 g)/4.00 g $mole^{-1}$
= 30.5 moles
thus, V = (30.5 moles)(22.4 liters $mole^{-1}$)
= 683 liters

REVIEW 3 IONS AND THE ELECTROLYSIS OF MOLTEN SALTS

The study of the electrolytic decomposition of molten salts is preparation for understanding concepts and words used in all branches of chemistry. The development here of ideas concerning the reduction and oxidation of ions in electrolysis reactions will serve as a review of Sections 3-1 and 3-2 of *Chemical Principles* and as an introduction to the study of ionic nomenclature and charge that will be continued in Review 6.

If you can correctly answer Items 50–60 at the end of the review, you are already familiar with this material and you need not work through the beginning items.

ITEM 1

Atoms are made up of protons, neutrons, and electrons. Although neutrons are electrically neutral, protons have a unit positive charge and electrons have a unit ________ charge.

ITEM 11

The conduction of electricity through any material requires the movement of charged particles. A metal conducts electricity because some of the electrons are not tightly held but are free to ________ through a metal crystal.

ITEM 21

Two chlorine atoms at the anode will form a molecule of chlorine gas and escape from the liquid. The reactions at the anode can be written

$$Cl^- \rightarrow Cl + e^-$$
$$2Cl \rightarrow Cl_2 \uparrow$$

The reaction at the cathode can be written ________.

ITEM 31

The electrolysis of molten cuprous oxide, Cu_2O, made up of Cu^+ ions and O^{2-} ions proceeds according to the equation $2Cu_2O \rightarrow 4Cu + O_2 \uparrow$. The reduction half-reaction is ________. The oxidation half-reaction is ________.

ITEM 41

To five significant figures, 1 $\mathscr{F}$ is 96,487 coulombs. One coulomb will produce ________ g of silver metal (at. wt: 107.868).

ITEM 51

The equation for the complete reaction (with the same number of electrons on both sides) is ________.

ITEM 2

Since an atom has equal numbers of protons and electrons, it is electrically neutral. If an electron is added to an atom, the atom becomes an *ion* and has a unit __________.

ITEM 12

Solid salts will not conduct electricity: All the electrons are tightly held to the ions, and the ions themselves are tightly held in the crystal. But when salts are melted, the liquid conducts electricity because the ions __________.

ITEM 22

Two electrons must be removed from Cl^- ions to form Cl_2, and at the same time two electrons must be added to __________ Na^+ ions to form __________.

ITEM 32

The salt silver bromide, AgBr, is made up of Ag^+ ions and Br^- ions. The __________ charged silver ion is called the __________ and the __________ charged bromide ion is called the __________.

ITEM 42

This quantity is important because the coulomb is now defined as the amount of electricity that will deposit 0.0011180 g of silver from a solution of silver nitrate. This is just the amount of electricity that will deposit 1/96,487 of a __________ of silver from a molten silver salt.

ITEM 52

Calculate the number of grams of iron (at. wt: 55.9) and chlorine gas (at. wt Cl: 35.5) produced by 0.500 A for 1.00 h.

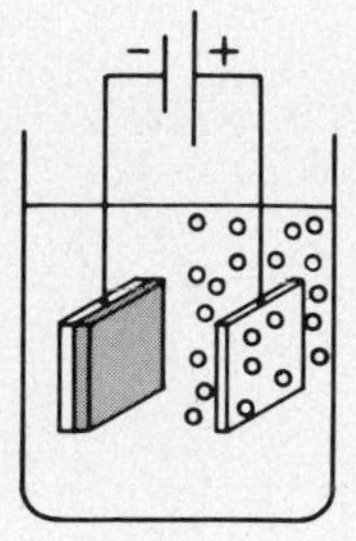

$Fe^{3+} + 3e^- \rightarrow Fe$
$2Cl^- \rightarrow Cl_2\uparrow + 2e^-$

ANSWER 1 negative

ANSWER 11 move

ANSWER 21 $Na^+ + e^- \rightarrow Na$

ANSWER 31 $Cu^+ + e^- \rightarrow Cu$
(or $4Cu^+ + 4e^- \rightarrow 4Cu$)
$2O^{2-} \rightarrow O_2\uparrow + 4e^-$

ANSWER 41

$$\frac{107.868 \text{ g mole}^{-1}}{96{,}487 \text{ coulombs mole}^{-1}} = 0.0011180$$

ANSWER 51

$$\begin{array}{l} 2Fe^{3+} + 6e^- \rightarrow 2Fe \\ 6Cl^- \rightarrow 3Cl_2 + 6e^- \\ \hline 2Fe^{3+} + 6Cl^- + 6e^- \rightarrow 2Fe + 3Cl_2\uparrow + 6e^- \end{array}$$

ITEM 3

Some atoms can add more than one electron. (You will study in Review 6, and in Chapter 6 and later chapters of *Chemical Principles*, how to predict which atoms can add electrons and how many.) The negative charge on any ion formed by addition of electrons to an atom is equal to ________.

ITEM 13

If two electrodes (rods of carbon or platinum) are immersed in molten sodium chloride and an electric potential is applied across the rods, one electrode will have excess electrons and a ________ charge. It will attract ions with a ________ charge.

ITEM 23

Thus, the total reaction for this process is written as the addition

$$2Na^+ + 2e^- \rightarrow 2Na$$
$$\underline{2Cl^- \qquad \rightarrow Cl_2\uparrow + 2e^-}$$

____

ITEM 33

When silver bromide is electrolyzed, silver metal collects at the ________, and the gas Br_2 is given off at the ________.

ITEM 43

The charge of one mole of electrons constitutes the quantity of electricity known as the ________. It is equivalent to ________ coulombs.

ITEM 53

Although solid crystals of ionic salts will not conduct electricity, molten salts and water solutions of ionic salts are conductors of electricity. In both cases the liquid conducts a current because ________.

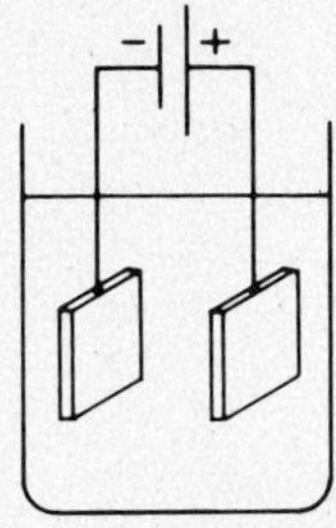

ANSWER 2 negative charge

ANSWER 12 are free to move through the liquid. They are no longer held in the crystal.

ANSWER 22 two
two Na atoms

ANSWER 32 positively
cation
negatively
anion

ANSWER 42 mole (or g-atom)

ANSWER 52

$$\frac{0.500 \text{ coulomb sec}^{-1} (3600 \text{ sec h}^{-1}) (1.00 \text{ h})}{96{,}500 \text{ coulombs } \mathscr{F}^{-1}}$$

$= 0.0187\ \mathscr{F}$

Fe: $0.0187\ \mathscr{F}\ (0.333 \text{ mole Fe } \mathscr{F}^{-1})$
$= 0.00623$ mole Fe
$0.00623 \text{ mole } (55.9 \text{ g mole}^{-1} \text{ Fe})$
$= 0.348$ g

Cl_2: $0.0187\ \mathscr{F}\ (0.500 \text{ mole } Cl_2\ \mathscr{F}^{-1})$
$= 0.00935$ mole Cl_2
$0.00935 \text{ mole } (71.0 \text{ g mole}^{-1}) = 0.664$ g

ITEM 4

For example, if an electron is added to a fluorine atom, F, the fluoride ion has a charge of 1− and is written F^-. If two electrons are added to an oxygen atom, O, the ion has a charge of ________ and is written ________.

ITEM 14

The negative electrode is called the *cathode.* The ions of molten sodium chloride, Na^+Cl^-, that will be attracted to the cathode are the ________ ions.

ITEM 24

The overall reaction is the decomposition of the salt NaCl into its elements. This process is called *electrolytic decomposition* or *electrolysis.* Write equations similar to the ones in Item 23 for the electrolysis of calcium chloride, $CaCl_2$ or $Ca^{2+}(Cl^-)_2$.

ITEM 34

The reduction half-reaction for the electrolysis of AgBr is ________, and the oxidation half-reaction is ________.

ITEM 44

An electrical current of 1 ampere (A) is equivalent to a current of 1 coulomb sec^{-1}. The number of faradays equivalent to a current of 1.00 A for 1.00 hour (h) is ________.

ITEM 54

What you have learned about the electrolysis of molten salts is valid for water solutions of salts, except that water itself may be oxidized or reduced. Therefore, it is more difficult to predict the reaction products. Nevertheless, there still must be a reduction reaction for every ________ reaction.

ANSWER 3 the number of electrons added

Cathode

ANSWER 13 negative
positive

Electrolysis

ANSWER 23

$2Na^+ + 2Cl^- + 2e^- \rightarrow 2Na + Cl_2\uparrow + 2e^-$
(or $2Na^+ + 2Cl^- \rightarrow 2Na + Cl_2\uparrow$)

ANSWER 33 cathode
anode

ANSWER 43 faraday
96,500

ANSWER 53 the *ions* are free to move to the electrodes

ITEM 5

Conversely, in some reactions electrons can be removed from an atom. If one electron is removed from an atom, the atom becomes an ion having a unit ________.

ITEM 15

Since positive ions are attracted to the ________ charged electrode, called the ________, positive ions are known as *cations.*

ITEM 25

When molten $CaCl_2$ is electrolyzed, the calcium metal collects at the ________ charged electrode, which is called the ________, and the chlorine is given off at the ________ electrode, which is called the ________.

ITEM 35

When the two reactions are added together so the $2e^-$ in each equation cancel, the equation for the total reaction is ________.

ITEM 45

The equation for the reduction of Ag^+ is ________.
The number of atoms of silver deposited by a current of 1.00 A for 1.00 h can be calculated as follows:
Number of $\mathscr{F}$ used is ________.
Number of Ag atoms deposited by 1 $\mathscr{F}$ is ________.
Number of Ag atoms deposited is ________.

ITEM 55

The electrolysis cell always consists of two electrodes immersed in the liquid. One has an excess of electrons pumped in by an electric generator and is called the ________. The other electrode is positively charged and is called the ________.

Cation

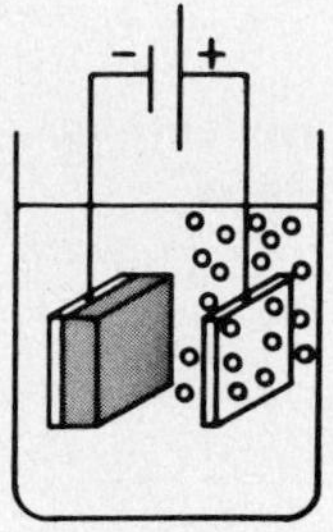

ANSWER 4 2−
O^{2-} or $O^{=}$

ANSWER 14 Na^+

ANSWER 24

$$\begin{array}{l} Ca^{2+} + 2e^- \rightarrow Ca \\ 2Cl^- \qquad\quad \rightarrow Cl_2\uparrow + 2e^- \\ \hline Ca^{2+} + 2Cl^- + 2e^- \rightarrow Ca + Cl_2\uparrow + 2e^- \end{array}$$

(or $Ca^{2+} + 2Cl^- \rightarrow Ca + Cl_2\uparrow$)

ANSWER 34 $Ag^+ + e^- \rightarrow Ag$
$2Br^- \rightarrow Br_2\uparrow + 2e^-$

ANSWER 44

$$\frac{1.00 \text{ coulomb sec}^{-1} (3600 \text{ sec h}^{-1})}{96{,}500 \text{ coulombs } \mathscr{F}^{-1}}$$

$$= 0.0373 \ \mathscr{F} \text{ h}^{-1}$$

ANSWER 54 oxidation

ITEM 6

Removal of one electron from a sodium atom, Na, forms an ion with the symbol ________, and if three electrons are removed from an aluminum atom, Al, the symbol for the ion is written

________.

ITEM 16

The positive electrode is called the *anode.* The ions of molten sodium chloride that are attracted to the anode are the ________ ions.

ITEM 26

When electrons are added to an atom or ion, the process is known as *reduction.* In the reaction

$$Ca^{2+} + 2Cl^- \rightarrow Ca + Cl_2 \uparrow$$

the ion that is reduced is ________.

ITEM 36

The equation for the half reaction $Ag^+ + e^- \rightarrow Ag$ indicates that $1e^-$ is added to 1 Ag^+ ion to produce 1 Ag atom. And it also indicates that when 1 *mole* of electrons is added to 1 mole of ________, 1 ________ of ________ is produced. (See Review 1 if this concept is unfamiliar.)

ITEM 46

The equation for the reduction of Li^+ to Li metal is ________. Calculate the number of grams of Li (at. wt: 6.94) deposited by a current of 0.500 A for 1.50 h.

ITEM 56

The positive ions of a salt are attracted to the ________ charged electrode and are called ________. The negative ions of a salt are attracted to the ________ charged electrode and are called ________.

Anode

Reduction

ANSWER 5 positive charge

ANSWER 15 negatively
cathode

ANSWER 25 negatively
cathode
positively charged
anode

ANSWER 35 $2Ag^+ + 2Br^- \rightarrow 2Ag + Br_2\uparrow$

ANSWER 45

$Ag^+ + 1e^- \rightarrow Ag$
from Item 44, 1.00 A for 1.00 h = 0.0373 $\mathscr{F}$
6.02×10^{23}
0.0373 $\mathscr{F}$ (6.02×10^{23} Ag atoms $\mathscr{F}^{-1}$)
= 2.25×10^{22} Ag atoms

ANSWER 55 cathode
anode

ITEM 7

Salts are compounds made up of ions. The negative ions can be any ions except hydroxide, OH^-, and the positive ions can be any except H^+. Since salts are electrically neutral, the total number of positive charges of the ions in a salt must be ________ the total number of negative charges.

ITEM 17

Negative ions are attracted to the ________ charged electrode, called the ________. Therefore, negative ions often are called *anions.*

ITEM 27

The equation for the reduction of a calcium ion is ________.

ITEM 37

The charge of 1 mole of electrons is given the name *faraday,* $\mathscr{F}$. Therefore, the number of electrons contained in the quantity of electricity known as the faraday is ________.

ITEM 47

The equation for the oxidation of chloride ion to chlorine gas is ________. The production of 1.00 mole of Cl_2 requires that 2.00 $\mathscr{F}$ of electrons be donated from Cl^- ions to the electric circuit. This is equivalent to ________ coulombs.

ITEM 57

Reduction is the process of ________, and oxidation is the process of ________.

Salt

ANSWER 6 Na^+
Al^{3+} (or Al^{+++})

ANSWER 16 Cl^-

Anion

ANSWER 26 Ca^{2+}

The faraday

ANSWER 36 Ag^+ ions
mole
Ag atoms

ANSWER 46

$Li^+ + e^- \rightarrow Li$

$$\frac{0.500 \text{ coulomb sec}^{-1} (3600 \text{ sec h}^{-1})(1.5 \text{ h})}{96{,}500 \text{ coulombs } \mathscr{F}^{-1}}$$

$$= 0.0280\ \mathscr{F}$$

$$0.0280\ \mathscr{F}\ (1.00 \text{ mole Li } \mathscr{F}^{-1})(6.94 \text{ g mole}^{-1} \text{ Li}) = 0.194 \text{ g}$$

ANSWER 56 negatively
cations
positively
anions

ITEM 8

Compounds that contain the OH^- ion usually are called *bases.* (A more general concept of a base will be discussed in Review 4.) Although the compound sodium hydroxide is made up of Na^+ and OH^- ions, NaOH is called a ________. Sodium chloride, NaCl, is a well-known example of a ________.

ITEM 18

When these names are applied to the ions of sodium chloride, the Cl^- ion is called the ________, and the Na^+ ion is called the ________ because Cl^- is attracted to the ________ and Na^+ is attracted to the ________.

ITEM 28

The opposite process, removal of electrons from an atom or ion, is known as *oxidation.* In the reaction

$$2Na^+ + 2Cl^- \rightarrow 2Na + Cl_2 \uparrow$$

the ________ ions are oxidized to ________.

ITEM 38

Electrical charge is usually measured in *coulombs.* The charge of 1 $\mathscr{F}$ is 96,500 coulombs, so the electrical charge of 1 mole of electrons is known to be ________.

ITEM 48

The equation for the reduction of Ca^{2+} to Ca is ________. The production of 3.00 moles of calcium metal requires ________ coulombs.

ITEM 58

The half-reaction for the oxidation of oxide ions, O^{2-}, to oxygen gas, O_2, is ________. The production of 1.00 mole of O_2 requires ________ $\mathscr{F}$ or ________ coulombs of electricity.

Oxidation

ANSWER 7 equal to

ANSWER 17 positively
anode

ANSWER 27 $Ca^{2+} + 2e^- \rightarrow Ca$

ANSWER 37 6.02×10^{23}

ANSWER 47

$2Cl^- \rightarrow Cl_2 \uparrow + 2e^-$
$2.00\ \mathscr{F}\ (96{,}500 \text{ coulombs } \mathscr{F}^{-1}) = 193{,}000$

ANSWER 57 addition of electrons
to atoms or ions
donation of electrons
from atoms or ions

ITEM 9

No compounds contain a free hydrogen ion, H^+ but compounds that dissolve in water and donate H^+ ions to water molecules are known as *acids.* Thus, hydrogen chloride, HCl, is an acid because ________.

ITEM 19

When a sodium ion is adjacent to the cathode, an electron jumps from the rod onto the ion. The addition of an electron to Na^+ converts it to ________, which becomes attached to the rod.

ITEM 29

The equation for the oxidation of Cl^- to Cl_2 is written ________.

ITEM 39

The charge, in coulombs, of one electron can be calculated from these values to be ________.

ITEM 49

The equation for the reduction of Al^{3+} to aluminum metal is ________. 0.500 $\mathscr{F}$ will produce ________ mole(s) of Al.

ITEM 59

The equation for the reduction half-reaction of a barium ion, Ba^{2+}, to barium metal is ________. Calculate the number of coulombs required to convert 0.357 g of $BaCl_2$ to Ba (at. wt: 137) and Cl_2.

ANSWER 8 base
salt

ANSWER 18 anion
cation
anode
cathode

ANSWER 28 Cl^-
Cl_2 (or Cl atoms that combine to form Cl_2 molecules)

ANSWER 38 96,500 coulombs

ANSWER 48

$Ca^{2+} + 2e^- \rightarrow Ca$

3.00 moles of Ca require 6.00 $\mathscr{F}$ of electrons

6.00 $\mathscr{F}$ (96,500 coulombs $\mathscr{F}^{-1}$)
= 579,000 coulombs

ANSWER 58

$2O^{2-} \rightarrow O_2 \uparrow + 4e^-$

4.00

4.00 $\mathscr{F}$ (96,500 coulombs $\mathscr{F}^{-1}$)
= 386,000 coulombs

ITEM 10

Acids and bases will be treated in Review 4. However, it is already necessary to know that when water solutions of acids and bases are mixed, the H^+ ions and the OH^- ions combine to form ________. The other ions in solution will form a ________ when the water is removed.

ITEM 20

When a chloride ion is adjacent to the anode, an electron from the ion is donated to the rod and conducted away. The removal of an electron from Cl^- converts it to ________.

ITEM 30

Each of these equations that include the number of electrons lost or gained by an atom is known as a *half-reaction* because each equation tells half the story. In a complete reaction there must be a reduction reaction for every ________ reaction.

ITEM 40

When 1 mole of AgBr is electrolyzed, ________ mole(s) of electrons is (are) required to reduce all Ag^+ to Ag. Given that the atomic weight of silver is 107.868, you know that the number of coulombs of electricity required to reduce 107.868 g of silver ions is ________.

ITEM 50

The electrolysis of ferric chloride, $FeCl_3$, produces iron metal and chlorine gas. The reduction reaction is Fe^{3+} + ________. The oxidation reaction is ________.

ITEM 60

Calculate the number of hours needed to maintain a current of 0.500 A in order to convert 5.00 g of Al_2O_3 to pure aluminum (at. wt: 27.0).

$Al \rightarrow Al^{3+} + 3e^-$

ANSWER 9

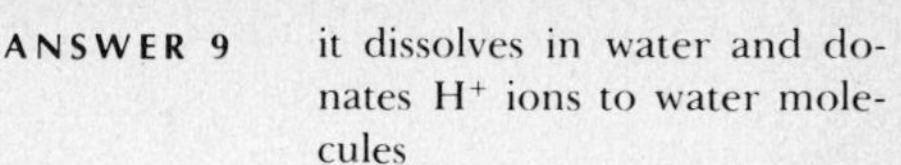

it dissolves in water and donates H^+ ions to water molecules

ANSWER 19 an atom of Na

ANSWER 29 $2Cl^- \rightarrow Cl_2\uparrow + 2e^-$

ANSWER 39

$$\frac{96{,}500 \text{ coulombs } \mathscr{F}^{-1}}{6.02 \times 10^{23} \text{ electrons } \mathscr{F}^{-1}} = 1.60 \times 10^{-19} \text{ coulombs electron}^{-1}$$

ANSWER 49 $Al^{3+} + 3e^- \rightarrow Al$

1 $\mathscr{F}$ will produce $\frac{1}{3}$ mole Al, so $\frac{1}{2}$ $\mathscr{F}$ will produce $\frac{1}{6}$ mole Al

ANSWER 59

$Ba^{2+} + 2e^- \rightarrow Ba$

0.357 g/208 g mole^{-1} $BaCl_2$ = 0.00172 mole $BaCl_2$

1 mole $BaCl_2$ gives 1 mole Ba

0.00172 mole (2 $\mathscr{F}$ mole^{-1} Ba)(96,500 coulombs $\mathscr{F}^{-1}$) = 332 coulombs

ANSWER 10 H_2O
salt

(*Item 11 is on page 90.*)

ANSWER 20 an atom of Cl

(*Item 21 is on page 90.*)

ANSWER 30 oxidation

(*Item 31 is on page 90.*)

ANSWER 40 1
96,500

(*Item 41 is on page 90.*)

ANSWER 50 $3e^- \rightarrow Fe$
$2Cl^- \rightarrow Cl_2\uparrow + 2e^-$

(*Item 51 is on page 90.*)

ANSWER 60

$Al \rightarrow Al^{3+} + 3e^-$
5.00 g/102 g $mole^{-1}$ Al_2O_3 = 0.0490 mole Al_2O_3
0.0490 mole Al_2O_3 gives 0.0980 mole Al
0.0980 mole (3 $\mathscr{F}$ $mole^{-1}$ Al)(96,500 coulombs $\mathscr{F}^{-1}$) = 28,400 coulombs
28,400 coulombs/0.5 coulomb sec^{-1} = 56,800 sec
56,800 sec/3600 sec h^{-1} = 15.8 h

REVIEW 4 STOICHIOMETRY

Perhaps the most practical of your chemical studies concerns stoichiometry. Whenever you synthesize a compound, measure the rate of a reaction, or carry out any one of dozens of kinds of chemical experiments, you will have to know how much of one reagent to mix with another and you will want to know how much product to expect. Stoichiometry is the application of the laws of chemical combination and is therefore an essential tool in determining the quantities of reactants used and products formed.

The first element of stoichiometry is the balanced chemical equation. Section 4–1 will show you how to interpret a chemical equation in terms of moles of reagents. Since most reactions are carried out in solution, Sections 4–2, 4–3, and 4–4 will describe solutions of ions and show you how to do calculations involving various kinds of concentrations of dissolved compounds. Sections 4–3 and 4–4 are specific for acids and bases: Section 4–3 will introduce the Brønsted theory of acid–base reactions and Section 4–4 will review the concentration terms used in stoichiometric calculations. You may be surprised to learn that heat is also a reagent in chemical reactions. Section 4–5 will demonstrate why and how this is so.

4–1 CHEMICAL EQUATIONS

Stoichiometry is based on chemical equations that are balanced with respect to mass, atoms, and charge. (Balancing of simple equations is discussed in Section 4–3 of *Chemical Principles.*) Once you have the equation for a reaction and understand what that equation means, the calculation of relative amounts of reactants and products is only a matter of using what you already know about gram-atoms, moles, and ideal gas laws. In this section, which reviews Sections 4–1 and 4–2 of *Chemical Principles*, you will see how these fit together in stoichiometric calculations.

If you can calculate easily and logically the answers to Items 25–29 you are already prepared for Section 4–2.

ITEM 1

If a sample contains 100 hydrogen molecules, the reaction of these according to the equation $2H_2 + O_2 \rightarrow 2H_2O$ requires __________ molecule(s) of oxygen and will form __________ molecule(s) of water.

ITEM 6

The coefficients in front of the formulas in a balanced chemical equation give the relative number of __________ or of __________ of each compound involved in the reaction.

ITEM 11

To produce 7 moles of CO_2 in the reaction $CH_4 + 2O_2 \rightarrow CO_2 + 2H_2O$, we would start with __________ moles of methane and __________ moles of O_2.

ITEM 16

To form 0.164 mole of ammonia by the reaction $N_2 + 3H_2 \rightarrow 2NH_3$ requires __________ of nitrogen and __________ of hydrogen.

ITEM 21

In the following items look up atomic weights in the periodic table in the inside front cover.

What weight of Na_2CO_3 would be needed to react with 0.124 mole of HCl according to the equation $Na_2CO_3 + 2HCl \rightarrow 2NaCl + H_2O + CO_2$?

Moles of Na_2CO_3 needed = __________.

Weight, in grams, of Na_2CO_3 needed = __________.

ITEM 26

Calculate the number of moles of HCl produced by the reaction of 1.5×10^{23} molecules of H_2 with 5.00 liters of Cl_2 at STP.

Moles of H_2 present = __________.

Moles of Cl_2 present = __________.

Moles of HCl produced = __________.

$H_2 + Cl_2 \rightarrow 2HCl$

ITEM 2

Although the equation for a chemical reaction shows how many individual molecules react to form product molecules, it is more convenient in doing chemical calculations to let each chemical symbol in the equation represent 1 __________ of the element and each formula represent 1 __________ of the compound.

ITEM 7

The combustion of methane, CH_4, leads to the formation of water and carbon dioxide. The balanced equation for this reaction is $CH_4 +$ __________ $O_2 \rightarrow$ __________ $CO_2 +$ __________.

ITEM 12

An air sample containing 0.240 mole of oxygen is capable of oxidizing __________ mole(s) of methane to CO_2 and H_2O.

ITEM 17

To use a balanced equation for deducing how much product can be formed in a given reaction, we must first know the number of __________ of the reactants. Then the equation will show how many __________ of the products are formed.

ITEM 22

Using the same type of steps as in the previous items, calculate the weights of Zn and HCl needed to produce 0.763 mole of hydrogen, according to the reaction $Zn + 2HCl \rightarrow ZnCl_2 + H_2$.

ITEM 27

How many grams of $AgNO_3$ are required to react completely with 0.624 mole of $BaCl_2$ by the reaction $2AgNO_3 + BaCl_2 \rightarrow 2AgCl + Ba(NO_3)_2$?

ANSWER 1

50
100

ANSWER 6

molecules
moles
(On the molecular level, the qualifications discussed in Section 4-2 of *Chemical Principles* must be observed.)

$CH_4 + 2O_2 \rightarrow CO_2 + 2H_2O$

ANSWER 11

7
14

ANSWER 16

$\frac{1}{2}(0.164 \text{ mole}) = 0.082 \text{ mole}$
$\frac{3}{2}(0.164 \text{ mole}) = 0.246 \text{ mole}$

ANSWER 21

$\frac{1}{2}(0.124 \text{ mole}) = 0.0620 \text{ mole}$
$0.0620 \text{ mole } (106 \text{ g mole}^{-1}) = 6.57 \text{ g}$

ANSWER 26

$$\frac{1.5 \times 10^{23} \text{ molecules}}{6.0 \times 10^{23} \text{ molecules mole}^{-1}} = 0.25 \text{ mole}$$

$$\frac{5.00 \text{ liters}}{22.4 \text{ liters mole}^{-1}} = 0.223 \text{ mole}$$

$2(0.223 \text{ mole}) = 0.446 \text{ mole}$

ITEM 3

Thus, the equation $2H_2 + O_2 \rightarrow 2H_2O$ states that 2 moles of hydrogen react with ________ and form ________.

ITEM 8

This equation (Answer 7) shows that the oxidation of one molecule of methane requires ________ of oxygen and that this reaction produces ________ and ________.

ITEM 13

To form 0.7 mole of H_2S by the laboratory preparation method represented by the equation $FeS + H_2SO_4 \rightarrow FeSO_4 + H_2S$, we would start with ________ mole(s) of FeS and ________ mole(s) of H_2SO_4.

ITEM 18

In the reaction represented by the equation $3Fe + 4H_2O \rightarrow Fe_3O_4 + 4H_2$, how many moles, and what weight, of hydrogen will be produced by the reaction of 4.76 moles of iron? Moles of H_2 formed = ________. Weight, in grams, of H_2 formed (at. wt H: 1.01) = ________.

ITEM 23

If 1.00 g, or ________ mole(s), of Zn is to react according to the equation $Zn + 2HCl \rightarrow ZnCl_2 + H_2$, at least ________ mole(s) or ________ g of HCl must be present.

ITEM 28

Oxygen reacts with ammonia according to the equation $4NH_3 + 7O_2 \rightarrow 4NO_2 + 6H_2O$. If 0.10 ml (i.e., 1.0×10^{-4} liter) of O_2 at STP is allowed to react in 1.04 g of NH_3 how many *molecules* of NO_2 are produced?

ANSWER 2 g-atom
mole

ANSWER 7 2
1
$2H_2O$
($CH_4 + 2O_2 \rightarrow CO_2 + 2H_2O$)

ANSWER 12 0.120

ANSWER 17 moles
moles

ANSWER 22 Moles of Zn needed = 0.763
Weight of Zn needed =
0.763 mole (65.4 g mole^{-1}) =
49.9 g
Moles of HCl needed =
$2 \times 0.763 = 1.526$
Weight of HCl needed =
1.526 moles (36.5 g
mole^{-1}) = 55.7 g

ANSWER 27 1.248 moles of $AgNO_3$ required
1.248 moles (170 g mole^{-1})
= 212 g

ITEM 4

The balanced equation for the reaction of hydrogen with acetylene to give ethane is $2H_2 + C_2H_2 \rightarrow C_2H_6$. In terms of molecules, this equation indicates that __________.

ITEM 9

To deal with samples more nearly the size used in the laboratory, we would say that for every mole of methane oxidized __________ mole(s) of oxygen are required and there are produced __________ and __________.

ITEM 14

The coefficients of the formulas of the compounds in a balanced equation indicate __________.

ITEM 19

The weight and volume at STP of chlorine needed for complete reaction with 10 g of H_2, according to the equation $H_2 + Cl_2 \rightarrow 2HCl$, can be calculated as follows:

$$\text{Moles of } H_2 = \frac{10 \text{ g}}{2.0 \text{ g mole}^{-1}} = \_\_\_\_\_\_\_\_\_\_.$$

Moles of Cl_2 needed = __________.
Weight of Cl_2 needed (at. wt Cl: 35.5) = __________.
Volume, in liters at STP, of Cl_2 needed = __________.

ITEM 24

Suppose that 2 moles of Zn are added to a solution containing 6 moles of HCl. Then __________ mole(s) of $ZnCl_2$ are produced and __________ mole(s) of __________ remain unreacted.

ITEM 29

How many grams of water and carbon dioxide are produced by the combustion of 1.00 liter at STP of methane, CH_4, in excess oxygen?

$Zn + 2HCl \rightarrow ZnCl_2 + H_2$

ANSWER 3 1 mole of oxygen
2 moles of water

ANSWER 8 two molecules
one molecule of CO_2
two molecules of H_2O

ANSWER 13 0.7
0.7

ANSWER 18 $\frac{4}{3}(4.76 \text{ moles}) = 6.35 \text{ moles}$
$6.35 \text{ moles } (2.02 \text{ g mole}^{-1}) = 12.8 \text{ g}$

ANSWER 23 $1.00 \text{ g}/(65.4 \text{ g mole}^{-1}) = 0.0153 \text{ mole}$
$2(0.0153 \text{ mole}) = 0.0306 \text{ mole}$
$0.0306 \text{ mole } (36.5 \text{ g mole}^{-1}) = 1.12 \text{ g}$

ANSWER 28

$$\text{moles } NH_3 = \frac{1.04 \text{ g}}{17.0 \text{ g mole}^{-1}} = 0.0612$$

$$\text{moles } O_2 = \frac{1.0 \times 10^{-4} \text{ liter}}{22.4 \text{ liters mole}^{-1}} = 4.5 \times 10^{-6}$$

Moles NH_3 needed to react with $O_2 = \frac{4}{7}(4.5 \times 10^{-6} \text{ mole}) = 2.6 \times 10^{-6}$. Thus, NH_3 is in excess. All O_2 will be used, and molecules of NO_2 produced $= \frac{4}{7}(4.5 \times 10^{-6} \text{ mole})(6.02 \times 10^{23} \text{ molecules mole}^{-1}) = 1.5 \times 10^{18}$ molecules.

ITEM 5

In terms of moles, the equation $2H_2 + C_2H_2 \rightarrow C_2H_6$ suggests that ________.

ITEM 10

If a sample containing 10 moles of methane were burned, ________ moles of oxygen would be required; ________ of CO_2 and ________ of H_2O would be produced.

ITEM 15

The formation of x moles of ammonia according to the reaction $N_2 + 3H_2 \rightarrow 2NH_3$ requires the use of ________ moles of nitrogen and ________ moles of hydrogen.

ITEM 20

The weight and volume at STP of N_2 needed to produce 150 g of NH_3 according to the equation $N_2 + 3H_2 \rightarrow 2NH_3$ is calculated as follows:

Moles of NH_3 to be produced $= 150 \text{ g}/17.0 \text{ g mole}^{-1}$
$= 8.82$ moles.

Moles of N_2 needed = ________.

Weight, in grams, of N_2 needed (at. wt N: 14.0) = ________.

Volume, in liters, of N_2 needed = ________.

ITEM 25

A 0.10-mole sample of Zn is allowed to react with 9.1 g of HCl. The reagent that is present in excess is ________. The reaction will produce ________ mole(s) of $ZnCl_2$ and ________ mole(s) of H_2. Left unreacted is ________ g of ________.

$CH_4 + 2O_2 \rightarrow CO_2 + 2H_2O$

$Zn + 2HCl \rightarrow ZnCl_2 + H_2$

ANSWER 4 two molecules of hydrogen combine with one molecule of acetylene, C_2H_2, to produce one molecule of ethane, C_2H_6

ANSWER 9 2
1 mole of CO_2
2 moles of H_2O

ANSWER 14 the relative numbers of moles (or molecules) of the reactants used and products formed

ANSWER 19 5.0
5.0
5.0 moles (71 g mole^{-1})
$= 360$ g
5.0 moles (22.4 liters mole^{-1})
$= 110$ liters

ANSWER 24 2
2
HCl

ANSWER 29

$CH_4 + 2O_2 \rightarrow CO_2 + 2H_2O$
moles methane
$= 1.00$ liter/22.4 liters mole^{-1} $= 0.0446$
moles H_2O produced
$= 2 \times 0.0446 = 0.0892$
weight H_2O produced
$= 0.0892$ mole $\times$ 18.0 g mole^{-1} $= 1.61$ g
moles CO_2 produced $= 0.0446$
weight CO_2 produced
$= 0.0446$ mole $\times$ 44.0 g mole^{-1} $= 1.96$ g

(*Turn to page 127.*)

ANSWER 5 2 moles of hydrogen combine with 1 mole of acetylene to produce 1 mole of ethane

(*Item 6 is on page 114.*)

ANSWER 10 20
10 moles
20 moles

(*Item 11 is on page 114.*)

ANSWER 15 $x/2$
$3x/2$

(*Item 16 is on page 114.*)

ANSWER 20 $\frac{1}{2} \times 8.82 = 4.41$
4.41 moles (28.0 g mole^{-1}) = 123 g
4.41 moles (22.4 liters mole^{-1}) = 98.8 liters

(*Item 21 is on page 114.*)

ANSWER 25 HCl (9.1 g/36.5 g mole^{-1} = 0.25 mole is present; only 0.20 mole is needed)
0.10
0.10
36.5 g mole^{-1} (0.25 − 0.20) mole = 1.8 g
HCl

(*Item 26 is on page 114.*)

4–2 IONS IN SOLUTION

Because it is often convenient to carry out chemical reactions in liquid solution, you must be able to calculate weight relations between reactants and products, one or more of which may be in solution. Quantities of solutions are more easily measured by volume than by weight. Therefore, it is necessary to relate the volume and concentration of a solution to the number of moles of dissolved compound. This section will supplement Section 4–4 of *Chemical Principles* in describing solutions of ions and showing how to do the necessary concentration and stoichiometry calculations.

Items 54–60 represent the different types of problems encountered in solution stoichiometry. If you can work these correctly, you are probably ready for Section 4–3.

ITEM 1

Combining two or more substances produces a mixture that can be *homogeneous* or *heterogeneous*. A solution is a mixture that has the same composition throughout and therefore can be described as

__________.

ITEM 11

The amount of solute in a unit volume of solution is called the *concentration*. The concentration of HCl in a solution containing 6.5 g HCl per liter of solution is __________ g $liter^{-1}$ or __________ g ml^{-1}.

ITEM 21

The number of moles (n) of solute in any given amount of a solution is obtained by multiplying V, the volume in liters, by the number of moles per liter (i.e., the molarity, M). Expressed in symbols, $n =$

__________.

ITEM 31

When NaCl is dissolved in water, each crystal, made of Na^+ and Cl^- ions, is broken, or dissociated, so the ions go into solution and are surrounded by

__________ molecules.

ITEM 41

Solution stoichiometry involves the weight relations between reactants and products for reactions that occur in solution. For a stoichiometric calculation it is necessary to have a balanced __________.

ITEM 51

In 100 ml of 0.05-molar $CuSO_4$ there is __________ mole of Cu^{2+}.

Solution

Concentration

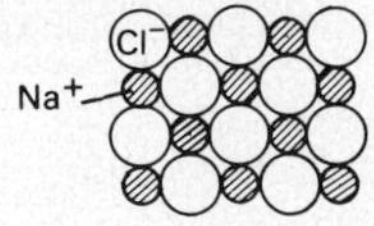

ITEM 2

If a liter of water to which a few crystals of salt are added is stirred well, the salt may dissolve completely and form a mixture that is homogeneous. This mixture can be called a ________.

ITEM 12

For many problems, the most useful concentration is the number of moles of solute per liter of solution. A solution containing 1.00 mole of solute in 500 ml of solution has a concentration of ________ mole(s) per liter.

ITEM 22

A sample of a 0.10-molar solution having a volume of 0.50 liter contains ________ mole(s) of solute.

ITEM 32

We can represent this process as

$$\text{NaCl(crystal)} \xrightarrow{H_2O} Na^+(aq) + Cl^-(aq),$$

in which (*aq*) is the abbreviation for "aqueous." $Na^+(aq)$ then represents a *hydrated* sodium ion, an ion that is surrounded by ________ molecules.

ITEM 42

A chemical equation is an attempt to show the amounts of reagents involved in a reaction. If a reaction occurs in an aqueous solution, it is likely that the actual reactants are not molecules but ________ that can be formed by the solution process.

ITEM 52

If excess Zn is added to 100 ml of 0.05-molar $CuSO_4$, how many gram-atoms of Cu will be formed?

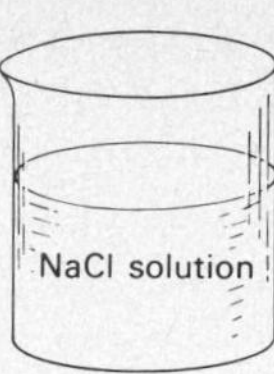

Hydrated

ANSWER 1 homogeneous

ANSWER 11 6.5

$$\frac{6.5\ \text{g liter}^{-1}}{1000\ \text{ml liter}^{-1}} = 0.0065$$

ANSWER 21 *MV* (or *VM*)

ANSWER 31 water (or solvent)

ANSWER 41 equation

ANSWER 51

$$\frac{100\ \text{ml}}{1000\ \text{ml liter}^{-1}} \times 0.05\ \text{mole liter}^{-1} = 0.005$$

ITEM 3

However, if a kilogram of salt is added to a liter of water, only a part of the salt will dissolve. A visible, physical dividing surface exists between the solid salt and the liquid; the entire mixture is ________.

ITEM 13

A solution containing 6.50 g of $CaCl_2$ (mol wt: 111) in 500 ml of solution has a concentration of ________ mole(s) per liter.

ITEM 23

To prepare 500 ml of a 0.30-molar solution, enough solvent to make 500 ml is added to ________ mole(s) of solute.

ITEM 33

Many other substances form ionic solutions in water. Some, like NaCl(crystal), are already composed of ________ before they are dissolved; others, like HCl(gas), are composed of neutral molecules before they are added to water.

ITEM 43

For instance, if $AgNO_3$ and KCl are dissolved in water, the first processes that occur in the solution are

$$AgNO_3(\text{crystal}) \xrightarrow{H_2O} \underline{\qquad\qquad}.$$

$$KCl(\text{crystal}) \xrightarrow{H_2O} \underline{\qquad\qquad}.$$

ITEM 53

If 0.13 g Zn (i.e., ________ g-atom) is added to 100 ml of 0.050-molar $CuSO_4$, how many gram-atoms (and how many grams) of Cu will precipitate?

ANSWER 2 solution

ANSWER 12 1.00 mole/0.500 liter
$= 2.00$ moles liter^{-1}

ANSWER 22

$n = MV$

0.10 mole liter^{-1} × 0.50 liter = 0.050 mole

ANSWER 32 water

ANSWER 42 (hydrated) ions

ANSWER 52

$$\frac{100 \text{ ml}}{1000 \text{ ml liter}^{-1}} \times 0.05 \text{ mole liter}^{-1}$$

$$= 0.005 \text{ mole} = 0.005 \text{ g-atom}$$

ITEM 4

The entire mixture of Item 3 is composed of solid salt and a liquid containing salt dissolved in water. The liquid part, however, is __________ and is correctly called a __________.

ITEM 14

The concentration in moles of solute per liter of solution is called the *molarity* (M) of the solution. What is the molarity of a solution containing 0.45 mole of solute in 3.0 liters of solution?

ITEM 24

If you want to add 0.30 mole of NaCl to a reaction mixture by adding 0.50-molar NaCl solution, what volume of solution must be added? __________. (It is convenient to recall that $n = MV$, which you can remember by the units involved, and to rearrange this general expression to find the quantity you want.)

ITEM 34

For ionic solids, such as NaCl, we use the term *gram formula weight* to represent the mass of 6.02×10^{23} Cl^- ions and __________ Na^+ ions. Thus, gram formula weight is sometimes used instead of __________ when simple molecules do not exist in the crystal.

ITEM 44

However, the reaction of interest between $AgNO_3$ and KCl is the formation of AgCl, a compound which does not dissolve appreciably in water. The equation for this reaction is written as

$$Ag^+(aq) + Cl^-(aq) \rightarrow AgCl(s)$$

in which AgCl(s) (for *solid*) is written to indicate that, for the purpose of stoichiometric calculations, the compound AgCl is *not* __________.

ITEM 54

When Cl_2 is added to a solution of $Fe(NO_3)_2$, the reaction that occurs is represented by the equation $2Fe^{2+} + Cl_2 \rightarrow 2Fe^{3+} + 2Cl^-$. What is the minimum weight of Cl_2 that must be added to 200 ml of 0.750-molar $Fe(NO_3)_2$ solution to oxidize completely the Fe^{2+} to Fe^{3+}?

Molarity

ANSWER 3 heterogeneous

ANSWER 13

$$\frac{6.50 \text{ g}}{111 \text{ g mole}^{-1}} = 0.0586 \text{ mole}$$

$$\frac{0.0586 \text{ mole}}{500 \text{ ml}} \left(\frac{1000 \text{ ml}}{\text{liter}}\right)$$

$$= 0.117 \text{ mole liter}^{-1}$$

ANSWER 23

0.30 mole liter^{-1} × 0.50 liter = 0.15 mole
(i.e., $n = MV = 0.30 \times 0.50 = 0.15$ mole)

ANSWER 33 ions

ANSWER 43 $Ag^+(aq) + NO_3^-(aq)$
$K^+(aq) + Cl^-(aq)$

ANSWER 53

$$\frac{0.13 \text{ g}}{65.4 \text{ g g-atom}^{-1}} = 0.0020 \text{ g-atom Zn.}$$

(0.100 liter)(0.050 mole liter^{-1})
= 0.005 mole Cu^{2+}
0.0020 g-atom Cu will precipitate since Cu^{2+} is present in excess.
0.0020 g-atom (63.5 g g-atom^{-1}) = 0.13 g Cu.

ITEM 5

In general, that component of a solution present in the largest amount is known as the *solvent*. The solvent can be a gas, a liquid, or a solid, but this section will treat only solutions in which the solvent is a liquid and will emphasize *aqueous* solutions, in which the solvent is __________.

ITEM 15

A 0.26-molar H_2SO_4 solution, often written as 0.26*M* H_2SO_4, contains __________ mole(s) of H_2SO_4 per liter of solution.

ITEM 25

If a solution is diluted (i.e., if pure solvent is added to it), the volume of solution changes, but the amount of solute __________.

ITEM 35

However, we have already stated that the term __________ represents 6.02×10^{23} *anything*. Thus, for simplicity we generally talk of a mole of NaCl and thereby mean __________.

ITEM 45

The complete ionic equation for the reaction of $AgNO_3$ and KCl in solution might be written as

$$Ag^{+}(aq) + NO_3^{-}(aq) + K^{+}(aq) + Cl^{-}(aq) \rightarrow \_\_\_\_\_\_.$$

ITEM 55

What is the molarity of Al^{3+} in a solution made by dissolving 12.7 g of $Al_2(SO_4)_3$ in sufficient water to make 100 ml of solution?

Solvent

ANSWER 4 homogeneous
solution

ANSWER 14 $\frac{0.45 \text{ mole}}{3.0 \text{ liters}} = 0.15 \text{ mole liter}^{-1}$

ANSWER 24

$$V = \frac{n}{M} = \frac{0.30 \text{ mole}}{0.50 \text{ mole liter}^{-1}} = 0.60 \text{ liter}$$

ANSWER 34 6.02×10^{23}
gram molecular weight

ANSWER 44 soluble in water
(or dissociated)

ANSWER 54

Moles Fe^{2+} present
$= 0.200 \text{ liter} \times 0.750 \text{ mole liter}^{-1}$
$= 0.150$ mole.
Moles Cl_2 needed $= \frac{1}{2} \times 0.150$ mole
$= 0.0750$ mole.
Weight $Cl_2 = 0.0750 \text{ mole} \times 70.9 \text{ g mole}^{-1}$
$= 5.32$ g.

ITEM 6

A component of a solution present in lesser amount than the major component is called a *solute.* It can be added to the solvent as a gas, a __________, or a __________.

ITEM 16

A solution is made by dissolving 5.0 g of KOH (mol wt: 56) in enough water to make 0.50 liter of solution. What is the molarity of this solution?

ITEM 26

In 0.750 liter of a 0.500-molar solution, there is (are) __________ mole(s) of solute. If sufficient solvent is added to make the volume 1.00 liter, there will be __________ mole(s) of solute. In the new solution the concentration is __________ mole(s) per liter.

ITEM 36

The ionic solid K_2SO_4 dissolves in water by dissociating into ions as

$$K_2SO_4(\text{crystal}) \xrightarrow{H_2O} 2K^+(aq) + SO_4^{2-}(aq)$$

Assuming that K_2SO_4 dissociates completely in water, each gram formula weight (or mole) of K_2SO_4 produces __________ *gram ions* (or moles) of K^+ and __________ gram ion (or mole) of SO_4^{2-}.

ITEM 46

However, the NO_3^- and K^+ ions are unchanged during the reaction, and it is customary to omit the unchanged ions. The designations (*s*) and (*aq*) are also sometimes omitted, although you should understand that ions in water solution are always hydrated. Thus, the simplest equation for the reaction that occurs in a solution of $AgNO_3$ and KCl is __________.

ITEM 56

When 25.0 ml of a 0.215-molar solution of K_2SO_4 are diluted to 125 ml, what is the molarity of the resulting solution?

Solute

ANSWER 5 water

ANSWER 15 0.26

ANSWER 25 does not change

ANSWER 35 mole
6.02×10^{23} Na^+ ions
and 6.02×10^{23} Cl^- ions

ANSWER 45 $AgCl(s) + NO_3^-(aq) + K^+(aq)$

ANSWER 55

Molecular weight of $Al_2(SO_4)_3$ is 342.
Thus, 12.7 g = 12.7/342 = 0.0371 mole.

$$\text{Molarity of } Al^{3+} = 2 \times \frac{0.0371 \text{ mole}}{0.100 \text{ liter}} = 0.742.$$

ITEM 7

Thus, a solution is a homogeneous mixture of a solvent and one or more ________.

ITEM 17

A solution is made by dissolving 6.4 g of bromine in enough carbon tetrachloride to make 75 ml of solution. What is the molarity of Br_2 in this solution? (Use the periodic table for atomic weights.)

ITEM 27

In a volume V_1 liters of a solution whose molarity is M_1, there are ________ moles of solute. After dilution to a volume V_2 the molarity is M_2. Since the number of moles of solute has not changed, $V_1M_1 =$ ________. This is a convenient expression for doing dilution problems.

ITEM 37

Concentrations of ions in solution are often expressed in terms of molarity. Thus, a 0.62-molar solution of K_2SO_4 contains (assuming complete dissociation) ________ mole(s) of K^+ and ________ mole(s) of SO_4^{2-} per liter.

ITEM 47

In 250 ml of 0.10-molar $AgNO_3$ there is (are) ________ mole(s) of Ag^+. In 400 ml of 0.10-molar NaCl there is (are) ________ mole(s) of Cl^-.

ITEM 57

What is the molarity of NO_3^- in a solution containing 0.600 mole of $Al(NO_3)_3$ in 500 ml of solution?

ANSWER 6 liquid
solid

ANSWER 16

$$\frac{(5.0/56)\text{ mole}}{0.50\text{ liter}} = 0.18\text{ mole liter}^{-1}$$

ANSWER 26

$0.500\text{ mole liter}^{-1}\ (0.750\text{ liter}) = 0.375\text{ mole}$

0.375

0.375

ANSWER 36 2
1

ANSWER 46 $Ag^+ + Cl^- \rightarrow AgCl$

ANSWER 56

$$M = \frac{0.215M \times 0.0250\text{ liter}}{0.125\text{ liter}} = 0.0430\text{ molar.}$$

ITEM 8

In an aqueous salt solution, water is the __________ and a salt is the __________.

ITEM 18

A solution of HNO_3 and water contains 70.0% HNO_3 by weight (i.e., 70.0% of the weight of the solution is HNO_3 and 30.0% is water). 1.00 ml of this solution weighs 1.42 g. In 1 liter of this solution, which weighs __________ g, the weight of HNO_3 is __________ g. The molarity of HNO_3 in this solution is therefore __________.

ITEM 28

In 50 ml of a 0.50-molar solution there is (are) __________ mole(s) of solute. After dilution to 75 ml there is (are) still __________ mole(s) of solute, but now the concentration of the solution is calculated to be __________ molar.

ITEM 38

For many purposes it is unimportant to know whether in an ionic solution there is *complete* dissociation of molecules into ions. Therefore, assume in the following items complete dissociation, except where otherwise indicated.

300 ml of a 0.500-molar solution of $Fe_2(SO_4)_3$ contains __________ mole(s) of Fe^{3+} and __________ mole(s) of SO_4^{2-}.

ITEM 48

If 250 ml of 0.10-molar $AgNO_3$ and 400 ml of 0.10-molar NaCl are mixed, __________ mole(s) of AgCl will form.

ITEM 58

450 ml of 0.10-molar HNO_3 are needed for a reaction. The 450 ml must be made with 70% HNO_3 solution having a density of 1.4 g ml^{-1}. What volume of 70% HNO_3 must be used?

$M_1V_1 = M_2V_2$

ANSWER 7 solutes

ANSWER 17 $\frac{(6.4/160)\text{ mole}}{75\text{ ml}}\left(\frac{1000\text{ ml}}{\text{liter}}\right)$

$= 0.53\text{ mole liter}^{-1}$

ANSWER 27 V_1M_1
V_2M_2

ANSWER 37 $2 \times 0.62 = 1.24$
0.62

ANSWER 47

$$Ag^+\ n = \frac{250\text{ ml}}{1000\text{ ml liter}^{-1}} \times 0.10\text{ mole liter}^{-1}$$

$$= 0.025\text{ mole}$$

$$Cl^-\ n = \frac{400\text{ ml}}{1000\text{ ml liter}^{-1}} \times 0.10\text{ mole liter}^{-1}$$

$$= 0.040\text{ mole}$$

ANSWER 57 $\frac{3(0.600\text{ mole})}{0.500\text{ liter}} = 3.60\text{ molar.}$

ITEM 9

Volumes of solutions used in chemical studies are usually measured in liters, milliliters (ml), or cubic centimeters (cm^3). One ml is equal to one thousandth of a __________, and 1 ml is almost equal to 1 cm^3.

ITEM 19

100 ml of an aqueous solution of acetic acid, CH_3COOH, weigh 100 g and contain 1.5% by weight of the acid. What is the molarity of the acetic acid?

ITEM 29

A 1.5-molar solution can be obtained by diluting 500 ml of a 2.0-molar solution to a total volume of __________ liter(s).

ITEM 39

A solution made by dissolving 0.15 mole of NaCl and 0.60 mole of KCl in enough water to make 1.00 liter has a Cl^- concentration of __________ mole(s) per liter.

ITEM 49

If Zn metal is added to an aqueous solution of $Cu(NO_3)_2$, copper metal appears. That is, copper *precipitates* from solution. The net equation for the reaction is written $Zn + Cu^{2+} \rightarrow Zn^{2+} + Cu$, thus indicating that for each atom of Zn that goes into solution __________ atom(s) of Cu precipitate(s).

ITEM 59

What volume of a 0.318-molar Na_2S solution is required to precipitate the Sb^{3+} in 25.0 ml of a 0.124-molar solution of $SbCl_3$? The equation for the reaction is $2Sb^{3+} + 3S^{2-} \rightarrow Sb_2S_3$.

ANSWER 8 solvent
solute

ANSWER 18

$$1.42 \text{ g ml}^{-1} \left(\frac{1000 \text{ ml}}{\text{liter}}\right)$$

$$= 1420 \text{ g liter}^{-1}$$

$$0.700 \times 1420 \text{ g} = 994 \text{ g}$$

$$\text{Molarity} = \frac{994 \text{ g liter}^{-1}}{63.0 \text{ g mole}^{-1}}$$

$$= 15.8 \text{ moles liter}^{-1}$$

ANSWER 28

$(0.50 \text{ mole liter}^{-1})\ 0.050 \text{ liter} = 0.025 \text{ mole}$
0.025

$$M_2 = \frac{M_1V_1}{V_2} = \frac{(0.50 \text{ mole liter}^{-1})(0.050 \text{ liter})}{0.075 \text{ liter}}$$

$$= 0.33 \text{ mole liter}^{-1}$$

(or $M = 0.025$ mole/0.075 liter = 0.33)

ANSWER 38

$$\left(\frac{300}{1000}\right) \text{liter } (0.500 \text{ mole liter}^{-1})(2) = 0.300 \text{ mole}$$

$$\left(\frac{300}{1000}\right) \text{liter } (0.500 \text{ mole liter}^{-1})(3) = 0.450 \text{ mole}$$

ANSWER 48 0.025 (since initial moles of $Ag^+ = 0.025$ and moles of $Cl^- = 0.040$)

ANSWER 58

Weight HNO_3 per liter of 70% solution
$= 0.70 \times 1.4 \text{ g ml}^{-1} \times 1000 \text{ ml liter}^{-1}$
$= 980 \text{ g liter}^{-1}$.

$$\text{Molarity of 70\% } HNO_3 = \frac{980 \text{ g liter}^{-1}}{63.0 \text{ g mole}^{-1}}$$

$$= 15.6 \text{ moles liter}^{-1}.$$

Volume of 70% HNO_3 needed

$$= \frac{0.45 \text{ liter} \times 0.10M}{15.6M}$$

$$= 0.0029 \text{ liter} = 2.9 \text{ ml.}$$

ITEM 10

A solution is made by dissolving 1.00 g of NaCl in sufficient water to make 10.0 ml of solution. Each ml of this solution contains __________ g of NaCl.

ITEM 20

The molarity of a solution is the number of __________ of solute per liter of __________.

ITEM 30

A solution of H_2SO_4 and H_2O that is 86% H_2SO_4 by weight has a density of 1.78 g ml^{-1} (i.e., 1.00 ml weighs 1.78 g). What volume of this 86% H_2SO_4 solution must be used to make 500 ml of 6.0-molar H_2SO_4?

Weight of H_2SO_4 per liter of solution = __________.

Molarity of 86% H_2SO_4 solution = __________.

Volume of 86% H_2SO_4 solution needed = __________.

ITEM 40

To what final volume should you dilute 250 ml of 0.15-molar $CaCl_2$ if the final concentration of Cl^- must be 0.10 molar?

ITEM 50

Each mole of Zn that goes into solution produces __________ mole(s) of Cu.

ITEM 60

How many moles of Ag_3PO_4 will be formed when 36.9 ml of 0.256-molar $AgNO_3$ solution are added to 27.4 ml of 0.188-molar Na_3PO_4 solution? The reaction is $3Ag^+ + PO_4^{3-} \rightarrow Ag_3PO_4$.

ANSWER 9 liter

ANSWER 19

wt liter^{-1} = 0.015(1000 g liter^{-1}) = 15 g liter^{-1}

$$\text{Molarity} = \frac{15 \text{ g liter}^{-1}}{60 \text{ g mole}^{-1}} = 0.25 \text{ mole liter}^{-1}$$

ANSWER 29

$$V_2 = \frac{M_1V_1}{M_2} = \frac{2.0M \times 0.500 \text{ liter}}{1.5M} = 0.67 \text{ liter}$$

ANSWER 39

0.15 mole liter^{-1} + 0.60 mole liter^{-1} = 0.75

ANSWER 49 one

ANSWER 59

Moles of Sb^{3+} = 0.0250 liter × 0.124 mole liter^{-1}
= 0.00310 mole.
And moles S^{2-} required = 0.00310($\frac{3}{2}$)
= 0.00465.
Thus, the volume of solution required is

$$\frac{0.00465 \text{ mole}}{0.318 \text{ mole liter}^{-1}} = 0.0146 \text{ liter} = 14.6 \text{ ml.}$$

ANSWER 10 0.100

(*Item 11 is on page 128.*)

ANSWER 20 moles
solution

(*Item 21 is on page 128.*)

ANSWER 30

wt H_2SO_4 per liter

$$= 0.86\ (1.78\text{ g ml}^{-1})\left(\frac{1000\text{ ml}}{\text{liter}}\right)$$

$$= 1530\text{ g liter}^{-1}$$

$$M = \frac{1530\text{ g liter}^{-1}}{98\text{ g mole}^{-1}} = 15.6\text{ moles liter}^{-1}$$

$$V = \frac{0.50\text{ liter} \times 6.0M}{15.6M} = 0.19\text{ liter}$$

(*Item 31 is on page 128.*)

ANSWER 40

$$V_2 = \frac{M_1V_1}{M_2} = \frac{(0.30M\text{ Cl}^-)\ 250\text{ ml}}{0.10M\text{ Cl}^-} = 750\text{ ml}$$

(*Item 41 is on page 128.*)

ANSWER 50 1

(*Item 51 is on page 128.*)

ANSWER 60

Moles Ag^+ present
$= 0.256$ mole liter^{-1} (0.0369 liter)
$= 0.00945$ mole.
Moles PO_4^{3-} present
$= 0.188$ mole liter^{-1} (0.0274 liter)
$= 0.00515$ mole.
Thus, PO_4^{3-} is present in excess of the stoichiometric amount, and (0.00945 mole)/3 = 0.00315 mole Ag_3PO_4 will be formed.

Although the stoichiometry of acid and base reactions is fundamentally the same as the stoichiometry of ionic salt reactions examined in this section, we will interrupt briefly our discussion to study acid and base reactions. Then we will return to stoichiometry as it is specifically applied to acids and bases.

4-3 ACID–BASE REACTIONS

The behavior of many chemical compounds can best be correlated by classifying them as acids or bases. Since your study of acids and bases in *Chemical Principles* was in terms of aqueous solutions, aqueous systems will be emphasized in this introduction to acid–base reactions. The definition of acids as hydrogen ion donors and of bases as hydrogen ion acceptors is the definition suggested by Johannes Brønsted.

ITEM 1

Positively charged hydrogen ions, H^+, can be produced in the gas phase by removing an ________ from a hydrogen atom.

ITEM 11

Since the oxygen atom of a water molecule has a pair of electrons available for sharing, it can accept a proton. In so doing, it is acting as a ________.

ITEM 21

The conjugate base of H_2O is ________. The conjugate acid of OH^- is ________.

ITEM 31

When acid–base reactions, which according to Brønsted theory involve ________ transfers, occur in aqueous solution, the solvent (________) is not inert; it enters into the reactions.

ITEM 41

The hydrogen phosphate ion, HPO_4^{2-}, is a weak acid; its conjugate base ________ is a ________ base.

ITEM 51

From a comparison of water solutions of CH_3ONa, NH_3, and NaCl, it is possible to say that, of the three bases CH_3O^-, NH_3, Cl^-, the strongest is ________ and the weakest is ________.

ITEM 2

Since a hydrogen atom consists of a proton and an electron, the species indicated by H^+ in chemical studies is really nothing more than a ________.

ITEM 12

Similarly, the hydroxide ion, OH^-, is a base because it can accept a ________ as, for example, in the reaction $H_3O^+ + OH^- \rightarrow$ ________.

ITEM 22

Removal of a proton from an acid produces the conjugate base of the acid. Addition of a proton to a base produces the ________.

ITEM 32

The ionization of an acid in water can be represented as $HA + H_2O \leftrightarrows$ ________ $+ A^-$ in which the two bases that compete for the proton are ________ and ________.

ITEM 42

In a 0.01-molar solution of HCN in water, about 0.02% of the HCN is ionized. In a 0.01-molar solution of acetic acid, $CH_3\overset{\overset{\displaystyle O}{\|}}{C}{-}OH$, about 5% of the acid is ionized. The acid that more readily donates a proton to water is ________ and therefore the stronger of the two acids is ________.

ITEM 52

The ionization of an acid in solution always requires the presence of a base. In aqueous solutions this base is ________.

$CN + H_2O \rightleftharpoons H_3O^+ + CN^-$

$$H_3\overset{\overset{\displaystyle O}{\|}}{C}OH + H_2O \rightleftharpoons H_3O^+ + CH_3\overset{\overset{\displaystyle O}{\|}}{C}O^-$$

ANSWER 1 electron

ANSWER 11 base

ANSWER 21 OH^-
H_2O

ANSWER 31 proton
H_2O

ANSWER 41 PO_4^{3-}
strong

ANSWER 51 CH_3O^-
Cl^-

ITEM 3

An exact description of H^+ in solution is still not possible, but it is certain that the ion is never free but is always associated with molecules of the ________.

ITEM 13

In the gas phase, neutral molecules of hydrogen nitrate, HNO_3, exist. The reaction that takes place when neutral molecules of gaseous HNO_3 are dissolved in water is $HNO_3 + H_2O \rightarrow$ ________ + NO_3^-.

ITEM 23

In a reaction of the type $HA + B \rightarrow A^- + HB^+$, we say that HA is acting as a(n) ________ and B as a(n) ________.

ITEM 33

Whether this general reaction goes to the left or to the right depends on the individual acid. For instance, hydrochloric acid is almost 100% ionized: The reaction $HCl + H_2O \leftrightarrows$ ________ goes almost entirely to the ________.

ITEM 43

Just as water serves not only as a solvent for the ionization of an acid but also as the necessary base, water can serve not only as solvent but also, when a base is added, as the ________.

ITEM 53

However, any base, if sufficiently strong, can effect the ionization of an acid. Next to water, the most common base-solvent for acids is ammonia. The ionization of HCl in NH_3 is written $HCl + NH_3 \leftrightarrows$ ________. The two bases competing for the proton are ________ and ________.

ANSWER 2 proton

ANSWER 12 proton (or hydrogen ion)
$2H_2O$

ANSWER 22 conjugate acid of the base

ANSWER 32 H_3O^+
A^-
H_2O

ANSWER 42 $CH_3\overset{\overset{\displaystyle O}{\|}}{C}OH$, acetic acid
$CH_3\overset{\overset{\displaystyle O}{\|}}{C}OH$, acetic acid

ANSWER 52 water

ITEM 4

For convenience, let us assume that the hydrogen is always bonded to one other atom by sharing a pair of electrons with that atom. Then, if H^+ ions were formed, they would react immediately with adjacent molecules that have ________ available for ________.

ITEM 14

If ammonia gas is dissolved in water, the reaction at the right occurs to an appreciable extent. In this reaction, H_2O is the ________ and NH_3 is the ________.

ITEM 24

When this reaction ($HA + B \rightarrow A^- + HB^+$) goes in the opposite direction ($HB^+ + A^- \rightarrow HA + B$), we say that ________ is acting as an acid and ________ as a base.

ITEM 34

In contrast, hydrocyanic acid, HCN, ionizes to the extent of less than 1% in solutions more concentrated than about 10^{-6} molar. The reaction $HCN + H_2O \leftrightarrows$ ________ goes almost entirely to the ________.

ITEM 44

For instance, the ionization of ammonia is written

$$NH_3 + H_2O \leftrightarrows NH_4^+ + OH^-$$

Again, the essential reaction is the transfer of a proton between the bases ________ and ________.

ITEM 54

Even compounds that react as acids in water may serve as bases for the ionization of other acids. For instance, acetic acid can donate one proton (the one attached to oxygen) to water.

$$CH_3\overset{\overset{\displaystyle O}{\|}}{C}{-}OH + H_2O \leftrightarrows CH_3\overset{\overset{\displaystyle O}{\|}}{C}{-}O^- + H_3O^+$$

CH_3COOH is acting as a(n) ________ and H_2O as a(n) ________.

$NH_3 + H_2O \rightarrow NH_4^+ + OH^-$

ANSWER 3 solvent

ANSWER 13 H_3O^+

ANSWER 23 acid
base

ANSWER 33 $H_3O^+ + Cl^-$
right

ANSWER 43 acid

ANSWER 53 $NH_4^+ + Cl^-$
Cl^-
NH_3

ITEM 5

In solution, protons can be transferred from a molecule or an ion in which they share a __________ to another molecule, or ion, that has a free __________.

ITEM 15

When the ionic compound NaOH is dissolved in water, the solution contains only the ions Na^+ and __________: There are no molecules of NaOH.

ITEM 25

Any acid–base reaction can be written as a reversible reaction: conjugate acid of base A + base B $\leftrightharpoons$ base A + conjugate acid of base B. The essential reaction is the transfer of a __________ from one base to another.

ITEM 35

The comparison of a solution of HCl in water with a solution of HCN at the same concentration shows that HCl donates a proton to water much more readily than does HCN. HCl is thus a *stronger* acid than HCN. A strong acid has a __________ tendency to donate a proton to a base than has a weak acid.

ITEM 45

In this reaction of NH_3 with H_2O, NH_4^+ is the conjugate __________ of __________ and OH^- is the conjugate __________ of __________.

ITEM 55

However, if HCl is dissolved in acetic acid, acetic acid can accept a proton.

$$HCl + CH_3\overset{\overset{O}{\|}}{C}-OH \leftrightharpoons CH_3\overset{\overset{^+OH}{\|}}{C}-OH + Cl^-$$

In this system, the two bases competing for the transferable proton are __________ and __________. HCl is acting as a(n) __________ and CH_3COOH as the solvent-__________.

ANSWER 4 electrons
sharing (or bonding)

ANSWER 14 acid (proton donor)
base (proton acceptor)

ANSWER 24 HB^+
A^-

ANSWER 34 $H_3O^+ + CN^-$
left

ANSWER 44 NH_3
OH^-

ANSWER 54 acid
base

ITEM 6

In solution, free H^+ ions do not occur, but they can be ________ from one molecule to another.

ITEM 16

When the molecular compound HCl dissolves in water, the reaction $HCl + H_2O \rightarrow$ ________ proceeds to such an extent that few, if any, HCl molecules remain.

ITEM 26

A molecule of sulfuric acid, H_2SO_4, can donate both its hydrogens to water molecules. The following steps can be written for the reaction of H_2SO_4 with water

$$H_2SO_4 + H_2O \leftrightarrows HSO_4^- + H_3O^+$$
$$HSO_4^- + H_2O \leftrightarrows \underline{\qquad\qquad}.$$

ITEM 36

It is apparent that the Cl^- ion has very little tendency to "hold on to" or to accept a proton: Cl^- is a weak base. On the other hand, the CN^- ion has a high affinity, or attraction, for a proton: CN^- is a relatively ________ base.

ITEM 46

When sodium methoxide, CH_3ONa, is dissolved in water, it dissociates into Na^+ and the base CH_3O^-, which has a strong affinity for protons. CH_3O^- reacts as follows

$$CH_3O^- + H_2O \leftrightarrows CH_3OH + \underline{\qquad\qquad}.$$

In this reaction the proton is transferred between the bases ________ and ________.

ITEM 56

Molecular bases, such as NH_3 and CH_3NH_2 (methylamine), require an acid for ionization. In aqueous solution, this acid is ________, as shown by the reaction of NH_3:

$$NH_3 + \underline{\qquad\qquad} \leftrightarrows \underline{\qquad\qquad}.$$

ANSWER 5 pair of electrons
pair of electrons

ANSWER 15 OH^-

ANSWER 25 proton

ANSWER 35 greater

ANSWER 45 acid
NH_3
base
H_2O

ANSWER 55 Cl^-

$$CH_3\overset{\overset{\displaystyle O}{\|}}{C}{-}OH$$

acid
base

ITEM 7

According to Brønsted's definitions of acids and bases, an acid is a substance that can donate a proton and a base is a substance that can ________.

ITEM 17

Therefore, the proton-transfer reaction that occurs when solutions of NaOH and HCl are mixed is $H_3O^+ + OH^- \rightarrow$ ________, and we see that the acid and the base in HCl and NaOH solutions are actually ________ and ________, respectively.

ITEM 27

In the first reaction, H_2SO_4 is the acid and ________ is its conjugate base. In the second reaction, HSO_4^- is the acid and ________ is its conjugate base.

ITEM 37

A strong acid readily gives up a proton; it follows that its conjugate base does not tend to ________ a proton and therefore is a ________ base.

ITEM 47

In the reaction $H_2O + CH_3O^- \leftrightarrows CH_3OH + OH^-$, CH_3OH is the conjugate ________ of ________.

ITEM 57

Again, however, the Brønsted theory does not require that H_2O be the solvent; any sufficiently strong acid can serve as the ionization medium for a base. Write the equation for the ionization of NH_3 in acetic acid

$$NH_3 + CH_3\overset{\overset{\displaystyle O}{\|}}{C}{-}OH \leftrightarrows \underline{\qquad\qquad}.$$

ANSWER 6 transferred

ANSWER 16 $H_3O^+ + Cl^-$

ANSWER 26 $SO_4^{2-} + H_3O^+$

ANSWER 36 strong

ANSWER 46 OH^-
CH_3O^-
OH^-

ANSWER 56 water
H_2O
$NH_4^+ + OH^-$

ITEM 8

When the gas hydrogen chloride, HCl, is dissolved in water, the reaction $HCl + H_2O \rightarrow Cl^- + H_3O^+$ occurs. In the reaction a proton is transferred from HCl to ________. Therefore, HCl is acting as a(n) ________ and H_2O as a(n) ________.

ITEM 18

When a base adds a proton, the species formed is an acid. This acid is known as the *conjugate acid* of the base. Thus, H_3O^+ is the conjugate acid of the base ________.

ITEM 28

Of the species H_2SO_4, HSO_4^-, and SO_4^{2-}, which can act most easily both as an acid and as a base?

ITEM 38

If an acid of any charge type, for example, HA^+, HA, or HA^- is weak, its conjugate base—________, ________, or ________, respectively—has a high proton affinity and is a ________ base.

ITEM 48

A comparison of an aqueous solution of CH_3O^- with that of NH_3 at the same concentration shows that NH_3 reacts to only a small extent, whereas the reaction of CH_3O^- with water goes to a large extent to the right. This comparison indicates that of the two bases, CH_3O^- and NH_3, ________ is the stronger.

ITEM 58

In any chemical system, then, according to the Brønsted definition of acids and bases, a molecule or ion acts as an acid if it ________ and as a base if it ________.

Conjugate acid

$CH_3O^- + H_2O \rightleftarrows CH_3OH + OH^-$
$NH_3 + H_2O \rightleftarrows NH_4^+ + OH^-$

ANSWER 7 accept a proton

ANSWER 17 $2H_2O$
H_3O^+
OH^-

ANSWER 27 HSO_4^-
SO_4^{2-}

ANSWER 37 accept
weak

ANSWER 47 acid
CH_3O^-

ANSWER 57 $NH_4^+ + CH_3\overset{\overset{\displaystyle O}{\|}}{C}—O^-$

ITEM 9

The H_3O^+ ion is called the *hydronium* (or *hydroxonium*) *ion.* It is formed by the transfer of a proton (H^+) from an acid molecule to a ________ molecule.

ITEM 19

Just as H_3O^+ is the conjugate ________ of the base H_2O, so H_2O is the conjugate ________ of the acid H_3O^+.

ITEM 29

Of the species H_3O^+, H_2O, OH^- (all very important in acid–base studies), which can act most easily both as an acid and as a base?

ITEM 39

Nitric acid, HNO_3, is a strong acid and therefore its conjugate base ________ is a ________ base.

ITEM 49

This type of reaction, in which a proton is donated from H_2O to a base, is an example of a hydrolysis reaction and results in the formation of ________ ions from H_2O molecules.

ANSWER 8 H_2O
acid
base

ANSWER 18 H_2O

ANSWER 28 HSO_4^-

ANSWER 38 A
A^-
A^{2-}
strong

ANSWER 48 CH_3O^-

ANSWER 58 donates a proton
accepts a proton

(*Turn to page 175.*)

ITEM 10

The water molecule may be drawn as $H : \ddot{\underset{\cdot\cdot}{O}} :$
$\quad H$

The dots represent electrons that are bonded or available for bonding. The diagram

$[H : \ddot{\underset{\cdot\cdot}{O}} : H]^+$
$\quad H$

represents __________.

ITEM 20

The conjugate base of the acid HCN is __________.
The conjugate base of the acid NH_4^+ is __________.

ITEM 30

Write equations for two reactions, one in which H_2O acts as a base and one in which H_2O acts as an acid. Use HCl as the acid in one reaction and NH_3 as the base in the other reaction.

ITEM 40

Nitrous acid, HNO_2, is a weak acid and therefore its conjugate base __________ is a __________ base.

ITEM 50

Whereas there is clear evidence of hydrolysis in solutions of CH_3ONa and NH_3, no reaction occurs when NaCl is added to water: neither Na^+ nor Cl^- is a strong enough __________ to remove a proton from __________.

ANSWER 9 water

ANSWER 19 acid
base

ANSWER 29 H_2O

ANSWER 39 NO_3^-
weak

ANSWER 49 OH^-

ANSWER 10 the hydronium ion

(*Item 11 is on page 152.*)

ANSWER 20 CN^-
NH_3

(*Item 21 is on page 152.*)

ANSWER 30

$HCl + H_2O \rightleftarrows H_3O^+ + Cl^-$ (H_2O is base)
$NH_3 + H_2O \rightleftarrows NH_4^+ + OH^-$ (H_2O is acid)

(*Item 31 is on page 152.*)

ANSWER 40 NO_2^-
strong

(*Item 41 is on page 152.*)

ANSWER 50 base
H_2O

(*Item 51 is on page 152.*)

4–4 ACID–BASE STOICHIOMETRY

Now that acids and bases have been defined and the proton-transfer reactions that are encountered most often have been described, we can use this background material to study the solution stoichiometry of acid–base reactions. The concepts of normality and equivalent weight, which are especially helpful in quantitative studies of acids and bases, will be a review of Section 4–5 of *Chemical Principles*.

In this section we assume that all acids and bases used as examples are completely ionized in solution. This is valid for our purpose here and has been done in order to avoid, temporarily, the fascinating problems of partial dissociation, which will be discussed in Chapter 5 of *Chemical Principles* and in Review 5 of this book.

Items 27–37 at the end of this section are representative of acid–base stoichiometry problems. If you can solve them, you can probably skip the rest of the section.

ITEM 1

When aqueous solutions of HCl and NaOH are mixed, the reaction can be written $HCl + NaOH \rightarrow NaCl + H_2O$. The actual "net" reaction is $H_3O^+ +$ _________ $\rightarrow$ _________.

ITEM 7

What volume of 0.25-molar NaOH solution would be necessary to provide 0.050 mole of OH^-?

ITEM 13

Since 1 equivalent of an acid can transfer 1 mole of protons to a base, 1 equivalent of H_2SO_4 is _________ of a mole of H_2SO_4.

ITEM 19

A 1-normal (*N*) solution of an acid or base contains one equivalent per liter of solution. One liter of a 0.5-normal solution of HCl contains _________ equivalent(s) of HCl.

ITEM 25

An equivalent of any base will react with _________ equivalent(s) of an acid in solution.

ITEM 31

How many milliliters of 1.32-normal KOH are required to react with 60 ml of 0.570-molar H_3PO_4?

ITEM 37

A sample of an acid weighing 1.55 g is dissolved in water. If 34.4 ml of 0.480-normal NaOH are required to react completely with the sample, what is the equivalent weight of the acid?

ITEM 2

Each mole of HCl furnishes ________ mole(s) of H_3O^+ in solution. Hence, ________ mole(s) of OH^- is (are) required for complete reaction.

ITEM 8

In 0.50 liter of 0.10-molar H_2SO_4 there is ________ mole of H_2SO_4 so ________ mole of H_3O^+ is available for reaction with a base.

ITEM 14

One molecule of the acid H_3PO_4 can transfer three protons to a base. One mole of H_3PO_4 is ________ equivalent(s) of H_3PO_4.

ITEM 20

A 0.6-normal solution contains ________ of acid or base per liter of solution.

ITEM 26

In 0.15 liter of 0.20-normal K_2CO_3 there is (are) ________ equivalent(s) that will react with ________ equivalent(s) of HCl, or with ________ equivalent(s) of H_2SO_4.

ITEM 32

The *equivalent weight*, E_w, of an acid is the weight of an acid (in grams) that can transfer an Avogadro's number of protons, that is, a "mole" of protons, to a base. The equivalent weight of an acid is therefore equal to the molecular weight divided by ________.

$H_3O^+ + OH^- \rightarrow 2H_2O$

Equivalent weight

ANSWER 1 OH^-
$2H_2O$

ANSWER 7 $V = \frac{n}{M} = \frac{0.050 \text{ mole}}{0.25 \text{ mole liter}^{-1}}$
$= 0.20$ liter

ANSWER 13 $\frac{1}{2}$

ANSWER 19 0.5

ANSWER 25 1

ANSWER 31

N of $H_3PO_4 = 3 \times 0.570 = 1.71$

$$V_2 = \frac{N_1V_1}{N_2}$$

$$= \frac{1.71 \text{ equiv liter}^{-1}(60 \text{ ml})}{1.32 \text{ equiv liter}^{-1}} = 78 \text{ ml}$$

ANSWER 37

equiv of NaOH = (0.0344 liter)(0.480 equiv liter^{-1})
= 0.0165
acid = 1.55 g/0.0165 equiv
= 94.0 g equiv^{-1}

ITEM 3

When aqueous solutions of H_2SO_4 and NaOH are mixed, the reaction may be written $H_2SO_4 + 2NaOH \rightarrow Na_2SO_4 + 2H_2O$ or, to show what is actually happening, $H_3O^+ +$ __________ $\rightarrow$ __________.

ITEM 9

Reaction with 0.10 mole H_3O^+ requires __________ mole(s) of OH^-.

ITEM 15

The amount of base that can accept 1 mole of protons is an equivalent of the base. One equivalent of NaOH is __________ mole(s) of NaOH.

ITEM 21

An N-normal solution contains __________ equivalent(s) per liter. The number of equivalents in V liters of an N-normal solution is __________.

ITEM 27

V_1 liters of N_1-normal acid solution contain __________ equivalents. V_2 liters of N_2-normal base contain __________ equivalents.

ITEM 33

The equivalent weight is often a unique property that can aid in the identification of an acid. The equivalent weight of HCl is __________ and of H_3PO_4 is __________. (See periodic table in the inside front cover for atomic weights.)

ANSWER 2 1
1

ANSWER 8 0.50 liter × 0.10 mole liter^{-1}
= 0.050 mole
0.10

ANSWER 14 3

ANSWER 20 0.6 equivalent

ANSWER 26

(0.15 liter)(0.20 equiv liter^{-1}) = 0.030 equiv
0.030
0.030

ANSWER 32 the number of transferable protons per molecule (which is equal to the number of equivalents per mole)

ITEM 4

Each mole of H_2SO_4 can produce __________ moles of H_3O^+ in solution; __________ moles of NaOH are required for complete reaction.

ITEM 10

Since some acid molecules can transfer more than one proton to a base, the number of moles of protons available can be __________ than the number of moles of acid.

ITEM 16

One mole of $Ca(OH)_2$ can furnish 2 moles of OH^- in solution. A mole of $Ca(OH)_2$ is __________ equivalent(s).

ITEM 22

In 2.5 liters of 0.20-normal NaOH there is (are) __________ equivalent(s) (often abbreviated to *equiv* or *eq*) of NaOH.

ITEM 28

Since each equivalent of any acid reacts with an equivalent of any base, the amounts required for complete reaction are related by the equation $V_1N_1 =$ __________.

ITEM 34

Acetic acid

$$CH_3\overset{\overset{\displaystyle O}{\|}}{C}—OH$$

can transfer one proton (the one attached to oxygen) to a base such as OH^-. The equivalent weight of acetic acid is __________.

ANSWER 3 OH^-
$2H_2O$

ANSWER 9 0.10

ANSWER 15 1

ANSWER 21 N
V liters $\times$ N equivalents liter^{-1}
$= VN$ equivalents

ANSWER 27 V_1N_1
V_2N_2

ANSWER 33

$$\frac{36.5 \text{ g mole}^{-1}}{1 \text{ equiv mole}^{-1}} = 36.5 \text{ g equiv}^{-1}$$

$$\frac{98.0 \text{ g mole}^{-1}}{3 \text{ equiv mole}^{-1}} = 32.7 \text{ g equiv}^{-1}$$

ITEM 5

In 0.50 liter of 0.10-molar HCl there is (are) ________ mole(s) of HCl or ________ mole(s) of H_3O^+.

ITEM 11

An *equivalent* of an acid is defined as the amount of acid that will transfer 1 mole of protons to a base. An equivalent of HCl is ________ mole(s) of HCl.

ITEM 17

The carbonate ion, CO_3^{2-}, is a base and can accept two protons to form H_2CO_3, which then decomposes to CO_2 and H_2O. One mole of Na_2CO_3 can furnish ________ mole(s) of CO_3^{2-} that can accept ________ mole(s) of protons.

ITEM 23

In the reaction $H_3O^+ + OH^- \rightarrow 2H_2O$, each mole of OH^- reacts with ________ mole(s) of H_3O^+.

ITEM 29

A 0.500-liter sample of 0.100-normal acid contains ________ equivalent. If this sample of acid is mixed with a 0.200-normal base solution, the volume of base required for complete reaction will be ________ liter(s).

ITEM 35

How many grams of

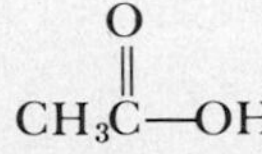

would be used to make up a solution containing 0.15 equiv in 100 ml of solution? What would be the normality of this solution?

Equivalent

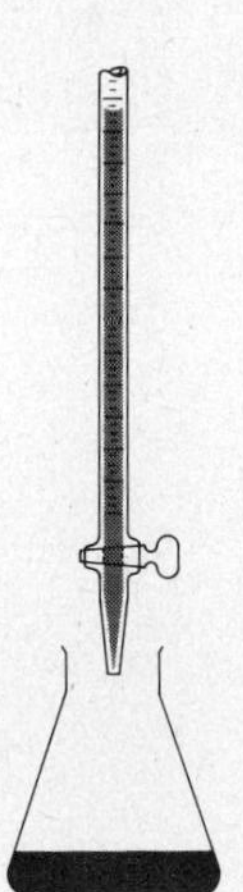

ANSWER 4 2
2

ANSWER 10 greater

ANSWER 16 2

ANSWER 22

(2.5 liters)(0.20 equiv liter^{-1}) = 0.50 equiv

ANSWER 28 V_2N_2

ANSWER 34 60.0

ITEM 6

For reaction with this 0.050 mole of H_3O^+, __________ mole of OH^- is required.

ITEM 12

Just as the word "mole" is used for gram molecular weight, the word "equivalent" is used generally instead of "gram equivalent weight." An equivalent is the weight of an acid, expressed in __________, that can transfer 6.02×10^{23} protons to a base.

ITEM 18

Therefore, 1 mole of CO_3^{2-} can furnish __________ equivalent(s) of base.

ITEM 24

An equivalent of acid will produce __________ mole(s) of H_3O^+ in solution. An equivalent of hydroxide base will produce __________ mole(s) of OH^- in solution.

ITEM 30

A 25.0-ml sample of $Ba(OH)_2$ solution required 37.2 ml of 0.110-normal sulfuric acid solution for complete reaction. The normality of the hydroxide solution was __________; the 25.0-ml sample contained __________ mole(s) of $Ba(OH)_2$.

ITEM 36

In 0.0520 liter of 0.204-normal NaOH, there is (are) __________ equivalent(s) of OH^- that will react with __________ equivalent(s) of

$$CH_3\overset{\overset{\displaystyle O}{\|}}{C}{-}OH$$

or __________ g of $CH_3\underset{\underset{\displaystyle O}{\|}}{C}{-}OH$

ANSWER 5

$n = MV = (0.50 \text{ liter})(0.10 \text{ mole liter}^{-1})$
$= 0.050 \text{ mole}$
0.050

ANSWER 11 1

ANSWER 17 1
2

ANSWER 23 1

ANSWER 29

$(0.500 \text{ liter})(0.100 \text{ equiv liter}^{-1}) = 0.0500 \text{ equiv}$

$$V_2 = \frac{V_1N_1}{N_2} = \frac{0.0500 \text{ equiv}}{0.200 \text{ equiv liter}^{-1}} = 0.250 \text{ liter}$$

ANSWER 35

$(60.0 \text{ g equiv}^{-1})(0.15 \text{ equiv}/100 \text{ ml}) = 9.0 \text{ g}/100 \text{ ml}$
$(0.15 \text{ equiv}/100 \text{ ml}) = 1.5 \text{ equiv liter}^{-1}$
Hence, solution is 1.5 normal.

ANSWER 6 0.050

(*Item 7 is on page 176.*)

ANSWER 12 grams

(*Item 13 is on page 176.*)

ANSWER 18 2

(*Item 19 is on page 176.*)

ANSWER 24 1
1

(*Item 25 is on page 176.*)

ANSWER 30

$$N_1 = \frac{N_2 V_2}{V_1}$$

$$= \frac{(37.2 \text{ ml})(0.110 \text{ equiv liter}^{-1})}{25.0 \text{ ml}} = 0.164$$

equivalents of $Ba(OH)_2 = VN$
$= (0.0250 \text{ liter})(0.164 \text{ equiv liter}^{-1})$
$= 0.00410$
moles of $Ba(OH)_2 = 0.00410/2 = 0.00205$

(*Item 31 is on page 176.*)

ANSWER 36

$(0.0520 \text{ liter})(0.204 \text{ equiv liter}^{-1}) = 0.0106$ equiv
0.0106
$(0.0106 \text{ equiv})(60.0 \text{ g equiv}^{-1}) = 0.636$ g

(*Item 37 is on page 176.*)

4–5 HEAT EFFECTS IN CHEMICAL REACTIONS

The heat effects that accompany chemical reactions have much practical and theoretical significance. The study of these effects has practical applications in the burning of fuels for heating purposes, the combustion of gasoline and other petroleum products for transportation, the use of chemical fuels and oxidants for rockets, and the sustenance of life itself through the controlled release of energy by chemical reactions occurring in plants and animals. In addition, the study of heat effects has led to advances in our understanding of chemical and biochemical reactions, chemical equilibrium, and molecular structure, and, in some measure, has contributed to all topics of current interest in chemistry.

As you have learned in Section 4–6 of *Chemical Principles,* for a given set of conditions, the quantity of heat evolved or absorbed in a chemical reaction is just as definite as is the amount of a product formed. In this section we will enlarge upon the discussion in Section 4–6 of *Chemical Principles* to help you to understand some of the methods for calculating the amount of energy evolved or absorbed in chemical reactions. On completion of the section you will be able to solve problems of this type: The heat of combustion of C(graphite) is −94.052 and C(diamond) is −94.505. Calculate ΔH for the reaction C(graphite) $\rightarrow$ C(diamond). Items 51–57 will provide a test of your knowledge.

ITEM 1

Every chemical compound is characterized by a particular arrangement of atoms in space, and every different arrangement has associated with it a certain amount of potential energy. Therefore, when one compound reacts to form another compound, the new arrangement of atoms will have either a greater or a lesser amount of ________ than the original.

ITEM 11

The equation $A + B \rightarrow C$ does not indicate whether the reaction is exothermic. Using q, rewrite the equation so it represents an exothermic reaction.

ITEM 21

In previous examples, water was formed as a liquid. The heat liberated when 1 mole of water vapor is formed from gaseous oxygen and hydrogen is 57.798 kcal. The thermochemical equation showing the formation of 1 mole of water vapor is ________.

ITEM 31

Another way of saying the same thing is that the net heat evolved or absorbed in the reaction $A \rightarrow C$ is (*independent of, dependent on*) the number and kind of intermediate steps.

ITEM 41

Insert horizontal lines and reagent names, and fill in the blank in the energy diagram at right for the reaction

$$H_2O(l) \rightarrow H_2O(g) \qquad \Delta H = +10.519 \text{ kcal}$$

ITEM 51

Using the data given in the thermochemical equations at the right, tabulate the heats of formation of the compounds below.

Compound	ΔH_f kcal
$H_2O(l)$	
$H_2O(g)$	
$NH_3(g)$	

H

$\Delta H =$ ______ kcal

reagents

$$H_2(g) + \tfrac{1}{2}O_2(g) \rightarrow H_2O(l) \quad \Delta H = -68.32 \text{ kcal}$$

$$2H_2(g) + O_2(g) \rightarrow 2H_2O(g) \quad \Delta H = -115.60 \text{ kcal}$$

$$N_2(g) + 3H_2(g) \rightarrow 2NH_3(g) \quad \Delta H = -22.08 \text{ kcal}$$

ITEM 2

The sum of the potential energy, arising from the __________ of atoms in a sample of a compound, and the kinetic energy, arising from their motion, is known as the *heat content* or *enthalpy, H,* of the sample. Although the absolute value of H is not usually known for a sample of a compound, it is the change in H, which can often be measured, that is important.

ITEM 12

Materials appearing on the left side of a chemical equation are called *reactants.* In an endothermic reaction, heat can be regarded as a __________. For such reactions, the amount of heat (symbol __________) would be written on the __________ side of the equation.

ITEM 22

The heat of combustion of gaseous pentane, C_5H_{12}, is 11.6 kcal g^{-1} when the products are gaseous CO_2 and liquid H_2O. The heat liberated in the combustion of 1 *mole* of pentane under these conditions is __________. The thermochemical equation for the combustion is __________.

ITEM 32

The heat evolved or absorbed in the reaction A $\rightarrow$ C depends only on the relative heat contents of __________ and __________, and not on whether the reaction is stepwise or direct.

ITEM 42

Fill in the following table:

Reaction	Heat (evolved or absorbed)	ΔH (+ or −)
Endothermic		
Exothermic		

ITEM 52

The heat of formation of a compound, which is given the symbol __________, is defined as the value of ΔH for the reaction in which __________ is formed from __________ at __________ °C and a pressure of __________.

Heat content
Enthalpy

ANSWER 1 potential energy

ANSWER 11 $A + B \rightarrow C + q$

ANSWER 21

$H_2(g) + \frac{1}{2}O_2(g) \rightarrow H_2O(g) + 57.798$ kcal

ANSWER 31 independent of

ANSWER 41

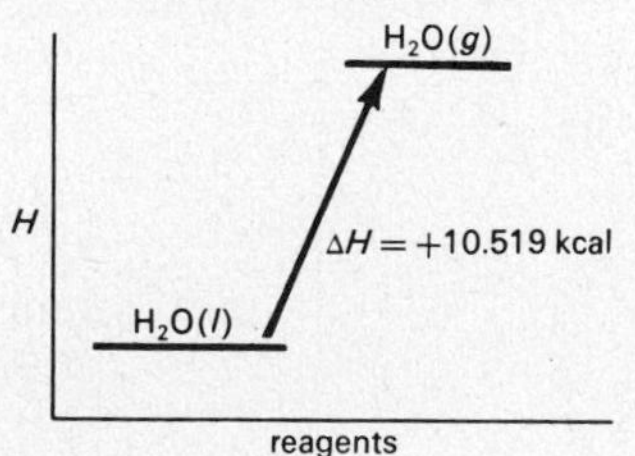

ANSWER 51

$H_2O(l)$ $\Delta H_f = -68.32$ kcal
$H_2O(g)$ $\Delta H_f = -57.80$ kcal
$NH_3(g)$ $\Delta H_f = -11.04$ kcal

ITEM 3

In the diagram at the right are shown the relative heat contents of equal amounts of compounds A and B. Compound B has a higher heat content than compound A. Therefore, energy must be absorbed by compound ________ in order that compound ________ be formed from it.

ITEM 13

The units in which q is given—the heat units most commonly used in thermochemistry—are *kilocalories.* Since the prefix "kilo" implies 1000, and since 1 calorie is defined as the amount of heat required to raise the temperature of 1 g of water from 14.5° C to 15.5° C, 1 kilocalorie (kcal) is the amount of heat required to raise the temperature of ________ from 14.5°C to 15.5°C.

ITEM 23

When 2 moles of gaseous hydrogen iodide (HI) are formed from 1 mole of $H_2(g)$ and 1 mole of $I_2(s)$, 12.40 kcal of heat are absorbed. Write the thermochemical equation for this reaction.

ITEM 33

If the reaction $A \rightarrow C$ is exothermic, the heat content of C must be ________ than that of A.

ITEM 43

One convention for writing thermochemical equations is to write the heat as if it were a reactant or product. Another convention is to write the equation and then to indicate the value of ΔH. Write both kinds of equations for the reaction $CO(g) + \frac{1}{2}O_2(g) \rightarrow CO_2(g)$, which is exothermic by 67.636 kcal.

ITEM 53

The formation of 1 mole of $NaOH(s)$ under standard state conditions from the elements $Na(s)$, $O_2(g)$, and $H_2(g)$ liberates 112.2 kcal of heat. Write an equation and ΔH for the formation of NaOH.

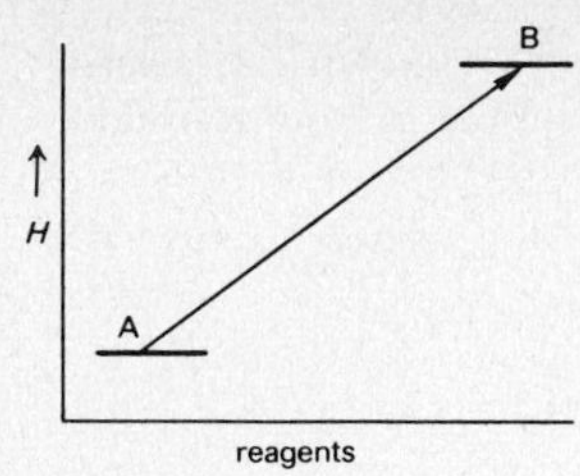

ANSWER 2 arrangement

ANSWER 12 reactant
q
left

ANSWER 22

Heat = (11.6 kcal g^{-1})(72.1 g $mole^{-1}$)
= 836 kcal $mole^{-1}$

$$C_5H_{12}(g) + 8O_2(g) \rightarrow 5CO_2(g) + 6H_2O(l) + 836 \text{ kcal}$$

ANSWER 32 A
C

ANSWER 42

Reaction	Heat	ΔH
Endothermic	Absorbed	+
Exothermic	Evolved	−

ANSWER 52

ΔH_f
1 mole of the compound
its elements in their commonly occurring state
25
1 atm

ITEM 4

Conversely, if compound B reacts to form compound A, and A has a lower heat content, it is certain that energy will be __________ in the process.

ITEM 14

When 2 moles of hydrogen are burned, they react with 1 mole of oxygen to form 2 moles of water. If the water is formed as liquid water, 136.634 kcal of heat are liberated for every mole of oxygen reacting; and the reaction, including the heat, can be written as $2H_2(g) + O_2(g) \rightarrow$ __________.

ITEM 24

Calculate the number of kcal of heat required for the formation of 10.0 g of gaseous hydrogen iodide.

ITEM 34

Indicate the relative heat contents of A and C on the diagram at the right by using the information that the reaction A → C is exothermic.

ITEM 44

On the energy diagram at the right, represent the reaction

$$C(s) + 2H_2(g) \rightarrow CH_4(g) \qquad \Delta H = -17.9 \text{ kcal}$$

ITEM 54

The formation of gaseous chlorine atoms from $Cl_2(g)$ under standard state conditions requires 29.0 kcal g-atom^{-1}. Write an equation and ΔH for the formation of Cl atoms.

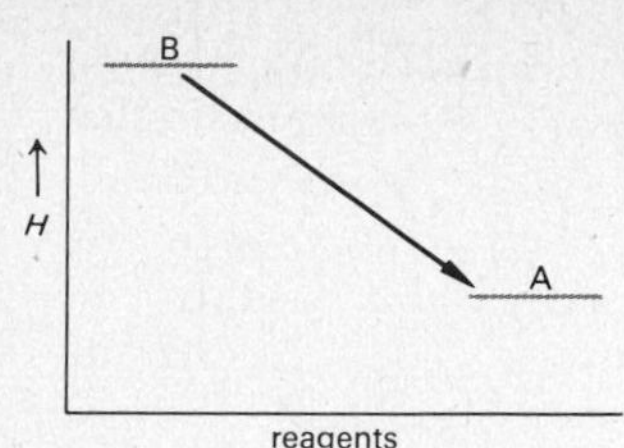

ANSWER 3 A
B

ANSWER 13 1000 g of water
(very close to 1 liter)

ANSWER 23

$12.40 \text{ kcal} + H_2(g) + I_2(s) \rightarrow 2HI(g)$

ANSWER 33 less

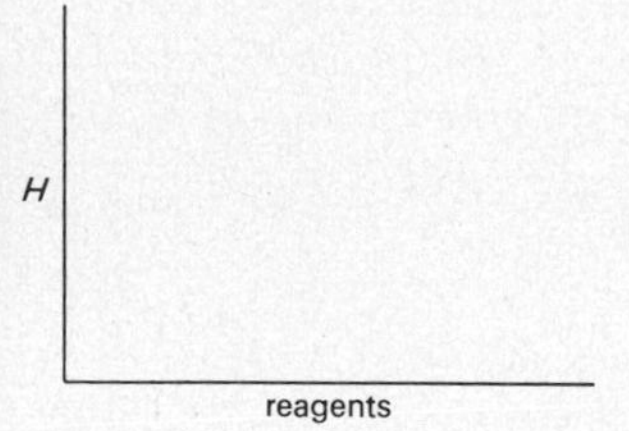

ANSWER 43

$CO(g) + \frac{1}{2}O_2(g) \rightarrow CO_2(g) + 67.636 \text{ kcal}$

$CO(g) + \frac{1}{2}O_2(g) \rightarrow CO_2(g) \qquad \Delta H = -67.636 \text{ kcal}$

ANSWER 53

$Na(s) + \frac{1}{2}O_2(g) + \frac{1}{2}H_2(g) \rightarrow NaOH(s)$

$\Delta H = -112.2 \text{ kcal}$

ITEM 5

The energy liberated or absorbed in chemical reactions usually is in the form of heat; therefore, the temperature of a reacting system changes as the reaction proceeds. If heat is liberated by the reaction, the energy warms the reactants and products and the temperature of the system ________. If the temperature falls, it is clear that ________ is being ________ by the reaction.

ITEM 15

The amount of heat liberated or absorbed depends on the physical state of the reagents. Therefore, thermochemical equations indicate physical states. The symbol (g) indicates that the substance is present in the ________ state; the symbol (l) indicates the ________ state; and the symbol (s) indicates the ________ state.

ITEM 25

It is often useful to determine the heat absorbed or liberated in a reaction that can be thought of as the combination of two or more reactions. Thus, when 1 mole of liquid water is formed from $H_2(g)$ and $O_2(g)$, 68.317 kcal of heat are ________; in the further conversion of 1 mole of liquid water to water vapor, 10.519 kcal are ________.

ITEM 35

Another way of expressing the change in heat content is frequently used in thermochemistry: The *heat of reaction*, ΔH, is defined as H(products) $-$ H(reactants). [The Greek letter Δ (delta) is commonly used to indicate increments or differences.] If H(products) is greater than H(reactants), will the value of ΔH be positive or negative?

ITEM 45

Calculate the amount of heat liberated by the formation of 5.00 liters of CH_4 at 25°C and 1 atm according to the equation

$$C(s) + 2H_2(g) \rightarrow CH_4(g) \qquad \Delta H = -17.9 \text{ kcal}$$

ITEM 55

The heat of combustion of gaseous pentane, C_5H_{12}, to gaseous CO_2 and liquid H_2O is 836 kcal mole^{-1}. Write an equation, including ΔH, for this combustion.

ANSWER 4 liberated (given off)

ANSWER 14 $2H_2O(l) + 136.634$ kcal

ANSWER 24

$$\frac{10.0 \text{ g}}{128 \text{ g mole}^{-1}} = 0.0781 \text{ mole}$$

$6.20 \text{ kcal} + \frac{1}{2}H_2(g) + \frac{1}{2}I_2(s) \rightarrow HI(g)$
$0.0781 \text{ mole } (6.20 \text{ kcal mole}^{-1}) = 0.484 \text{ kcal}$

ANSWER 34

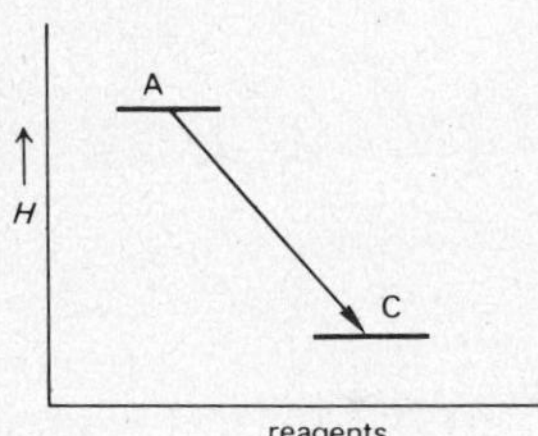

Heat of reaction

ANSWER 44

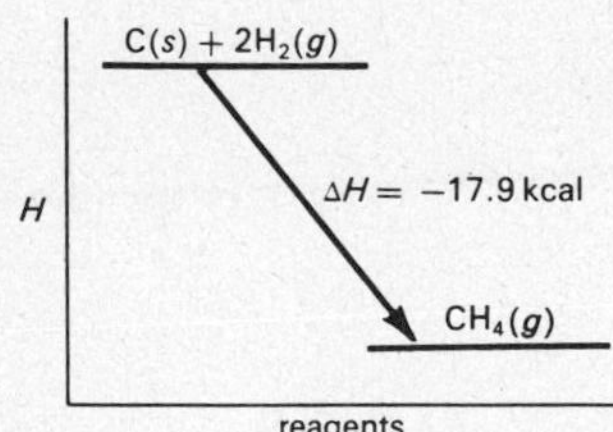

ANSWER 54 $\frac{1}{2}Cl_2(g) \rightarrow Cl(g)$
$\Delta H = +29.0$ kcal

ITEM 6

Experimentally, the heat change is determined by observing a system isolated from external heating and cooling. Before the reactants are mixed, the amount of heat required to change the temperature of the system by 1° C is measured. Then the reactants are mixed, and the liberated or absorbed heat is determined by measuring the change in ________ of the entire system.

ITEM 16

It is also necessary to indicate reagents that are in solution. The notation $Ag^+(aq)$ indicates that the ion is dissolved in ________.

ITEM 26

From the data at the right, the total reaction by which 1 mole of $H_2O(g)$ is formed from $H_2(g)$ and $O_2(g)$ can be recognized as occurring with the ________ of ________ kcal.

ITEM 36

Then, for an endothermic reaction, in which heat is absorbed and H(products) is ________ than H(reactants), the value of ΔH has a ________ sign.

ITEM 46

It is well to note that, although we have defined ΔH, the heat of reaction, as ________, ΔH is precisely defined as the difference in heat contents of the reaction products and of the reactants at constant pressure and at a definite temperature, with every substance in a definite physical state.

ITEM 56

Calculate the amount of heat released by the combustion of 3.00 g of gaseous pentane (mol wt: 72.1).

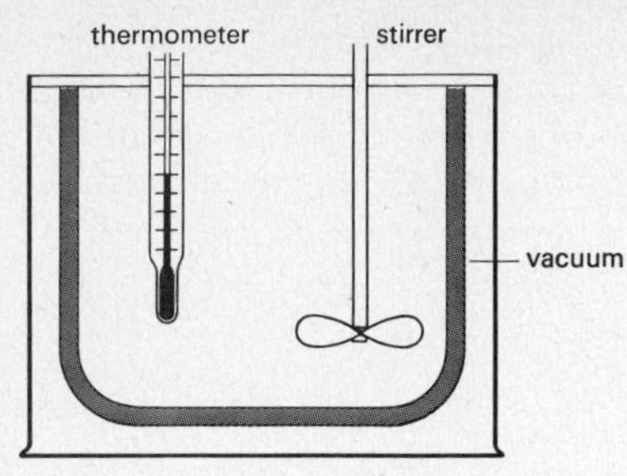

$$H_2(g) + \tfrac{1}{2}O_2(g) \rightarrow H_2O(l) + 68.317 \text{ kcal}$$
$$10.519 \text{ kcal} + H_2O(l) \rightarrow H_2O(g)$$

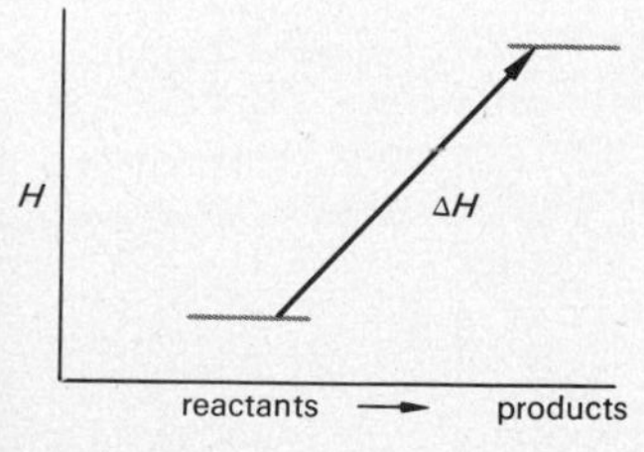

ANSWER 5 rises
heat
absorbed

ANSWER 15 gas
liquid
solid

ANSWER 25 liberated
absorbed

ANSWER 35 positive

ANSWER 45

$$n = \frac{PV}{RT} = \frac{1 \text{ atm } (5.00 \text{ liters})}{0.0821 \dfrac{\text{liter atm}}{\text{deg mole}} (298 \text{ deg})}$$
$$= 0.204 \text{ mole}$$
$$0.204 \text{ mole } (17.9 \text{ kcal mole}^{-1}) = 3.65 \text{ kcal}$$

ANSWER 55

$C_5H_{12}(g) + 8O_2(g) \rightarrow 5CO_2(g) + 6H_2O(l)$
$\Delta H = -836$ kcal

ITEM 7

The word "exothermic" comes from the Greek word *exo*, meaning "out of," and *therme*, meaning "heat." Therefore, *exothermic* is used to describe reactions in which heat is ________. As an exothermic reaction proceeds, the temperature of the system ________.

ITEM 17

The equation at the right is another example of a thermochemical equation. It differs from ordinary equations in that it not only includes ________ in the equation but also indicates the ________ of the reagents.

ITEM 27

This result is obtained in a convenient, systematic way if the thermochemical equations for the two steps are written one underneath the other and are then added as if they were algebraic equations. Carry out this procedure with the data at the right to obtain the thermochemical equation for the reaction

$$H_2(g) + \tfrac{1}{2}O_2(g) \rightarrow H_2O(g)$$

ITEM 37

Correspondingly, for an exothermic reaction, in which H(products) is ________ than H(reactants), the value of ΔH has a ________ sign.

ITEM 47

In whatever way thermochemical equations are written, equations of two or more reactions (run under the same conditions of pressure and temperaure) can be added or subtracted to give the equation for a new reaction. When two reactions are added, ΔH for the new total reaction will be the ________ of the ΔH's for the individual component reactions.

ITEM 57

The standard heat of formation of $H_2O(l)$ is −68.3 kcal mole^{-1}. The standard heat of formation of sulfur trioxide, $SO_3(g)$, is −94.5 kcal mole^{-1}. Calculate ΔH for the reaction

$$3H_2(g) + SO_3(g) \rightarrow S(s) + 3H_2O(l)$$

Exothermic

ANSWER 6 temperature

$3H_2(g) + 6C(s) + 11.7\ \text{kcal} \rightarrow C_6H_6(l)$

ANSWER 16 water

$H_2(g) + \frac{1}{2}O_2(g) \rightarrow H_2O(l) + 68.317\ \text{kcal}$
$10.519\ \text{kcal} + H_2O(l) \rightarrow H_2O(g)$

ANSWER 26 liberation
$68.317 - 10.519 = 57.798$

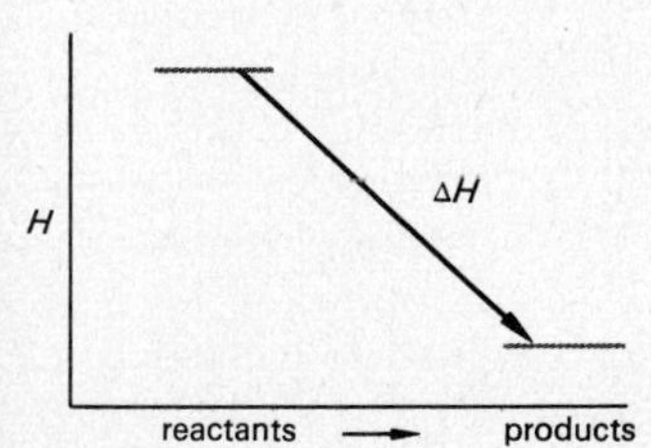

ANSWER 36 greater
plus (positive)

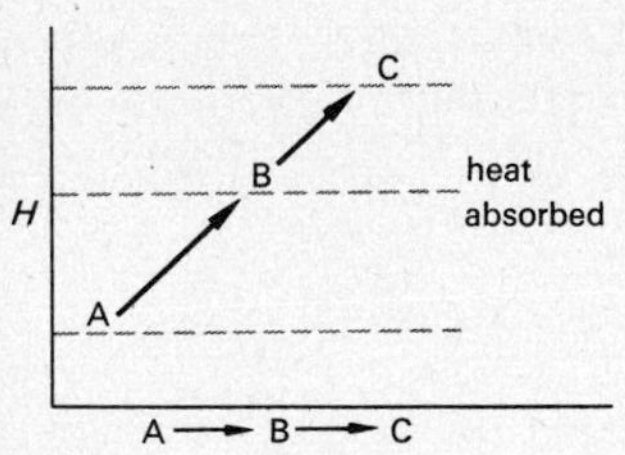

ANSWER 46 $H(\text{products}) - H(\text{reactants})$

ANSWER 56

$$\frac{3.00\ \text{g}}{72.1\ \text{g mole}^{-1}} = 0.0416\ \text{mole}$$

$0.0416\ \text{mole}\ (836\ \text{kcal mole}^{-1}) = 34.8\ \text{kcal}$

ITEM 8

The Greek prefix for "into" is *endo*; hence, reactions that absorb heat are described as __________ reactions. During such reactions, the temperature of the reaction system will __________.

ITEM 18

If the equation at the right were written to show the formation of 1 instead of 2 moles of water (in which case __________ as much heat would be evolved) the thermochemical equation would be __________.

ITEM 28

Similarly, one can subtract equations. Thus, the thermochemical equations for the combustion of C(graphite) and CO(g) are

$$\begin{aligned} &\text{C(graphite)} + O_2(g) \rightarrow CO_2(g) + 94.052 \text{ kcal} \\ &\text{CO}(g) + \tfrac{1}{2}O_2(g) \rightarrow CO_2(g) + 67.636 \text{ kcal} \end{aligned}$$

Subtraction of the second equation from the first (and moving terms with negative signs to the other side) yields __________.

ITEM 38

The thermochemical equation

$$H_2(g) + \tfrac{1}{2}O_2(g) \rightarrow H_2O(g) + 57.798 \text{ kcal}$$

shows that the heat content of H_2O is __________ than that of $H_2(g)$ + $\frac{1}{2}O_2(g)$. Complete the diagram at the right that represents this reaction by adding symbols for the reagents and the value for the heat of reaction. The sign of the value of ΔH must be __________.

ITEM 48

The reaction C(graphite) → C(diamond) is difficult to perform. For this reaction, however, ΔH can be calculated from the heats of combustion of C(graphite) and C(diamond) of −94.052 kcal mole^{-1} and −94.505 kcal mole^{-1}, respectively. Calculate ΔH.

$2H_2(g) + O_2(g) \rightarrow 2H_2O(l) + 136.634$ kcal

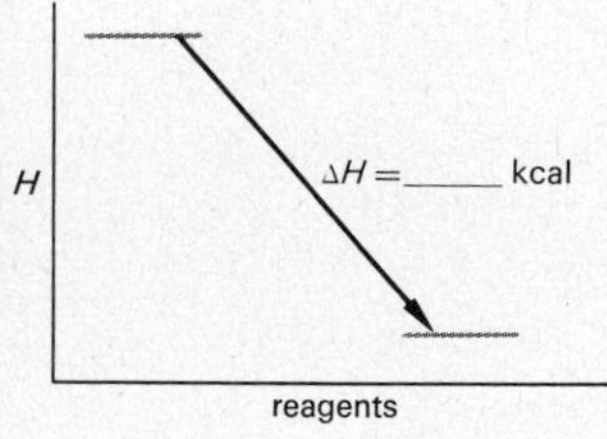

ANSWER 7 liberated
rises

ANSWER 17 heat
physical state

ANSWER 27

$$\begin{array}{l} H_2(g) + \tfrac{1}{2}O_2(g) \rightarrow H_2O(l) + 68.317 \text{ kcal} \\ 10.519 \text{ kcal} + H_2O(l) \rightarrow H_2O(g) \\ \hline H_2(g) + \tfrac{1}{2}O_2(g) \rightarrow H_2O(g) + 57.798 \text{ kcal} \end{array}$$

[The $H_2O(l)$ terms can be cancelled.]

ANSWER 37 less
minus (negative)

ANSWER 47 sum

ANSWER 57

$H_2(g) + \tfrac{1}{2}O_2(g) \rightarrow H_2O(l)$ $\Delta H = -68.3$ kcal
$S(s) + \tfrac{3}{2}O_2(g) \rightarrow SO_3(g)$ $\Delta H = -94.5$ kcal

To obtain 3 moles of water and 1 mole S, the second equation must be subtracted from the first equation, which has been multiplied by a factor of 3.

$$\begin{array}{ll} 3H_2(g) + \tfrac{3}{2}O_2(g) \rightarrow 3H_2O(l) & \Delta H = -204.9 \text{ kcal} \\ S(s) + \tfrac{3}{2}O_2(g) \rightarrow SO_3(g) & \Delta H = -94.5 \text{ kcal} \\ \hline 3H_2(g) + SO_3(g) \rightarrow S(s) + 3H_2O(l) & \Delta H = -110.4 \text{ kcal} \end{array}$$

(*Turn to page 214.*)

ITEM 9

Some reactions proceed with very little heat being liberated or absorbed. But temperature changes in most reactions can indicate the liberation of heat (__________ reactions) or the absorption of heat (__________ reactions.)

ITEM 19

The equation that answers Item 18 illustrates the third difference that is sometimes observed when the heat effects accompanying a reaction are shown, namely, that the coefficients are sometimes __________.

ITEM 29

In this way, the heat effects for difficult-to-study reactions can be deduced from data on experimentally convenient reactions, such as combustion reactions. In the example given, the heat evolved when C(graphite) is burned to CO(*g*) would be difficult to determine experimentally because during the reaction some CO might be further oxidized to __________.

ITEM 39

When ΔH is used to express the change in heat content, the thermochemical equation is written as

$$H_2(g) + \tfrac{1}{2}O_2(g) \rightarrow H_2O(g) \qquad \Delta H = -57.798 \text{ kcal}$$

When this kind of equation is used, it is important to remember that ΔH is defined as __________.

ITEM 49

Complete the energy diagram at the right to represent the relative energies of CO_2, [C(graphite) + O_2], and [C(diamond) + O_2]. The diagram should show that diamond has a __________ heat content than graphite.

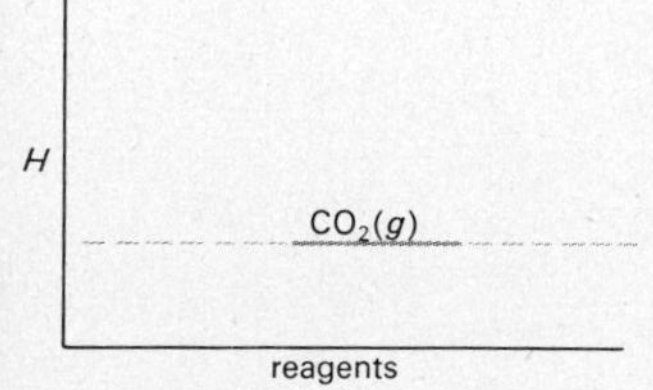

ANSWER 8 endothermic
fall

ANSWER 18

half

$H_2(g) + \frac{1}{2}O_2(g) \rightarrow H_2O(l) + 68.317$ kcal

ANSWER 28

C(graphite) $+ \frac{1}{2}O_2(g) \rightarrow CO(g) + 26.416$ kcal

ANSWER 38 less
negative

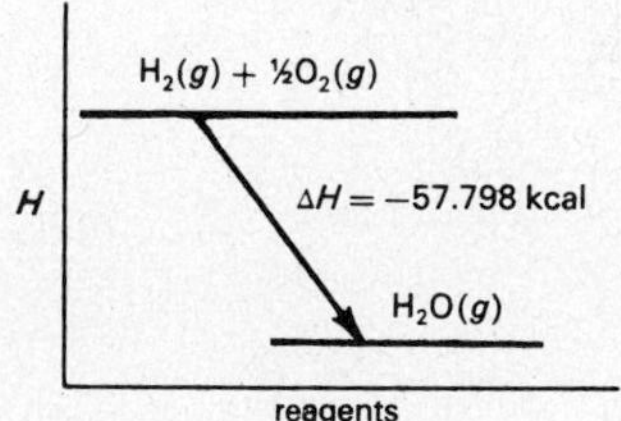

ANSWER 48

The combustion equations must be subtracted to yield the desired reaction.

C(graphite) $+ O_2(g) \rightarrow CO_2(g)$ $\quad \Delta H = -94.052$ kcal

C(diamond) $+ O_2(g) \rightarrow CO_2(g)$ $\quad \Delta H = -94.505$ kcal

C(graphite) $\rightarrow$ C(diamond) $\quad \Delta H = +0.453$ kcal

ITEM 10

The general term *reagent* can refer to either a *reactant* or a *product.* Heat can be regarded as a reagent: In an __________ reaction, heat is a product. If q is used to represent the amount of heat liberated, it would be written on the __________ side of the equation for the reaction.

ITEM 20

The *heat of combustion* of a compound is the heat change accompanying the complete combustion (burning in oxygen) of 1 mole of a compound (unless otherwise specified). Either of these equations

$$2H_2(g) + O_2(g) \rightarrow 2H_2O(l) + 136.634 \text{ kcal}$$
$$H_2(g) + \tfrac{1}{2}O_2(g) \rightarrow H_2O(l) + 68.317 \text{ kcal}$$

shows that the heat of combustion of hydrogen is __________ kcal mole^{-1}.

ITEM 30

The procedure of combining reactions to obtain data on a third reaction depends on the fact that the total heat liberated or absorbed in the successive reactions A → B and B → C is the same as in the single, direct reaction A → __________.

ITEM 40

The expression

$$H_2(g) + I_2(s) \rightarrow 2HI(g) \qquad \Delta H = +12.40 \text{ kcal}$$

states that the heat content of $H_2(g) + I_2(s)$ is __________ than that of $2HI(g)$ by __________ kcal. Is the reaction endothermic or exothermic?

ITEM 50

The *heat of formation,* often indicated as ΔH_f, of a compound is defined as the value of ΔH for the reaction in which 1 mole of that compound is formed from its elements in their commonly occurring state at 25° C and 1 atm pressure. Thus, the heat of formation of $HCl(g)$ is __________ kcal mole^{-1}.

$\underbrace{\text{Reactants} \rightarrow \text{Products}}_{\text{Reagents}}$

ANSWER 9 exothermic
endothermic

Heat of combustion

ANSWER 19 fractions

ANSWER 29 CO_2

ANSWER 39 $H(\text{products}) - H(\text{reactants})$

ANSWER 49 greater

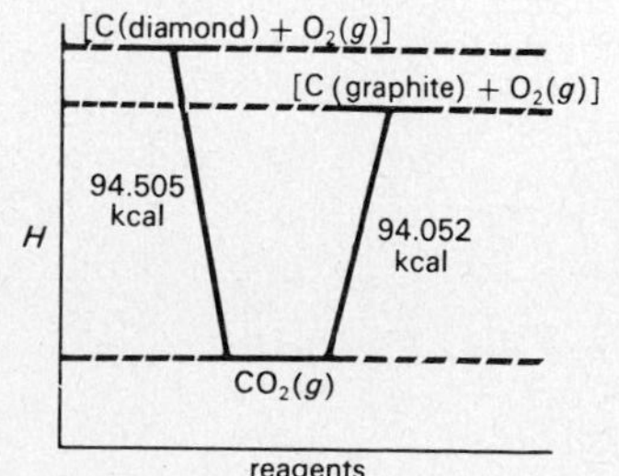

Heat of formation

$\frac{1}{2}H_2(g) + \frac{1}{2}Cl_2(g) \rightarrow HCl(g)$
$\Delta H = -22.1$ kcal

ANSWER 10 exothermic
right

(*Item 11 is on page 192.*)

ANSWER 20 68.317

(*Item 21 is on page 192.*)

ANSWER 30 C

(*Item 31 is on page 192.*)

ANSWER 40 less
12.40
endothermic

(*Item 41 is on page 192.*)

ANSWER 50 −22.1

(*Item 51 is on page 192.*)

This review has presented many ideas and required much calculation, but this quantitative chemistry is necessary as a basis for everyday laboratory work. Throughout these pages you have seen the importance of the mole concept and the balanced equation in determining the relative amounts of reactants and products, including heat, in a chemical reaction. You have seen how the concentration and volume of a solution are related to the number of moles of solute, and you have learned the theory and stoichiometric terms used most often in acid–base reactions. All this is essential to your further study of chemistry.

REVIEW 5 EQUILIBRIUM CALCULATIONS

Whenever a chemist desires to prepare a compound, he wants to use a reaction that "goes to completion," but chemical reactions proceed only until they reach a state of equilibrium. When the equilibrium state is reached, the system can contain mostly products, mostly reactants, or any intermediate proportion of products and reactants.

The position of equilibrium in a chemical system can sometimes be deduced from an expression relating the concentrations of all reagents to a constant for the reaction. For example, if $N_2O_4(g)$ is introduced into a container, dissociation to $NO_2(g)$ occurs according to the reaction equation $N_2O_4(g) \rightleftharpoons 2\,NO_2(g)$. After some definite fraction of N_2O_4 has been converted to NO_2, the concentrations of the two reagents will remain constant, and equilibrium is established. The concentrations of the reagents at equilibrium are related by an expression that has, for this example, the form $K = [NO_2]^2/[N_2O_4]$, in which K is the equilibrium constant for this reaction. This is an approximation that is valid for ideal behavior in much the same way that the expression $PV = nRT$ is really valid only for ideal gases.

5-1 HOMOGENEOUS GAS REACTIONS

This section deals with the equilibria established by gases in a container having a fixed volume and temperature. It is assumed that you know how to write the expression for the equilibrium constant for the reaction if the equation for the reaction is given and that you are able to solve quadratic equations of the form $ax^2 + bx + c = 0$.

A typical problem that you will learn to solve is the following: "Calculate the equilibrium concentrations when 0.60 mole of PCl_5 and 0.020 mole of PCl_3 are allowed to come to equilibrium, according to the reaction $PCl_5(g) \rightleftharpoons PCl_3(g) + Cl_2(g)$ in a 4.7-liter container at 250°C. The equilibrium constant at this temperature is 0.042." When you have finished this review you will be able to solve without difficulty this problem and those of Items 75 through 78.

ITEM 1

At 250°C phosphorus pentachloride dissociates according to the equation at the right. If 1.00 mole of PCl_5 were introduced into a container at 250°C, at a later time there would be (less, more) than 1.00 mole PCl_5 in the container.

ITEM 16

A reaction cell originally contained 1.50 moles of H_2 and 1.50 moles of O_2. After the cell was heated, it contained 0.34 mole of H_2O. Fill in the following table.

	Initially	Used up	Remaining
Moles H_2	1.50		
Moles O_2	1.50		

ITEM 31

The value for the equilibrium constant for the reaction $H_2(g) + I_2(g) \rightleftharpoons 2HI(g)$ at 426°C is 54. In one experiment the equilibrium concentrations of I_2 and HI were $[I_2] = 1.3 \times 10^{-3}$ and $[HI] = 1.6 \times 10^{-2}$. Calculate the equilibrium concentration of H_2.

ITEM 46

Solution of this quadratic equation gives a value of x of 0.016. Tabulate the number of moles and the concentrations at equilibrium.

	N_2O_4	NO_2
Moles	__________	__________
Conc., moles liter^{-1}	__________	__________

ITEM 61

Solution of the quadratic equation yields $x = 0.07$ mole. The number of moles at equilibrium of H_2 = __________, of I_2 = __________, and of HI = __________. Use these values to check the solution of the problem by calculating the value of the equilibrium constant from them.

$PCl_5(g) \rightleftharpoons PCl_3(g) + Cl_2(g)$

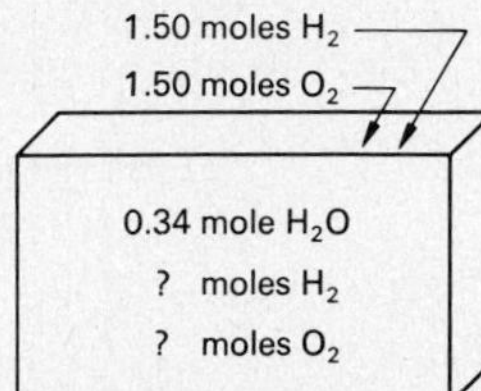

	$N_2O_4(g) \rightleftharpoons$	$2NO_2(g)$
Moles	$0.100 - x$	$2x$

$V = 2.0$ liters

	$2HI(g) \rightleftharpoons$	$H_2(g)$	$+\ I_2(g)$
Equil. moles	$3.00 - 2x$	x	$2.50 + x$

$V = 4.00$ liters

ITEM 2

Each mole of PCl_5 that dissociates produces __________ mole(s) of PCl_3 and __________ mole(s) of Cl_2.

ITEM 17

To perform the calculations in the preceding items, you had to know the __________ for the reaction.

ITEM 32

Once you know the equation for a reaction, the value of K (the symbol for the __________), and the initial concentrations of reagents, you can calculate the concentrations at equilibrium.

ITEM 47

The percentage of N_2O_4 that dissociates (in the system treated in the preceding items) is __________.

ITEM 62

Any component of a mixture of gases has a concentration and a pressure, known as its *partial pressure.* Since the concentration and partial pressure are simply related (assuming ideal gas behavior) by the expression $p = RT(n/V)$, in which R is a constant, T is the absolute temperature, and n/V is the concentration in moles per unit volume, equilibrium calculations for gas reactions can be done either in terms of concentrations or of __________.

$PCl_5(g) \rightleftharpoons PCl_3(g) + Cl_2(g)$

	At equilibrium $N_2O_4(g)$	$\rightleftharpoons 2NO_2(g)$
Moles	0.084	0.032
Conc., moles liter^{-1}	0.042	0.016

ANSWER 1 less

ANSWER 16

Used up	Remaining
0.34	1.16
0.17	1.33

ANSWER 31

$$54 = \frac{(1.6 \times 10^{-2})^2}{[H_2](1.3 \times 10^{-3})}$$

$$[H_2] = 3.6 \times 10^{-3}$$

ANSWER 46

0.084 0.032
0.042 0.016

ANSWER 61

0.07
2.57
2.86

$$K = \frac{(0.07/4)(2.57/4)}{(2.86/4)^2} = 0.022$$

ITEM 3

As the reaction proceeds, the number of moles of PCl_5 will _____crease, the number of moles of PCl_3 and Cl_2 will _____crease, and the total number of moles in the system will _____crease until a state of equilibrium is attained.

ITEM 18

It is convenient to write the initial and later amounts beneath the symbols in the equation, as at the right. Complete the entries.

ITEM 33

A 1.8-mole sample of PCl_5 was added to a 5.0-liter reaction chamber at 250°C. To calculate the amount of each reagent that was present after equilibrium had been established, it is first necessary to write the expression for K. In this case $K =$ __________.

ITEM 48

Equilibrium calculations of the types just treated can be made directly in terms of concentrations. As an example, let $x =$ number of moles per liter of N_2O_4 that dissociate, and tabulate below the equilibrium concentrations from the data at the right.

Initial $[N_2O_4] =$ __________ $[NO_2] =$ __________

Equil. $[N_2O_4] =$ __________ $[NO_2] =$ __________

ITEM 63

When partial pressures are used in equilibrium calculations, the equilibrium constant is written K_p. The equilibrium expression for the reaction $PCl_5(g) \rightleftharpoons PCl_3(g) + Cl_2(g)$ is

$$\_\_\_\_\_\_\_\_\_\_ = \frac{(p_{PCl_3})(p_{Cl_2})}{\_\_\_\_\_\_\_\_\_\_}$$

$PCl_5(g) \rightleftharpoons PCl_3(g) + Cl_2(g)$

	$PCl_5(g)$	$\rightleftharpoons$ $PCl_3(g)$	+ $Cl_2(g)$
Initial moles	0.83	0.19	0
Later moles	0.71		

$PCl_5(g) \rightleftharpoons PCl_3(g) + Cl_2(g)$

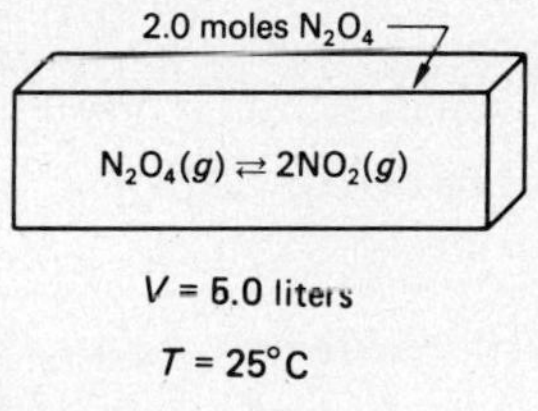

V = 5.0 liters

T = 25°C

ANSWER 2 1
1

ANSWER 17 equation

ANSWER 32 equilibrium constant

ANSWER 47 $\dfrac{(0.016)}{(0.100)}\, 100 = 16\%$

ANSWER 62 partial pressures

(*Chemical Principles* defers discussion of partial pressures in equilibrium calculations until Chapter 15.)

ITEM 4

A 1.00-mole sample of PCl_5 was introduced into a container as indicated at the right. If there is now only 0.75 mole of PCl_5, there must be ________ mole(s) of PCl_3 and ________ mole(s) of Cl_2.

ITEM 19

Complete the entries at the right.

ITEM 34

The second step is the expression of the equilibrium concentrations of the reagents in terms of some convenient unknown. Let x be the number of moles of PCl_5 that dissociate. Complete the table at the right.

ITEM 49

The next step in the calculation is, as usual, the substitution of the equilibrium concentrations, in terms of x, into the expression for K. Since $K = 0.0058$, the equation that permits the calculation of the number of ________ per ________ that dissociate is ________.

ITEM 64

The expression for K_p for the reaction $2HI(g) \rightleftharpoons H_2(g) + I_2(g)$ is ________.

1.00 mole PCl_5

0.75 mole PCl_5

? mole PCl_3

? mole Cl_2

	$2H_2O(g) \rightleftharpoons$	$2H_2(g)$ +	$O_2(g)$
Initial moles	5.8	0	0
Later moles	2.0		

	$PCl_5(g) \rightleftharpoons$	$PCl_3(g)$ +	$Cl_2(g)$
Initial moles	1.8	0	0
Equil. moles	$1.8 - x$		
Equil. conc.			

$V = 5.0$ liters

	$N_2O_4(g) \rightleftharpoons$	$2NO_2(g)$
Equil. conc., moles liter^{-1}	$0.40 - x$	$2x$

$K = 0.0058$ at 25°C

ANSWER 3 de
in
in

ANSWER 18 $PCl_3(g)$: $0.19 + 0.12 = 0.31$
$Cl_2(g)$: $0 + 0.12 = 0.12$

ANSWER 33 $\frac{[PCl_3][Cl_2]}{[PCl_5]}$

ANSWER 48 $\frac{2.0}{5.0} = 0.40$ 0

$0.40 - x$ $2x$

ANSWER 63 K_p
(p_{PCl_5})

ITEM 5

If 1.00 mole of PCl_5 were introduced and x moles now remain, the amount of PCl_3 present would be ________ mole(s).

ITEM 20

Complete the entries at the right.

ITEM 35

The third step is the substitution of the concentrations, in terms of x, and the value of K, 0.042, into the expression for the equilibrium constant. This gives the equation ________.

ITEM 50

Solution of this equation yields $x = 0.023$ mole liter^{-1}. Tabulate below the concentrations and number of moles at equilibrium.

	$N_2O_4(g)$	$\rightleftharpoons$	$2NO_2(g)$
Equil. conc., moles liter^{-1}	________		________
Equil. moles	________		________

ITEM 65

Note that K and K_p are equal for a reaction in which no change in the total number of molecules occurs as the reaction proceeds. For other reactions, K and K_p are different [but related by $K_p = K(\mathrm{RT})^{\Delta n}$, in which Δn is the increase in the number of molecules in the reaction]. Write the expressions for K and K_p for the reaction $2SO_2 + O_2 \rightleftharpoons 2SO_3$. Is K_p equal to K?

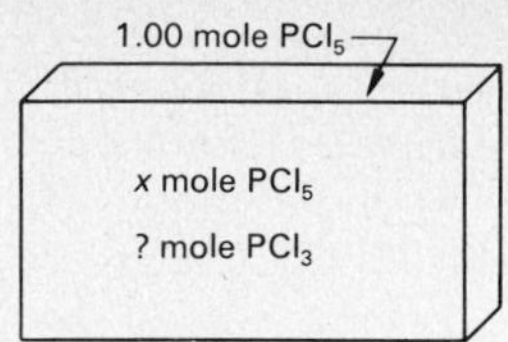

	$2CO_2(g) \rightleftharpoons$	$2CO(g)$	$+\ O_2(g)$
Initial moles	0.0	0.8	0.6
Later moles	0.2		

	$PCl_5(g)$	$\rightleftharpoons PCl_3(g)$	$+\ Cl_2(g)$
Equil. conc.	$\dfrac{1.8 - x}{5.0}$	$\dfrac{x}{5.0}$	$\dfrac{x}{5.0}$

$K = 0.042$ at 250°C

	$N_2O_4(g)$	$\rightleftharpoons 2NO_2(g)$
Equil. conc., moles liter^{-1}	$0.40 - x$	$2x$

$V = 5.0$ liters

ANSWER 4 0.25
0.25

ANSWER 19 $H_2(g)$: $5.8 - 2.0 = 3.8$
$O_2(g)$: $\frac{1}{2}(3.8) = 1.9$

ANSWER 34

$PCl_5(g) \rightleftharpoons$	$PCl_3(g)$	$+\ Cl_2(g)$
1.8	0	0
$1.8 - x$	x	x
$\dfrac{1.8 - x}{5.0}$	$\dfrac{x}{5.0}$	$\dfrac{x}{5.0}$

ANSWER 49 moles
liter

$$0.0058 = \frac{(2x)^2}{0.40 - x}$$

ANSWER 64

$$K_p = \frac{(p_{H_2})(p_{I_2})}{(p_{HI})^2}$$

ITEM 6

The *concentration* of a compound is the amount of that compound per unit volume, often expressed in terms of moles per liter. If a 1.00-mole sample of PCl_5 were introduced into a 5.0-liter vessel, the initial concentration (before any dissociation) would be ________ mole liter^{-1}. The concentration of PCl_5 will _____crease with time.

ITEM 21

The concentrations of the reagents in a reaction vessel can be calculated if one knows the number of ________ of the reagents and the ________ of the container.

ITEM 36

Solution of the quadratic equation (verify if you like) gives $x = 0.52$. Thus, the equilibrium amounts are

PCl_5: ________ mole(s)
PCl_3: ________ mole(s)
Cl_2: ________ mole(s)

ITEM 51

For problems requiring calculation of equilibrium amounts of reagents when the equilibrium constant and the initial amounts of the dissociating species are known, it is necessary to have a balanced ________ for the reaction, to be able to write the expression for the ________ and to express, in terms of some convenient unknown, the ________.

ITEM 66

When initial partial pressures and K_p for a reaction are known and equilibrium pressures must be calculated, the procedure is exactly the same as for calculations using concentrations. The first step is to write the expression for the ________ for the balanced ________.

$$PCl_5(g) \rightleftharpoons PCl_3(g) + Cl_2(g)$$

	$PCl_5(g)$	$PCl_3(g)$	$Cl_2(g)$
Equil. moles	$1.8 - x$	x	x

ANSWER 5 $1.00 - x$

ANSWER 20 $CO(g)$: $0.8 - 0.2 = 0.6$
$O_2(g)$: $0.6 - 0.1 = 0.5$

ANSWER 35

$$0.042 = \frac{(x/5.0)(x/5.0)}{(1.8 - x)/5.0}$$

$$= \frac{x^2}{5.0(1.8 - x)}$$

ANSWER 50 0.38 0.046
1.9 0.23

ANSWER 65

$$K = \frac{[SO_3]^2}{[SO_2]^2[O_2]}$$

$$K_p = \frac{(p_{SO_3})^2}{(p_{SO_2})^2(p_{O_2})}$$

No. $K_p = \dfrac{(RT)^2[SO_3]^2}{(RT)^2[SO_2]^2(RT)[O_2]}$

$$= \frac{1}{RT}\,\frac{[SO_3]^2}{[SO_2]^2[O_2]}$$

ITEM 7

A 1.00-mole sample of PCl_5 was put into a 10.0-liter container. Later, when 0.30 mole of PCl_5 remained, the number of moles of PCl_3 was ________ and the concentration of PCl_3 was ________ mole(s) liter^{-1}.

ITEM 22

Complete the entries at the right for hydrogen, nitrogen, and ammonia in a 5.0-liter container.

ITEM 37

The percentage dissociation is often useful information. In the example of Item 36, ________ mole(s) of the original 1.8 moles dissociated; therefore, the percent dissociation = 100(________/1.8) = ________%.

ITEM 52

Then substitution of the ________ and the ________ into the expression for the equilibrium constant yields an equation that can be solved for *x*.

ITEM 67

HI is put in a vessel held at 490°C at an initial pressure of 1.50 atm. Letting *x* represent the decrease in pressure of HI because of dissociation, tabulate at the right the partial pressures of each reagent at equilibrium.

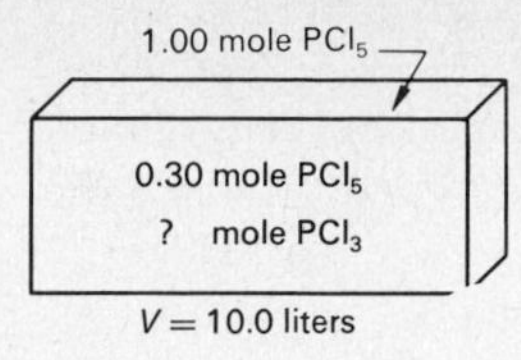

	$3H_2(g)$	$+ \; N_2(g)$	$\rightleftharpoons \; 2NH_3(g)$
Initial moles	0.6	0.4	0.0
Later moles	0.3	0.3	0.2
Later conc.			

$V = 5.0$ liters

At equilibrium

$PCl_5 = 1.3$ moles

$PCl_3 = 0.52$ mole

$Cl_2 = 0.52$ mole

	$2HI(g)$	$\rightleftharpoons \; H_2(g)$	$+ \; I_2(g)$
Initial p	1.50	0	0
Equil. p			

ANSWER 6 0.20
de

ANSWER 21 moles
volume

ANSWER 36 $1.8 - 0.5 = 1.3$
0.52
0.52

ANSWER 51 equation
equilibrium constant
equilibrium concentrations of the reagents

ANSWER 66 equilibrium constant
equation

ITEM 8

A container of volume *V* originally contained *n* moles of PCl_5. If *x* moles of PCl_3 are formed, there remain __________ mole(s) of PCl_5. The concentrations are then

PCl_5 = __________ moles $liter^{-1}$
PCl_3 = __________ moles $liter^{-1}$
Cl_2 = __________ moles $liter^{-1}$

ITEM 23

Complete the table at the right.

ITEM 38

In another experiment, a mixture of 1.00 mole of PCl_5 and 1.00 mole of PCl_3 was put into a 5.0-liter reaction vessel at 250°C. To calculate the amounts of the reagents present at equilibrium, it is first necessary to write the expression for the equilibrium constant: $K =$ __________.

ITEM 53

The equation for the nitrogen–hydrogen–ammonia reaction is $N_2(g) + 3H_2(g) \rightleftharpoons 2NH_3(g)$. The equilibrium expression is __________.

ITEM 68

At 490°C, K_p for the reaction $2HI(g) \rightleftharpoons H_2(g) + I_2(g)$ is 0.022. Write the expression for the equilibrium constant in terms of *x*, the decrease in pressure of HI.

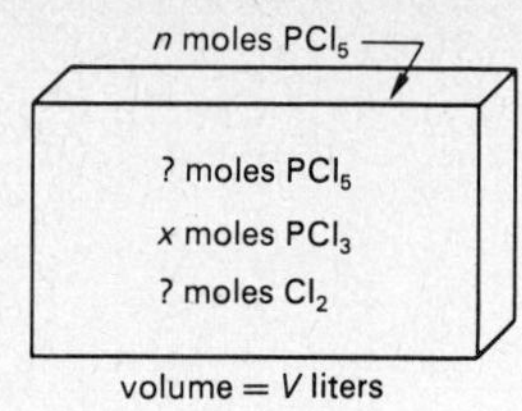

	$2HCl(g) \rightleftharpoons$	$H_2(g)$ +	$Cl_2(g)$
Initial moles	2.86	1.00	0
Later moles	1.30		
Later conc.			

$V = 12.0$ liters

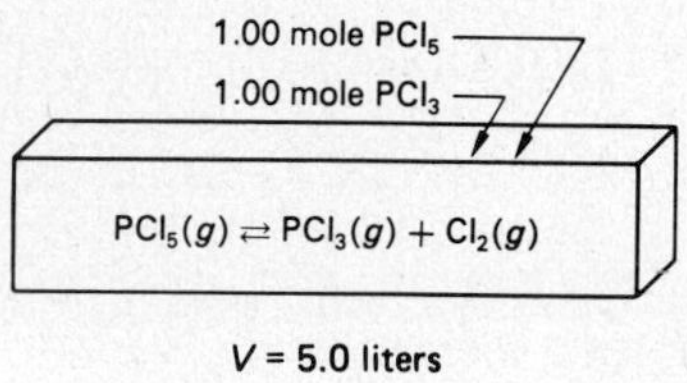

ANSWER 7 0.70
0.070

ANSWER 22 $H_2(g)$: 0.06
$N_2(g)$: 0.06
$NH_3(g)$: 0.04

ANSWER 37 0.52
$(100)(0.52/1.8) = 29$

ANSWER 52 equilibrium constant
concentrations, expressed
in terms of the unknown,

ANSWER 67 $1.50 - x$ $x/2$ $x/2$

ITEM 9

The concentration of a compound in moles per liter is often indicated by placing brackets around its formula. Thus, $[PCl_5] = 0.20$ means that the concentration of PCl_5 is ________.

ITEM 24

Complete the table at the right.

ITEM 39

The next step is to express the equilibrium concentrations in terms of a convenient unknown. Again, let x be the number of moles of PCl_5 that dissociate, and write expressions below for the number of moles and the concentrations at equilibrium in terms of x.

	$PCl_5(g)$ $\rightleftharpoons$	$PCl_3(g)$ +	$Cl_2(g)$
Initial moles	1.00	1.00	0
Later moles	________	________	________
Later conc., moles liter^{-1}	________	________	________

ITEM 54

At 500°C, 12.00 moles of NH_3 are sealed in a 2.00-liter reaction vessel. As a first step in the calculation of the equilibrium concentrations of NH_3, N_2, and H_2 in this system, let x = number of *moles per liter* of NH_3 that dissociate. Then the expressions for the equilibrium concentrations of each reagent are ________.

ITEM 69

Solution of the quadratic equation yields $x = 0.34$ atm. What are the partial pressures of HI, H_2, and I_2 at equilibrium?

Equil. p_{HI}: ________
Equil. p_{H_2}: ________
Equil. p_{I_2}: ________

24

	$3H_2(g)$ +	$N_2(g)$ ⇌	$2NH_3(g)$
Initial moles	0.00	0.35	0.92
Later moles			0.70
Later conc.			

$V = 4.0$ liters

39

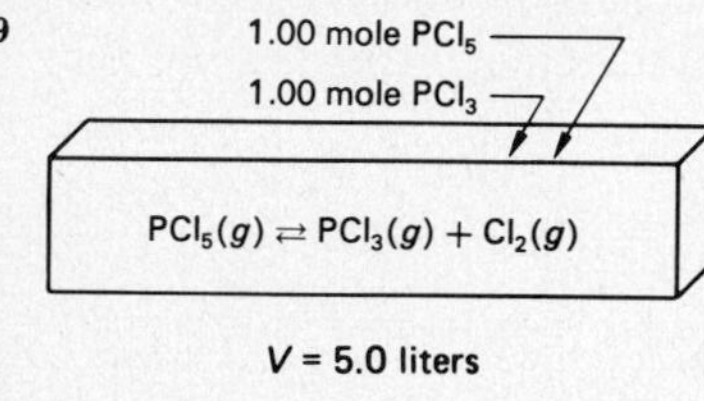

54 $N_2(g) + 3H_2(g) \rightleftharpoons 2NH_3(g)$

69

	$2HI(g)$ ⇌	$H_2(g)$ +	$I_2(g)$
Equil. p	$1.50 - x$	$x/2$	$x/2$

ANSWER 8

$n - x$
$(n - x)/V$
x/V
x/V

ANSWER 23

$2HCl(g)$ ⇌	$H_2(g)$ +	$Cl_2(g)$
2.86	1.00	0
1.30	1.00 + 1.56/2 = 1.78	0.78
0.108	0.148	0.065

ANSWER 38

$$\frac{[PCl_3][Cl_2]}{[PCl_5]}$$

ANSWER 53

$$K = \frac{[NH_3]^2}{[N_2][H_2]^3}$$

ANSWER 68

$$K_p = \frac{(p_{H_2})(p_{I_2})}{(p_{HI})^2}$$

$$0.022 = \frac{(x/2)(x/2)}{(1.50 - x)^2}$$

ITEM 10

In the system at the right, which resulted from the addition of PCl_5 to a 6.00-liter reaction chamber, $[PCl_5]$ is __________, $[PCl_3]$ is __________, and $[Cl_2]$ is __________.

ITEM 25

The concentration results from Item 24 can be indicated with the notation $[H_2]=$ __________, $[N_2]=$ __________, and $[NH_3]=$ __________.

ITEM 40

Using the expressions of Item 39 and the value of K at 250°C, 0.042, set up an equation that permits the calculation of x.

ITEM 55

For the reaction $N_2(g) + 3H_2(g) \rightleftharpoons 2NH_3(g)$, $K = 0.0579$ at 500°C. Write the equation from which x can be calculated.

ITEM 70

The expression for K_p for the reaction $PCl_5(g) \rightleftharpoons PCl_3(g) + Cl_2(g)$ is __________.

0.60 mole PCl_5
0.30 mole PCl_3
? mole Cl_2

$V = 6.00$ liters

ANSWER 9 0.20 mole liter^{-1}

ANSWER 24

$3H_2(g)$	$+ N_2(g) \rightleftharpoons$	$2NH_3(g)$
0.00	0.35	0.92
0.33	0.46	0.70
0.082	0.12	0.18

ANSWER 39

$1.00 - x$	$1.00 + x$	x
$\frac{1.00 - x}{5.0}$	$\frac{1.00 + x}{5.0}$	$\frac{x}{5.0}$

ANSWER 54

	$N_2(g)$	$+ 3H_2(g) \rightleftharpoons$	$2NH_3(g)$
Initial conc.	0	0	6.00
Final conc.	$\frac{1}{2}x$	$\frac{3}{2}x$	$6.00 - x$

ANSWER 69

HI: 1.16 atm
H_2: 0.17 atm
I_2: 0.17 atm

ITEM 11

If PCl_5 were introduced into a container at 250°C, some PCl_3 and some Cl_2 would form. The dissociation equation __________ shows that the number of moles of PCl_3 will be __________ the number of moles of Cl_2.

ITEM 26

The brackets around a chemical symbol indicate that the __________ of the compound is being given and that the units are __________ per __________.

ITEM 41

Solution of the quadratic equation gives $x = 0.15$. Complete the tabulation of the equilibrium amounts below.

	Moles initially	Moles at equilibrium
PCl_5	1.00	
PCl_3	1.00	
Cl_2	0.0	

ITEM 56

Solution of the fourth-power equation yields $x = 3.07$ moles liter^{-1}. What are the equilibrium concentrations of N_2, H_2, and NH_3?

ITEM 71

For the reaction $PCl_5(g) \rightleftharpoons PCl_3(g) + Cl_2(g)$, K_p at 250°C is 1.78. If a reaction vessel initially contains 2.00 atm of PCl_5 and 0.50 atm of Cl_2 at 250°C, the expression for the equilibrium constant in terms of x, the decrease in the pressure of PCl_5, is __________.

$$PCl_5(g) \rightleftharpoons PCl_3(g) + Cl_2(g)$$

	$PCl_5(g)$	$PCl_3(g)$	$Cl_2(g)$
Equil. moles	$1.00 - x$	$1.00 + x$	x

$$N_2(g) + 3H_2(g) \rightleftharpoons 2NH_3(g)$$

	$N_2(g)$	$3H_2(g)$	$2NH_3(g)$
Equil. conc.	$0.5x$	$1.5x$	$6.00 - x$

ANSWER 10

$$\frac{0.60 \text{ mole}}{6.00 \text{ liters}} = 0.10$$

$$\frac{0.30 \text{ mole}}{6.00 \text{ liters}} = 0.050$$

$$\frac{0.30 \text{ mole}}{6.00 \text{ liters}} = 0.050$$

ANSWER 25

0.082
0.12
0.18

ANSWER 40

$$0.042 = \frac{x(1.00 + x)}{5.0(1.00 - x)}$$

ANSWER 55

$$0.0579 = \frac{(6.00 - x)^2}{(0.500x)(1.50x)^3} = \frac{(6.00 - x)^2}{1.69x^4}$$

ANSWER 70

$$K_p = \frac{(p_{PCl_3})(p_{Cl_2})}{p_{PCl_5}}$$

ITEM 12

According to the equation at the right, the dissociation of 1 mole of H_2O produces __________ mole(s) of H_2 and __________ mole(s) of O_2.

ITEM 27

The expression for the equilibrium constant for the reaction

$$PCl_5(g) \rightleftharpoons PCl_3(g) + Cl_2(g)$$

has the form $K =$ __________.

ITEM 42

Check the calculations of the preceding items by calculating the value of K from the results of Item 41. (Recall that $K = 0.042$ was used to obtain these values.)

ITEM 57

What is the percentage dissociation of NH_3 in Item 56?

ITEM 72

Solve the quadratic equation for x. Remember that for $ax^2 + bx + c = 0$

$$x = \frac{-b \pm \sqrt{b^2 - 4ac}}{2a}$$

$2H_2O(g) \rightleftharpoons 2H_2(g) + O_2(g)$

At equilibrium
$[PCl_5] = \frac{0.85}{5.0} = 0.17$
$[PCl_3] = \frac{1.15}{5.0} = 0.23$
$[Cl_2] = \frac{0.15}{5.0} = 0.030$

ANSWER 11 $PCl_5(g) \rightleftharpoons PCl_3(g) + Cl_2(g)$
equal to

ANSWER 26 concentration
moles
liter

ANSWER 41 0.85
1.15
0.15

ANSWER 56 $[N_2] = 1.54$
$[H_2] = 4.60$
$[NH_3] = 2.93$

ANSWER 71 $1.78 = \frac{x(0.50 + x)}{2.00 - x}$

ITEM 13

If n moles of H_2O are heated until x moles have dissociated, the amounts of the three reagents present will be $H_2O =$ ________ moles, $H_2 =$ ________ moles, and $O_2 =$ ________ moles.

ITEM 28

In one experiment PCl_5 was heated at 250°C until its concentration remained constant. The equilibrium concentration of each reagent was that given at the right. Calculate the value of K at 250°C.

ITEM 43

For the system $N_2O_4(g) \rightleftharpoons 2NO_2(g)$ the applicable equilibrium expression is ________.

ITEM 58

The equation for the hydrogen–iodine–hydrogen iodide reaction is $2HI(g) \rightleftharpoons H_2(g) + I_2(g)$. The equilibrium expression for this reaction is ________.

ITEM 73

Tabulate the equilibrium partial pressures.

$p_{PCl_5} =$ ________
$p_{PCl_3} =$ ________
$p_{Cl_2} =$ ________

What is the percentage decomposition of PCl_5?

$2H_2O(g) \rightleftharpoons 2H_2(g) + O_2(g)$

ANSWER 12 1

$\frac{1}{2}$

ANSWER 27 $\frac{[PCl_3][Cl_2]}{[PCl_5]}$

At equilibrium

$[PCl_5] = 0.015$
$[PCl_3] = 0.025$
$[Cl_2] = 0.025$

$V = 5.0$ liters

ANSWER 42 $K = \frac{(0.23)(0.030)}{0.17} = 0.041$

(0.042 is not obtained because the value of x was rounded off to two significant figures for Item 41.)

ANSWER 57 $\frac{3.07}{6.00} \times 100 = 51.2\%$

	$PCl_5(g)$	$\rightleftharpoons$	$PCl_3(g)$	+	$Cl_2(g)$
Equil. p	$2.00 - x$		x		$0.50 + x$

ANSWER 72 $x^2 + 2.28x - 3.56 = 0$

$$x = \frac{-2.28 \pm \sqrt{5.20 + 14.24}}{2}$$

$x = 1.06$

ITEM 14

If n moles of H_2O were heated until the system contained, as a result of dissociation, y moles of O_2, the amounts present would be y moles of O_2, ________ moles of H_2, and ________ moles of H_2O.

ITEM 29

If the cell had been filled originally with pure PCl_5, its concentration before dissociation must have been ________, and the total number of moles of PCl_5 before dissociation must have been ________.

ITEM 44

An amount of 0.100 mole of N_2O_4 is placed in a 2.0-liter vessel at 25°C and allowed to equilibrate. The expressions for the equilibrium concentrations of the reagents in terms of x, the number of moles of N_2O_4 that dissociate, are

	N_2O_4	NO_2
Initial moles	________	________
Equil. moles	________	________
Equil. conc.	________	________

ITEM 59

A mixture of 3.00 moles of HI and 2.50 moles of I_2 is heated at 490°C in a 4.00-liter reaction vessel. Letting $x =$ number of moles of H_2 present at equilibrium, express the equilibrium concentration of each reagent in terms of x.

ITEM 74

Check the solution to Item 73 by substituting the calculated partial pressures into the equilibrium expression and solving for K_p.

$2H_2O(g) \rightleftharpoons 2H_2(g) + O_2(g)$

ANSWER 13 $n - x$
x
$x/2$

At equilibrium

$[PCl_5] = 0.015$
$[PCl_3] = 0.025$
$[Cl_2] = 0.025$

$V = 5.0$ liters

ANSWER 28 $$\frac{(0.025)(0.025)}{0.015} = 0.042$$

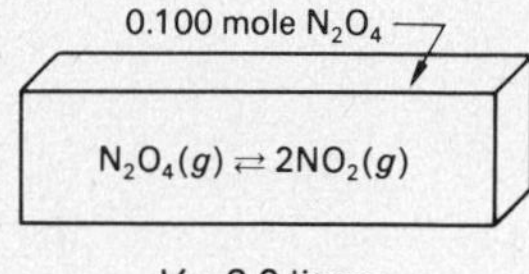

ANSWER 43 $$K = \frac{[NO_2]^2}{[N_2O_4]}$$

$2HI(g) \rightleftharpoons H_2(g) + I_2(g)$

ANSWER 58 $$K = \frac{[H_2][I_2]}{[HI]^2}$$

$PCl_5(g) \rightleftharpoons PCl_3(g) + Cl_2(g)$

$K_p = 1.78$ at 250°C

ANSWER 73 0.94
1.06
1.56
% decomposition

$$= \frac{(100)(1.06)}{2.00} = 53$$

ITEM 15

A reaction cell contained 0.52 mole of H_2 and 0.26 mole of O_2. After the cell was heated, it was found to contain 0.12 mole of H_2O. Fill in the following table.

	Initially	Used up	Remaining
Moles H_2	0.52		
Moles O_2	0.26		

ITEM 30

Write the expression for the equilibrium constant K for the reaction $H_2(g) + I_2(g) \rightleftharpoons 2HI(g)$.

ITEM 45

At 25°C, the value of K for the reaction $N_2O_4(g) \rightleftharpoons 2NO_2(g)$ is 0.0058. An equation from which x can be calculated is __________.

ITEM 60

The equilibrium constant for the reaction $2HI(g) \rightleftharpoons H_2(g) + I_2(g)$ at 490°C is 0.022. Write the equation from which a value of x can be obtained.

Items 75 through 78 are four problems to test your skill at doing equilibrium calculations. Set up each problem in an organized fashion and solve the quadratic equations. Remember that the forms of the concentration expressions in terms of x will vary with the definition of x, but that there is only one correct answer.

(*Turn to page 248.*)

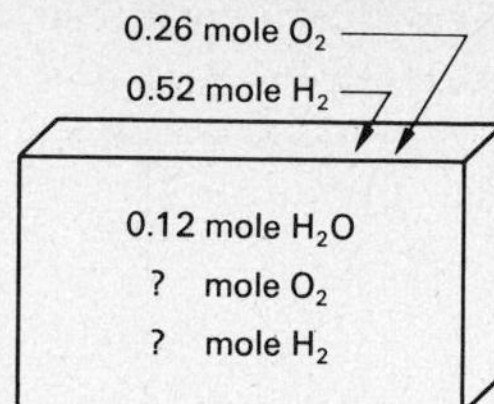

ANSWER 14 $2y$

$n - 2y$

ANSWER 29

$[0.015] + [0.025] = 0.040 \text{ mole liter}^{-1}$
$(0.040 \text{ mole liter}^{-1})(5.0 \text{ liters}) = 0.20 \text{ mole}$

ANSWER 44

	N_2O_4	NO_2
Initial moles	0.100	0
Equil. moles	$0.100 - x$	$2x$
Equil. conc., moles liter^{-1}	$\frac{0.100 - x}{2.0}$	$\frac{2x}{2.0}$

ANSWER 59

	$2HI(g)$ $\rightleftharpoons$	$H_2(g)$ +	$I_2(g)$
Initial moles	3.00	0	2.50
Equil. moles	$3.00 - 2x$	x	$2.50 + x$
Equil. conc.	$\frac{3.00 - 2x}{4.00}$	$\frac{x}{4.00}$	$\frac{2.50 + x}{4.00}$

ANSWER 74 $K_p = \frac{(1.06)(1.56)}{0.94} = 1.76$

ITEM 75

For the reaction $3H_2(g) + N_2(g) \rightleftharpoons 2NH_3(g)$, the equilibrium constant K_p at 500°C is 1.43×10^{-5}. If the equilibrium partial pressure of H_2 is 3.00 atm and N_2 is 1.30 atm, what is the partial pressure of NH_3 at equilibrium?

ANSWER 15

Used up	Remaining
0.12	0.40
0.06	0.20

(*Item 16 is on page 218.*)

ANSWER 30

$$K = \frac{[HI]^2}{[H_2][I_2]}$$

(*Item 31 is on page 218.*)

ANSWER 45

$$0.0058 = \frac{(2x/2.0)^2}{(0.100 - x)/2.0}$$

$$= \frac{2.0x^2}{0.100 - x}$$

(*Item 46 is on page 218.*)

ANSWER 60

$$0.022 = \frac{(x/4.00)[(2.50 + x)/4.00]}{[(3.00 - 2x)/4.00]^2}$$

$$= \frac{x(2.50 + x)}{(3.00 - 2x)^2}$$

(*Item 61 is on page 218.*)

ITEM 76

For the reaction $2HI(g) \rightleftharpoons H_2(g) + I_2(g)$, the equilibrium constant, K, at 490°C is 0.022. If 2.7 moles of HI are allowed to dissociate in a 4.3-liter reaction vessel at 490°C, what *concentrations* of HI, H_2, and I_2 are present at equilibrium?

ANSWER 75

$$K_p = \frac{(p_{NH_3})^2}{(p_{H_2})^3(p_{N_2})}$$

$$1.43 \times 10^{-5} = \frac{(p_{NH_3})^2}{(3.00)^3(1.30)}$$

$$(p_{NH_3})^2 = 5.02 \times 10^{-4}$$

$$p_{NH_3} = 0.0224 \text{ atm}$$

ITEM 77

The equilibrium constant, K, for the reaction $H_2(g) + CO_2(g) \rightleftharpoons H_2O(g) + CO(g)$ is 0.719 at 1000°K. If 3.00 moles of H_2O and 2.00 moles of CO are allowed to react in a 5.00-liter reaction vessel at 1000°K, what is the number of moles of each of the four reagents at equilibrium?

ANSWER 76

$$K = \frac{[H_2][I_2]}{[HI]^2}$$

If x = number of moles of HI that dissociate

	$2HI(g) \rightleftharpoons$	$H_2(g)$ +	$I_2(g)$
Initial moles	2.7	0	0
Equil. moles	$2.7 - x$	$x/2$	$x/2$
Equil. conc., moles liter^{-1}	$\left(\frac{2.7 - x}{4.3}\right)$	$\left(\frac{x/2}{4.3}\right)$	$\left(\frac{x/2}{4.3}\right)$

$$K = 0.022 = \frac{[(x/2)/(4.3)][(x/2)/(4.3)]}{[(2.7 - x)/4.3]^2}$$

$$= \frac{x^2/4}{(2.7 - x)^2}$$

$x = 0.62$ mole

Equil. [HI] = 0.49 $[H_2] = 0.072$ $[I_2] = 0.072$

Or if y = number of moles liter^{-1} of HI that dissociate

	$2HI(g) \rightleftharpoons$	$H_2(g)$ +	$I_2(g)$
Initial conc.	$\frac{2.7}{4.3} = 0.63$	0	0
Equil. conc.	$0.63 - y$	$y/2$	$y/2$

$$K = 0.022 = \frac{(y/2)(y/2)}{(0.63 - y)^2} = \frac{y^2/4}{(0.63 - y)^2}$$

$y = 0.144$ mole liter^{-1}

Equil. [HI] = 0.49 $[H_2] = 0.072$ $[I_2] = 0.072$

ITEM 78

The equilibrium constant, K, for the reaction $PCl_5(g) \rightleftharpoons PCl_3(g) + Cl_2(g)$ is 0.042 at 250°C. In experiment A, 1.00 mole of PCl_5 is allowed to dissociate in a 5.0-liter reaction vessel at 250°C. In experiment B, a mixture of 1.00 mole of PCl_5 and 0.50 mole of Cl_2 is allowed to react in the 5.0-liter reaction vessel. For the two experiments, calculate and compare the percentage decomposition of PCl_5.

ANSWER 77

$$K = \frac{[H_2O][CO]}{[H_2][CO_2]}$$

Let x = number of moles of H_2O that react

	$H_2(g)$ +	$CO_2(g)$ ⇌	$H_2O(g)$ +	$CO(g)$
Initial moles	0	0	3.00	2.00
Equil. moles	x	x	$3.00 - x$	$2.00 - x$
Equil. conc.	$x/5$	$x/5$	$(3.00 - x)/5$	$(2.00 - x)/5$

$$K = 0.719 = \frac{[(3.00 - x)/5][(2.00 - x)/5]}{(x/5)(x/5)}$$

$$= \frac{(3.00 - x)(2.00 - x)}{x^2}$$

$$0.281x^2 - 5.00x + 6.00 = 0$$

$x = 1.30$ moles

	$H_2(g)$ +	$CO_2(g)$ ⇌	$H_2O(g)$ +	$CO(g)$
Equil. moles	1.30	1.30	1.70	0.70

ANSWER 78

$$K = \frac{[PCl_3][Cl_2]}{[PCl_5]}$$

Let x = number of moles of PCl_5 that dissociate

Experiment A

	$PCl_5(g)$	$\rightleftharpoons$	$PCl_3(g)$	$+ Cl_2(g)$
Initial moles	1.00		0	0
Equil. moles	$1.00 - x$		x	x
Equil. conc.	$(1.00 - x)/5.0$		$x/5.0$	$x/5.0$

$$0.042 = \frac{(0.20x)(0.20x)}{0.20 - 0.20x}$$

$$0.040x^2 + 0.0084x - 0.0084 = 0$$

$x = 0.37$ mole

$$\% \text{ decomposition} = 100 \times \frac{0.37}{1.00} = 37$$

Experiment B

	$PCl_5(g)$	$\rightleftharpoons$	$PCl_3(g)$	$+$	$Cl_2(g)$
Initial moles	1.00		0		0.50
Equil. moles	$1.00 - x$		x		$0.50 + x$
Equil. conc.	$(1.00 - x)/5.0$		$x/5.0$		$(0.50 + x)/5.0$

$$0.042 = \frac{(0.20x)(0.10 + 0.20x)}{0.20 - 0.20x}$$

$$0.040x^2 + 0.028x - 0.0084 = 0$$

$x = 0.23$ mole

$$\% \text{ decomposition} = 100 \times \frac{0.23}{1.00} = 23$$

(*Turn to page 258.*)

In this section you have learned how to do the ideal behavior calculations associated with equilibria in homogeneous gas reactions. You have found that it is first necessary to have the balanced equation for the reaction and to be able to write the expression for the equilibrium constant, K. If you know K, as well as the initial concentrations of the reagents, you can calculate the equilibrium concentrations in terms of some convenient unknown. Sometimes the partial pressures of gases are used instead of concentrations, and you have learned the corresponding relationship between K_p and K.

In the next two sections the same principles will be applied to other types of equilibria.

5-2 THE IONIZATION OF ACIDS AND BASES

When certain acids and bases are dissolved in water they undergo only partial dissociation or ionization. Thus a solution of acetic acid in water contains

$$\begin{array}{l} CH_3\underset{\displaystyle \overset{\|}{O}}{C}OH \end{array}$$

molecules and

$$\begin{array}{l} CH_3\underset{\displaystyle \overset{\|}{O}}{C}O^- \end{array}$$

and H_3O^+ ions. Some inorganic acids and bases, too, are only partially ionized in water; hydrogen sulfide and ammonia are examples. Partially ionized acids and bases such as these are referred to collectively as weak acids and bases.

This review will help you with parts of Sections 5-2 through 5-9 of *Chemical Principles*. We will examine equilibria that exist between the various species in solutions of weak acids and bases and will present some simple types of concentration calculations. Items 81 through 85 are representative of the kinds of problems that you will learn to solve.

ITEM 1

According to the Brønsted definitions, an acid is a compound that donates a proton, and a base is a compound that __________.

ITEM 15

Substitution of the experimentally determined concentrations into the expression for the acidity constant gives the value

$$K_a = \frac{(\_\_\_\_\_\_)(\_\_\_\_\_\_)}{(\_\_\_\_\_\_)} = \_\_\_\_\_\_$$

ITEM 29

Hydrochloric acid, HCl, is a strong acid and dissociates completely in dilute aqueous solution. The hydrogen ion concentration of a 0.0010-molar solution of HCl is __________. The pH is __________.

ITEM 43

The value of K_a for HCN at 25°C is 4.0×10^{-10}. The equation to be solved for x is then __________.

ITEM 57

Since K_b for NH_3 is 1.74×10^{-5} and the equilibrium expression is __________, the equation that must be solved for x to determine the equilibrium concentrations of the reagents is __________.

ITEM 71

The extent of dissociation of 0.020-molar HNO_2 in 0.10-molar HCl is 0.45%, and the dissociation of 0.020-molar HNO_2 in 0.00010-molar HCl is 14%. This is an example of the *common ion effect*: An increase in the concentration of the common ion, the ion __________, causes a(n) _____crease in the extent of dissociation of the weak acid.

ITEM 85

The acid HSO_4^- partially dissociates according to the equation $HSO_4^- \rightleftharpoons H^+ + SO_4^{2-}$. K_a for HSO_4^- is 1.2×10^{-2} at 25°C. What is the value of $[H^+]$ in a solution made by adding 0.0050 mole of $NaHSO_4$ to enough water to make 100 ml of solution?

CH_3COOH	$\rightleftharpoons$	H^+	+	CH_3COO^-
$0.019M$		$0.00058M$		$0.00058M$

	HCN	$\rightleftharpoons$ H^+	+ CN^-
Equil. M	$0.020 - x$	x	x

$$NH_3 + H_2O \rightleftharpoons NH_4^+ + OH^-$$

ITEM 2

In an aqueous solution of an acid, the acid dissociates by donating a proton to __________.

ITEM 16

The weak base methylamine, CH_3NH_2, reacts with water according to the equation $CH_3NH_2 + H_2O \rightleftharpoons CH_3NH_3^+ + OH^-$. The expression for K_b is __________.

ITEM 30

In a 0.0028-molar solution of the strong acid nitric acid, HNO_3, $[H^+] =$ __________. The pH is ________.

ITEM 44

Solution of this equation yields $x = 2.8 \times 10^{-6}$. The concentration of H^+ at equilibrium is therefore __________. The concentration of undissociated HCN is essentially unchanged from the initial concentration.

ITEM 58

This equation may be simplified considerably by trying the assumption that x is small compared with both 0.0200 and 0.100. The simplified equation is __________.

ITEM 72

A solution of 0.080-molar HNO_2 is prepared. It has $[H^+] =$ __________, $[NO_2^-] =$ __________, and $[HNO_2] =$ __________.

$0.0028 = 2.8 \times 10^{-3}$

$$\log 0.0028 = -3 + \log 2.8$$
$$= -3 + 0.5$$
$$= -2.5$$

$HNO_2 \rightleftharpoons H^+ + NO_2^-$

$K_a = 4.5 \times 10^{-4}$

ANSWER 1 accepts a proton

ANSWER 15

$$\frac{(5.8 \times 10^{-4})(5.8 \times 10^{-4})}{1.9 \times 10^{-2}} = 1.8 \times 10^{-5}$$

ANSWER 29 $1.0 \times 10^{-3}M$

3

ANSWER 43

$$\frac{x^2}{0.020 - x} = 4.0 \times 10^{-10}$$

ANSWER 57

$$K_b = \frac{[NH_4^+][OH^-]}{[NH_3]}$$

$$1.74 \times 10^{-5} = \frac{x(x + 0.0200)}{0.100 - x}$$

ANSWER 71 H^+

de

ANSWER 85

$$K_a = \frac{[H^+][SO_4^{2-}]}{[HSO_4^-]}$$

Let x = number of moles per liter of HSO_4^- that dissociate.

$$1.2 \times 10^{-2} = \frac{x^2}{0.050 - x}$$

K_a is large and no simplification is possible.

$x = 0.019$

Therefore, $[H^+] = 0.019$.

(*Turn to page 290.*)

ITEM 3

For example, in an aqueous solution of the weak acid nitrous acid, HNO_2, the HNO_2 molecules donate protons to the water molecules to form NO_2^- and H_3O^+ ions. The equation for this ionization reaction is $HNO_2 + H_2O \rightleftharpoons$ ________.

ITEM 17

Since methylamine is a relatively weak base, it has ________ tendency to accept a proton from an acid than does a strong base. The weaker a base is, the smaller is the extent of proton addition. This corresponds to a ________ value of K_b.

ITEM 31

The fact that $[H^+][OH^-] = 1.0 \times 10^{-14}$ in aqueous solution at 25°C allows us to calculate either $[H^+]$ or $[OH^-]$ whenever the other is known. In a 0.0010-molar solution of HCl, the concentration of OH^- is ________.

ITEM 45

The extent of dissociation of HCN is so small that x is minute compared with 0.020, the molarity of HCN. We could simplify the equation of Item 43 by making the assumption that $0.020 - x \simeq 0.020$. The simplified equation is ________.

ITEM 59

Solution of the simplified equation gives a value for x of ________. The assumption that x is small compared with ________ is justified.

ITEM 73

The concentration of the H^+ ions in a dilute solution of HNO_2 will be decreased if more ________ ions are added because of the ________ ion effect.

$CH_3NH_2 + H_2O \rightleftharpoons CH_3NH_3^+ + OH^-$

$$4.0 \times 10^{-10} = \frac{x^2}{0.020 - x}$$

$x = 2.8 \times 10^{-6}$

$HNO_2 \rightleftharpoons H^+ + NO_2^-$

ANSWER 2 water

ANSWER 16 $\dfrac{[CH_3NH_3^+][OH^-]}{[CH_3NH_2]}$

ANSWER 30 2.8×10^{-3}
2.5

ANSWER 44 2.8×10^{-6} molar

ANSWER 58 $1.74 \times 10^{-5} = \dfrac{0.0200x}{0.100}$

ANSWER 72 Let x = number of moles per liter of HNO_2 dissociated.

$$4.5 \times 10^{-4} = \frac{x^2}{0.080 - x}$$

$$\simeq \frac{x^2}{0.080}$$

$x = 6.0 \times 10^{-3}$
$= [H^+] = [NO_2^-]$

$[HNO_2] = 0.074$

ITEM 4

The equilibrium constant expression for this reaction has the form

$$K_{eq} = \frac{[H_3O^+][NO_2^-]}{[HNO_2][H_2O]}$$

in which the brackets around the symbols denote __________.

ITEM 18

A 0.0100-molar solution of methylamine in water is 22.4% ionized. The concentrations needed for the calculation of K_b are CH_3NH_2, __________; $CH_3NH_3^+$, __________; and OH^-, __________.

ITEM 32

Sodium hydroxide, NaOH, dissociates completely in water solution. In a solution of 1.3×10^{-4}-molar NaOH, the hydroxide ion concentration is __________ and the hydrogen ion concentration is __________.

ITEM 46

Such simplifications are usually satisfactory when the value of x calculated from the simplified equation is less than about 10% of the value to which it is added or from which it is __________.

ITEM 60

In the solution 0.0200 molar in NaOH and 0.100 molar in NH_3, $[NH_4^+]$ = __________, $[OH^-]$ = __________, $[NH_3]$ = __________, and $[Na^+]$ = __________.

ITEM 74

Suppose that you want a solution in which the hydrogen ion concentration in 0.080-molar HNO_2 is 1.0×10^{-3} mole liter^{-1}. The number of moles per liter of HNO_2 dissociated at equilibrium will be __________.

$HNO_2 + H_2O \rightleftharpoons H_3O^+ + NO_2^-$

ANSWER 3 $H_3O^+ + NO_2^-$

ANSWER 17 less
low

$CH_3NH_2 + H_2O \rightleftharpoons CH_3NH_3^+ + OH^-$

ANSWER 31

$$\frac{K_w}{[H^+]} = \frac{1.0 \times 10^{-14}}{1.0 \times 10^{-3}}$$
$$= 1.0 \times 10^{-11} \text{ molar}$$

ANSWER 45

$$4.0 \times 10^{-10} = \frac{x^2}{0.020}$$

ANSWER 59 8.7×10^{-5}
0.0200 and 0.100

At equilibrium
$[NH_4^+] = x$
$[OH^-] = 0.0200 + x$
$[NH_3] = 0.100 - x$

ANSWER 73 NO_2^-
common

ITEM 5

However, because $[H_2O]$ remains essentially constant in dilute solutions of HNO_2 (those of concern to us here), we can write a new and simpler expression in which the constant terms are collected: $K_a = K_{eq}[H_2O] =$ ________.

ITEM 19

The value of K_b for CH_3NH_2 is

$$K_b = \frac{(\_\_\_\_\_\_)(\_\_\_\_\_\_)}{(\_\_\_\_\_\_)} = \_\_\_\_\_\_$$

ITEM 33

What is the value of $[H^+]$ for a solution prepared by adding enough water to 0.0070 mole of HNO_3 to make 350 ml of solution?

ITEM 47

The expression in terms of x that will yield the concentration of H^+ in a 0.85-molar solution of acetic acid, CH_3COOH, is ________.

ITEM 61

In a solution 0.100 molar in NH_3, the concentration of NH_4^+ is 1.3×10^{-3} molar (Item 53) and the percentage ionization of NH_3 is ________. In a solution 0.100 molar in NH_3 and 0.0200 molar in NaOH, the concentration of NH_4^+ is 8.7×10^{-5} molar and the percentage ionization is ________.

ITEM 75

You can control the hydrogen ion concentration by adding the appropriate amount of $NaNO_2$ to furnish NO_2^- ions. To calculate the concentration of $NaNO_2$ needed to produce a $[H^+]$ of 1.0×10^{-3} in a 0.080-molar HNO_2 solution, let x equal the required $NaNO_2$ molarity and write the equilibrium concentrations of NO_2^- ________, H^+ ________, and HNO_2 ________.

$$K_{eq} = \frac{[H_3O^+][NO_2^-]}{[HNO_2][H_2O]}$$

ANSWER 4 concentrations in moles per liter

$$CH_3NH_2 + H_2O \rightleftharpoons CH_3NH_3^+ + OH^-$$

$$K_b = \frac{[CH_3NH_3^+][OH^-]}{[CH_3NH_2]}$$

ANSWER 18

CH_3NH_2:

$0.0100 - (0.224 \times 0.0100) = 7.8 \times 10^{-3}M$

$CH_3NH_3^+$:

$0.224 \times 0.0100 = 2.24 \times 10^{-3}M$

OH^-:

$0.224 \times 0.0100 = 2.24 \times 10^{-3}M$

ANSWER 32 1.3×10^{-4} molar

$$\frac{K_w}{[OH^-]} = \frac{1.0 \times 10^{-14}}{1.3 \times 10^{-4}}$$

$$= 7.7 \times 10^{-11} \text{ molar}$$

$$CH_3COOH \rightleftharpoons H^+ + CH_3COO^-$$

$K_a = 1.8 \times 10^{-5}$ at 25°C

ANSWER 46 subtracted

ANSWER 60 8.7×10^{-5}
0.0201
0.100
0.0200

ANSWER 74 1.0×10^{-3}

ITEM 6

An equilibrium constant expression of the same form can be found directly by omitting the solvent in the ionization equation. Thus, $HNO_2 \rightleftharpoons H^+ + NO_2^-$ leads to the equilibrium expression ________. $[H_3O^+]$ represents the same concentration of hydrogen ions as does $[H^+]$, so the expressions of Items 5 and 6 are the same.

ITEM 20

Water has the capacity to act as either an acid or a base and, in fact, pure water is ionized to a small extent according to the equation $H_2O + H_2O \rightleftharpoons H_3O^+ + OH^-$. The constant for the dissociation of water, in which $[H_2O]$ has been combined with the equilibrium constant, is known as the *ion product* for water, K_w. The expression for K_w is ________.

ITEM 34

The calculation of $[H^+]$ and $[OH^-]$ for solutions of strong acids and bases of given concentration requires only the value of K_w at the appropriate temperature. For solutions of weak acids and bases, it is also necessary to know the value of K_a, the ________ constant, or K_b, the ________ constant.

ITEM 48

To simplify the solution of the equation for x, the trial approximation $0.85 - x \simeq 0.85$ can be made. When this is done a value for x of ________ is obtained. The approximation accordingly (is, is not) justified.

ITEM 62

The effect of the *common ion*, OH^-, on the ionization of NH_3 in H_2O is to ____crease the extent of ionization.

ITEM 76

Substitution of the expressions from Item 75 into the equilibrium expression for HNO_2 ($K_a = 4.5 \times 10^{-4}$) gives the equation ________.

ANSWER 5 $\frac{[H_3O^+][NO_2^-]}{[HNO_2]}$

ANSWER 19 $\frac{(2.24 \times 10^{-3})(2.24 \times 10^{-3})}{7.8 \times 10^{-3}}$

$= 6.4 \times 10^{-4}$

ANSWER 33 0.0070 mole in 350 ml equals 0.020 mole liter^{-1}. Therefore, $[H^+] = 0.020$.

ANSWER 47 $1.8 \times 10^{-5} = \frac{x^2}{0.85 - x}$

in which x = number of moles per liter of acetic acid that dissociate

ANSWER 61 $\frac{100(1.3 \times 10^{-3})}{1.00 \times 10^{-1}} = 1.3\%$

$\frac{100(8.7 \times 10^{-5})}{1.00 \times 10^{-1}} = 0.087\%$

ANSWER 75 $[NO_2^-] = x + 1.0 \times 10^{-3}$

$[H^+] = 1.0 \times 10^{-3}$

$[HNO_2] = 0.080 - 1.0 \times 10^{-3}$

$= 0.079$

ITEM 7

Ammonia is a weak base in aqueous solution. It ionizes according to the equation $NH_3 + H_2O \rightleftharpoons NH_4^+ + OH^-$. Accordingly, the expression for its basicity constant, in which the constant term $[H_2O]$ is absorbed into the equilibrium constant, is $K_b =$ ________.

ITEM 21

At 25°C, the concentration of H^+ in pure water is 1.01×10^{-7} molar. The numerical value of K_w at 25°C is ________.

ITEM 35

Once K_a or K_b for a compound has been determined, the value can be used to calculate the concentrations of all species at equilibrium in solutions of various concentrations. Of course, you must also have a balanced ________ for the dissociation.

ITEM 49

Based on the result of Item 48, the percentage dissociation of 0.85-molar acetic acid at 25°C is calculated to be ________.

ITEM 63

A solution is prepared in which the total concentration of HNO_2 is 0.020 molar and the concentration of HCl is 0.10 molar. To calculate the percentage dissociation of HNO_2 in this solution, first write the concentration expressions in x for HNO_2, ________; H^+, ________; and NO_2^-, ________.

ITEM 77

Solution of the equation yields $x =$ ________. Therefore, if the equilibrium concentration of H^+ in the 0.080-molar HNO_2 solution is 1.0×10^{-3}, the concentration of $NaNO_2$ is ________ and the total NO_2^- concentration is ________.

$K_w = [H^+][OH^-]$

$CH_3COOH \rightleftharpoons H^+ + CH_3COO^-$
0.85*M* 0.0039*M* 0.0039*M*

$HNO_2 \rightleftharpoons H^+ + NO_2^-$

ANSWER 6 $K_a = \dfrac{[H^+][NO_2^-]}{[HNO_2]}$

ANSWER 20 $K_w = [H^+][OH^-]$
(or $K_w = [H_3O^+][OH^-]$)

ANSWER 34 acidity
basicity

ANSWER 48 $1.8 \times 10^{-5} = x^2/0.85$
$x = 0.0039 = [H^+]$
is (0.0039 is less than 10% of 0.85.)

ANSWER 62 de

ANSWER 76

$$K_a = \frac{[H^+][NO_2^-]}{[HNO_2]}$$

$$4.5 \times 10^{-4} = \frac{(1.0 \times 10^{-3})(x + 1.0 \times 10^{-3})}{0.079}$$

ITEM 8

The numerical value of K_a or K_b indicates the degree of ionization and, therefore, the strength of the acid or base. A weak acid or base has ________ tendency to donate or accept a proton than does a strong acid or base.

ITEM 22

The product $[H^+][OH^-]$ for pure water equals 1.0×10^{-14} at 25°C. Even when the concentrations of H^+ and OH^- are not equal, it is still true that, to a good approximation, the product $[H^+][OH^-]$ is equal to ________ in *any* dilute aqueous solution at 25°C.

ITEM 36

Then it is necessary to express the ________ of the reagents at equilibrium in terms of some convenient unknown and to substitute these relations into the expression for the ________.

ITEM 50

Calculate the percentage dissociation of acetic acid at 25°C in a 0.0085-molar solution, which is more dilute than the solution of Item 49 by a factor of 100.

ITEM 64

K_a for HNO_2 at 25°C is 4.5×10^{-4}. The expression for K_a for HNO_2 is ________, and the equation in terms of x is ________.

ITEM 78

In solutions of HF, hydrofluoric acid ($K_a = 6.7 \times 10^{-4}$), the H^+ ion concentration can be adjusted by the addition of ________ ions and the F^- ion concentration can be adjusted by the addition of ________ ions.

$CH_3COOH \rightleftharpoons H^+ + CH_3COO^-$

$K_a = 1.8 \times 10^{-5}$ at 25°C

ANSWER 7 $\frac{[NH_4^+][OH^-]}{[NH_3]}$

ANSWER 21 $1.01 \times 10^{-7} \times 1.01 \times 10^{-7}$
$= 1.02 \times 10^{-14}$

ANSWER 35 equation

ANSWER 49 $100\left(\frac{0.0039}{0.85}\right) = 0.46\%$

ANSWER 63 Let x = number of moles per liter of HNO_2 that dissociate.
$[HNO_2] = 0.020 - x$
$[H^+] = 0.10 + x$
$[NO_2^-] = x$

ANSWER 77 0.035
0.035 mole liter^{-1}
$0.035 + 0.001$
$= 0.036$ mole liter^{-1}

ITEM 9

Therefore, the extent of dissociation and the value of K_a or K_b of a weak acid or base are ________ than those of a strong acid or base.

ITEM 23

A solution in which the concentrations of H^+ and OH^- are equal is called a *neutral* solution. At 25°C a neutral aqueous solution must have $[H^+] =$ ________ and $[OH^-] =$ ________.

ITEM 37

Dissociation of the weak acid nitrous acid occurs as follows: $HNO_2 \rightleftharpoons H^+ + NO_2^-$. The expression for K_a is ________.

ITEM 51

The preceding calculations illustrate a general principle: As the concentration of a weak acid or base is decreased, the degree of ionization ________, although the actual concentration of the H^+ or OH^-, respectively, ________.

ITEM 65

Solution of the equation by using the simplifications that can be justified later, ________ and ________, yields the equation ________.

ITEM 79

In a 0.012-molar HF solution to which NaF has been added, the concentration of H^+ is 1.0×10^{-4} molar. Set up the expression for the equilibrium constant in terms of x, the concentration of NaF.

$K_w = [H^+][OH^-]$
$= 1.0 \times 10^{-14}$ at 25°C

$\left[\begin{array}{c} O \\ \Vert \\ CH_3COH \end{array}\right]$	$[H^+]$	% dissociation
0.85	3.9×10^{-3}	0.46
0.0085	3.9×10^{-4}	4.6

$HF \rightleftharpoons H^+ + F^-$

$K_a = 6.7 \times 10^{-4}$

ANSWER 8 less

ANSWER 22 1.0×10^{-14}

ANSWER 36 concentrations
acidity or basicity constant
(or equilibrium constant)

ANSWER 50

$$1.8 \times 10^{-5} = \frac{x^2}{0.0085 - x} \simeq \frac{x^2}{0.0085}$$

$$x = 3.9 \times 10^{-4} = [H^+]$$

$$\% \text{ dissociation} = \frac{(3.9 \times 10^{-4})100}{8.5 \times 10^{-3}}$$

$$= 4.6\%$$

ANSWER 64

$$K_a = \frac{[H^+][NO_2^-]}{[HNO_2]}$$

$$4.5 \times 10^{-4} = \frac{(0.10 + x)(x)}{0.020 - x}$$

ANSWER 78 F^-
H^+

ITEM 10

Acids and bases that are classified as *strong* are completely ionized in dilute solutions. Conversely, acids and bases that are classified as ________ are incompletely ionized in dilute solutions.

ITEM 24

A common measure of acidity is the pH, defined as $pH = -\log [H^+]$. For a neutral solution the $pH = -\log$ (________) = ________.

ITEM 38

The equilibrium concentrations of species present in an aqueous nitrous acid solution can be expressed in terms of x, the number of moles per liter of HNO_2 that dissociate. Complete the entries at the right for a solution of HNO_2 made by adding 0.100 mole of HNO_2 to enough water to make 1.00 liter of solution.

ITEM 52

Ammonia, NH_3, ionizes in water according to the equation $NH_3 + H_2O \rightleftharpoons NH_4^+ + OH^-$. The expression for K_b for ammonia is ________.

ITEM 66

The value of x is ________. Was the simplification justified?

ITEM 80

Solution of the equation gives $x =$ ________. Therefore, the concentration of NaF in the HF–NaF solution must be ________.

The following five problems will be a review and a test of what you remember about acid–base ionization calculations. Set up the solution to each problem in a logical way and solve it completely.

$K_w = [H^+][OH^-]$
$= 1.0 \times 10^{-14}$ at 25°C

	HNO_2	$\rightleftharpoons$ H^+	+ NO_2^-
Equil. M	______	______	______

ANSWER 9 smaller

ANSWER 23 1.0×10^{-7}
1.0×10^{-7}

ANSWER 37 $K_a = \dfrac{[H^+][NO_2^-]}{[HNO_2]}$

ANSWER 51 increases
decreases

ANSWER 65 $0.10 + x \simeq 0.10$
$0.020 - x \simeq 0.020$

$$4.5 \times 10^{-4} = \frac{0.10x}{0.020}$$

ANSWER 79

$$K_a = \frac{[H^+][F^-]}{[HF]}$$

$$6.7 \times 10^{-4} = \frac{(1.0 \times 10^{-4})(x + 1.0 \times 10^{-4})}{0.012 - 1.0 \times 10^{-4}}$$

ITEM 11

Acetic acid, $CH_3\overset{O}{\overset{\|}{C}}OH$, is an acid in aqueous solution because some of the molecules donate to water the hydrogen atom that is attached to an oxygen atom. The equation for the dissociation of acetic acid can be written $CH_3COOH + H_2O \rightleftharpoons CH_3COO^- +$ ________ or, more simply, $CH_3COOH \rightleftharpoons$ ________.

ITEM 25

An *acidic* solution has a greater concentration of H^+ than of OH^-. Nevertheless, the product $[H^+][OH^-]$ must be equal to ________.

ITEM 39

The value of K_a for nitrous acid at 25°C is 4.5×10^{-4}. The value of x can be calculated from the equation ________.

ITEM 53

Let x be the number of moles per liter of NH_3 that are ionized at equilibrium, and calculate the concentrations of NH_3, NH_4^+, and OH^- in a 0.100-molar solution of NH_3 at 25°C.

ITEM 67

The percentage dissociation of 0.020-molar HNO_2 in 0.10-molar HCl at 25°C is ________.

ITEM 81

What is the concentration of NH_4Cl in an NH_3–NH_4Cl solution having an initial concentration of NH_3 of 0.0090 molar and a pH of 9.0? K_b for NH_3 is 1.8×10^{-5}.

$$K_a = \frac{[H^+][NO_2^-]}{[HNO_2]}$$

$NH_3 + H_2O \rightleftharpoons NH_4^+ + OH^-$

$K_b = 1.74 \times 10^{-5}$ at 25°C

ANSWER 10 weak

ANSWER 24 1.0×10^{-7}
7

ANSWER 38 HNO_2: $0.100 - x$
H^+: x
NO_2^-: x

ANSWER 52 $K_b = \frac{[NH_4^+][OH^-]}{[NH_3]}$

ANSWER 66 9.0×10^{-5}
Yes.

ANSWER 80 0.080
0.080 mole liter^{-1}

ITEM 12

The expression for the acidity constant of acetic acid is $K_a =$ __________.

ITEM 26

A solution in which the hydrogen ion concentration is 1.0×10^{-4} has a pH of __________.

ITEM 40

Solution of this equation gives $x = 0.0065$. The equilibrium concentrations are $[H^+] =$ __________, $[NO_2^-] =$ __________, and $[HNO_2] =$ __________. The percentage dissociation of HNO_2 in a 0.100-molar solution at 25°C is __________.

ITEM 54

The solutions considered thus far have contained only one acid or base in addition to water. Consider a solution that is 0.100 molar in NH_3 and 0.0200 molar in NaOH. The concentration of OH^- ions in 0.0200-molar NaOH, in which there is no NH_3, is __________. If ammonia is added, the OH^- concentration will _____crease.

ITEM 68

Consider a solution of 0.020-molar HNO_2 in 0.00010-molar HCl. Calculate the number of moles per liter of HNO_2 that dissociate. Try the simplifying assumptions.

ITEM 82

Formic acid, $H\overset{\overset{\displaystyle O}{\|}}{C}OH$, partially dissociates in water according to the equation $HCOOH \rightleftharpoons H^+ + HCOO^-$. At 25°C, a 0.025-molar solution of formic acid is 8.1% dissociated. Calculate the concentration of each of the ions and molecules, except H_2O, in the solution.

$CH_3COOH \rightleftharpoons H^+ + CH_3COO^-$

ANSWER 11 H_3O^+

$H^+ + CH_3COO^-$

ANSWER 25 1.0×10^{-14} (at 25°C)

At equilibrium

$[H^+] = x$

$[NO_2^-] = x$

$[HNO_2] = 0.100 - x$

ANSWER 39 $4.5 \times 10^{-4} = \dfrac{(x)(x)}{0.100 - x}$

$NH_3 + H_2O \rightleftharpoons NH_4^+ + OH^-$

ANSWER 53

$$1.74 \times 10^{-5} = \frac{x^2}{0.100 - x} \simeq \frac{x^2}{0.100}$$

$x = 0.0013 = [NH_4^+] = [OH^-]$

$[NH_3] = 0.100 - 0.001 = 0.099$

$HNO_2 \rightleftharpoons H^+ + NO_2^-$

$K_a = 4.5 \times 10^{-4}$

ANSWER 67 $\dfrac{100(9.0 \times 10^{-5})}{2.0 \times 10^{-2}} = 0.45\%$

ANSWER 81

$[H^+] = 1.0 \times 10^{-9}$

$$[OH^-] = \frac{1.0 \times 10^{-14}}{1.0 \times 10^{-9}} = 1.0 \times 10^{-5}$$

$$K_b = \frac{[NH_4^+][OH^-]}{[NH_3]}$$

Let x = necessary $[NH_4Cl]$.

$$1.8 \times 10^{-5} = \frac{(x + 1.0 \times 10^{-5})(1.0 \times 10^{-5})}{9.0 \times 10^{-3} - 1.0 \times 10^{-5}}$$

$x = 0.016 = [NH_4Cl]$.

ITEM 13

It is possible to calculate the equilibrium constant of a weak acid or base if the total concentration of the acid or base and its percentage ionization at the given concentration are known. If 100 ml of solution are prepared with 0.0020 mole of acetic acid, the total concentration of acetic acid is __________.

ITEM 27

Since, in any acidic solution, the hydrogen ion concentration is greater than __________ mole liter^{-1}, the pH of an acidic solution is __________ than 7.

ITEM 41

The equation for the dissociation of hydrocyanic acid, HCN, in aqueous solution is $HCN \rightleftharpoons$ __________. The expression for K_a is $K_a =$ __________.

ITEM 55

No matter what the OH^- concentration in the solution is, the value of __________ remains constant with constant temperature.

ITEM 69

Are the assumptions justified in Item 68?

ITEM 83

The value of $[H^+]$ for a solution of HCN in water is 1.0×10^{-5}. Calculate [HCN] in this solution. (K_a for HCN at 25°C is 4.0×10^{-10}.)

ANSWER 12 $\dfrac{[H^+][CH_3COO^-]}{[CH_3COOH]}$

ANSWER 26 4

ANSWER 40 H^+: 0.0065 mole liter^{-1}
NO_2^-: 0.0065 mole liter^{-1}
HNO_2: 0.093 mole liter^{-1}

$$\left(\frac{0.0065}{0.100}\right)100 = 6.5\%$$

ANSWER 54 0.0200 molar
in

ANSWER 68 Let x = number of moles per liter of HNO_2 that dissociate.

$$4.5 \times 10^{-4} = \frac{(0.00010 + x)x}{0.020 - x}$$

$$4.5 \times 10^{-4} = \frac{0.00010x}{0.020}$$

$$x = 0.09$$

ANSWER 82

$$[H^+] = [H\overset{\overset{\displaystyle O}{\|}}{C}O^-] = 0.081 \times 0.025 = 0.0020$$

$$[H\overset{\overset{\displaystyle O}{\|}}{C}OH] = 0.025 - 0.002 = 0.023$$

$$K_w = [H^+][OH^-]$$

$$[OH^-] = \frac{1.0 \times 10^{-14}}{2.0 \times 10^{-3}} = 5.0 \times 10^{-12}$$

ITEM 14

When acetic acid of this concentration is at equilibrium at 25°C, 2.9% of the acid is dissociated. Since the total acetic acid concentration is 0.020 molar, concentrations of the individual species are

$[H^+] =$ ____________
$[CH_3COO^-] =$ ____________
$[CH_3COOH] =$ ____________

ITEM 28

A *basic* solution is one in which the concentration of OH^- ions is greater than the concentration of ____________ ions. Therefore, the pH of a basic solution is ____________ than 7.

ITEM 42

When the molarity of a solution is specified, it is the total concentration of solute—dissociated plus undissociated—that is given. Thus, the total concentration of HCN in a 0.020-molar solution is ____________. If x is the moles per liter of HCN that dissociate, $[H^+] =$ ____________, $[CN^-] =$ ____________, and $[HCN] =$ ____________.

ITEM 56

In a solution that is 0.100 molar in NH_3 and 0.0200 molar in NaOH, the concentrations of NH_4^+, NH_3, and OH^- in terms of x (the moles per liter of NH_3 ionizing) are $[NH_4^+]$ ____________ and $[NH_3]$ ____________. The $[OH^-]$ resulting from NH_3 reaction is ____________, and the total $[OH^-]$ is ____________.

ITEM 70

Using only the assumption $0.020 - x \simeq 0.020$ does not save any work; recalculate the value of x without simplification. What is the percentage dissociation?

ITEM 84

How many moles of sodium acetate, $CH_3\overset{O}{\overset{\|}{C}}ONa$, are present in 500 ml of a sodium acetate–0.075-molar acetic acid solution that has a hydrogen ion concentration of 0.00010 molar? K_a for acetic acid is 1.8×10^{-5} at 25°C.

$CH_3COOH \rightleftharpoons H^+ + CH_3COO^-$

ANSWER 13 0.020 mole liter^{-1} (0.020*M*)

ANSWER 27 1.0×10^{-7}
less

$HCN \rightleftharpoons H^+ + CN^-$

ANSWER 41 $H^+ + CN^-$

$$\frac{[H^+][CN^-]}{[HCN]}$$

$NH_3 + H_2O \rightleftharpoons NH_4^+ + OH^-$

$K_b = 1.74 \times 10^{-5}$ at 25°C

ANSWER 55 K_b for ammonia

$$x = \frac{-b \pm \sqrt{b^2 - 4ac}}{2a}$$

ANSWER 69 No, 0.09 is greater
than 0.020 and 0.00010.

ANSWER 83

$$K_a = \frac{[H^+][CN^-]}{[HCN]}$$

$$[H^+] = [CN^-] = 1.0 \times 10^{-5}$$

$$4.0 \times 10^{-10} = \frac{(1.0 \times 10^{-5})^2}{[HCN]}$$

$$[HCN] = \frac{1.0 \times 10^{-10}}{4.0 \times 10^{-10}} = 0.25$$

ANSWER 14

H^+: $0.029 \times 0.020M = 0.00058M$
CH_3COO^-: $0.029 \times 0.020M = 0.00058M$
CH_3COOH: $0.020M - 0.00058M = 0.019M$

(Item 15 is on page 260.)

ANSWER 28 H^+
greater

(Item 29 is on page 260.)

ANSWER 42 0.020 molar
x
x
$0.020 - x$

(Item 43 is on page 260.)

ANSWER 56 x
$0.100 - x$
x
$x + 0.0200$

(Item 57 is on page 260.)

ANSWER 70

$$4.5 \times 10^{-4} = \frac{(0.00010 + x)x}{0.020 - x}$$

$$x^2 + (5.5 \times 10^{-4})x - 9.0 \times 10^{-6} = 0$$

$$x = 0.0027$$

$$\% = \frac{100 \times 0.0027}{0.020} = 14\%$$

(Item 71 is on page 260.)

ANSWER 84

$$K_a = \frac{[H^+][CH_3COO^-]}{[CH_3COOH]}$$

Let $x = [CH_3COO^-]$ from CH_3COONa.

$$1.8 \times 10^{-5} = \frac{(1.0 \times 10^{-4})(x + 1.0 \times 10^{-4})}{0.075 - 1.0 \times 10^{-4}}$$

$x = 1.3 \times 10^{-2} = 0.013$.
Number of moles $= VM = 0.50 \times 0.013$
$= 0.0065$ mole

(Item 85 is on page 260.)

The basic calculations pertaining to equilibria involving the ionization of acids and bases in aqueous solution are now part of your chemical knowledge. You have learned that the expression for the acidity or basicity constant, K_a or K_b, shows the relationship between the concentration of an acid or base and its ionization products.

The degree to which an acid or base ionizes in water depends upon the value of K_a or K_b, the concentration of the solution, and on the concentration of any added "common" ion. You have learned that the H^+ and OH^- concentrations in aqueous solution are related by the equation $K_w = [H^+][OH^-]$, in which the ion product for water, K_w, is 1.0×10^{-14} at 25°C.

5–3 HETEROGENEOUS EQUILIBRIUM AND SOLUBILITY PRODUCT

The process of solution of many simple ionic compounds in water can be viewed as a simple transfer of the ions from the solid to the aqueous phase that is accompanied by the simultaneous hydration of the ions. If, for a certain amount of the solid and water, all the solid dissolves, the concentrations of the ions in solution depend directly on the relative amounts of water and solid. However, if all of the solid does not dissolve, the solution becomes saturated and additional solid does not increase the concentration of the ions in solution. In such a case there is an equilibrium between the ions in solution and the ions of the solid.

This review examines the equilibria existing in saturated solutions of relatively insoluble salts. It will help you to understand Section 5–10 of *Chemical Principles.* (The simple treatment given here is not applicable to salts that dissolve to more than 0.01 mole liter^{-1} because of complications arising from interactions between the ions.) When temperatures are not specified, room temperature is assumed. When you have finished the review, you will be able to solve easily problems such as those in Items 57 through 60.

ITEM 1

The equation for the solution of AgCl in water can be written $AgCl(s) \rightleftharpoons Ag^+(aq) + Cl^-(aq)$. However, the notations (*s*) for solid and (*aq*) for aqueous are often omitted, with the understanding that AgCl represents the undissolved silver chloride, and Ag^+ and Cl^- are the hydrated silver and chloride ions. The simplified equation is ________.

ITEM 11

The equation for the solution of strontium chromate is $SrCrO_4(s) \rightleftharpoons Sr^{2+} + CrO_4^{2-}$, so $K_{sp} =$ ________.

ITEM 21

Solution of this equation yields $x = 3.5 \times 10^{-4}$. Thus, in a saturated calcium fluoride solution at 20°C, $[Ca^{2+}] =$ ________, and $[F^-] =$ ________ mole liter^{-1}.

ITEM 31

Because we can assume that NaF is completely dissociated in this solution, the $[F^-]$ contributed by the sodium fluoride is ________, and that contributed by the calcium fluoride is ________. The total $[F^-]$ is ________.

ITEM 41

Solution of the original equation in *x*, which is $7.7 \times 10^{-13} = (1.0 \times 10^{-6} + x)(x)$, gives $x =$ ________.

ITEM 51

If $[Pb^{2+}]$ is 1.0×10^{-5} in a solution containing Pb^{2+} and Cl^-, what is the maximum possible concentration of Cl^-?

0.010 M
NaF
$[Ca^{2+}] = x$
CaF_2

$$x = \frac{-b \pm \sqrt{b^2 - 4ac}}{2a}$$

$$PbCl_2 \rightleftharpoons Pb^{2+} + 2Cl^-$$

$$K_{sp} = 1.7 \times 10^{-5}$$

ITEM 2

The expression for the equilibrium constant for this reaction can be written $K_{sp} = [Ag^+][Cl^-]$. A term containing [AgCl] does not appear in the expression because AgCl is a pure solid phase with essentially fixed concentration. The brackets indicate that __________.

ITEM 12

In a saturated solution of pure strontium chromate at 15°C, the concentration of CrO_4^{2-} is 0.006 molar. The numerical value of K_{sp} for strontium chromate at 15°C is __________.

ITEM 22

The solubility product constant for barium sulfate at 20°C is 9.9×10^{-11}. Calculate the solubility of barium sulfate in water at 20°C.

ITEM 32

The solubility product expression for calcium fluoride has the form __________. The value of K_{sp} is 1.7×10^{-10} at 20°C. The expression that must be solved to obtain x is __________.

ITEM 42

In Items 39 through 41, $[Ag^+]$ = __________, $[Br^-]$ = __________, and the solubility of AgBr in 1.0×10^{-6}-molar silver nitrate solution is __________.

ITEM 52

The solubility product constant for strontium sulfate is 2.8×10^{-7}. $SrSO_4$ is precipitated from a Sr^{2+} solution by adding enough H_2SO_4 to make the final $[SO_4^{2-}]$ equal to 0.0010 molar. Calculate the molarity of Sr^{2+} remaining in solution.

$BaSO_4 \rightleftharpoons Ba^{2+} + SO_4{}^{2-}$

$[Ca^{2+}] = x$
$[F^-] = 0.010 + 2x$
CaF_2

$[Ag^+] = 1.0 \times 10^{-6} + x$
$[Br^-] = x$
AgBr

$SrSO_4 \rightleftharpoons Sr^{2+} + SO_4{}^{2-}$

ANSWER 1 $AgCl \rightleftharpoons Ag^+ + Cl^-$

ANSWER 11 $[Sr^{2+}][CrO_4^{2-}]$

ANSWER 21 3.5×10^{-4}
7.0×10^{-4}

ANSWER 31 0.010
$2x$
$0.010 + 2x$

ANSWER 41 $x^2 + (1.0 \times 10^{-6})x$
$- 7.7 \times 10^{-13} = 0$
$x = 5.1 \times 10^{-7}$

ANSWER 51 Let $x = [Cl^-]$.
$K_{sp} = [Pb^{2+}][Cl^-]^2$
$1.7 \times 10^{-5} = (1.0 \times 10^{-5})(x)^2$
$x = 1.3$ moles liter^{-1}

ITEM 3

The expression $K_{sp} = [Ag^+][Cl^-]$ states that the product of the concentrations of Ag^+ and Cl^- in a __________ solution of silver chloride is a constant. This constant is called the *solubility product constant.*

ITEM 13

The equation for the solution of strontium fluoride is $SrF_2 \rightleftharpoons Sr^{2+} + 2F^-$. Accordingly, K_{sp}= __________. The solubility of strontium fluoride at 25°C is 0.0073 g in 100 ml of solution. At 25°C the numerical value for K_{sp} for strontium fluoride is __________.

ITEM 23

The preceding calculations have been limited to solutions of a pure salt. Consider, however, the solubility of AgCl in a 0.010-molar sodium chloride solution. In a 0.010-molar salt solution containing no silver chloride, $[Cl^-]$ = __________.

ITEM 33

Solution of the equation in Answer 32 gives $x = 1.7 \times 10^{-6}$. Therefore, $[Ca^{2+}]$ is __________, $[F^-]$ is __________, and the solubility of CaF_2 in 0.010-molar sodium fluoride is __________.

ITEM 43

For a solution of lead iodate, $Pb(IO_3)_2(s) \rightleftharpoons Pb^{2+} + 2IO_3^-$, and $K_{sp} = 3.2 \times 10^{-13}$. Calculate the molarity of IO_3^- in 0.0030-molar $Pb(NO_3)_2$ saturated with lead iodate. (The lead nitrate is completely soluble and dissociated.)

ITEM 53

Calculate the maximum concentration of barium nitrate, $Ba(NO_3)_2$, possible in a 7.5×10^{-4}-molar solution of potassium chromate, K_2CrO_4. The solubility product constant of $BaCrO_4$ is 2.4×10^{-10}. Assume 100% dissociation of $Ba(NO_3)_2$ and K_2CrO_4, both of which are quite soluble compared with $BaCrO_4$.

ANSWER 2 the concentration is being specified in units of moles per liter

ANSWER 12 $K_{sp} = [Sr^{2+}][CrO_4^{2-}]$
$= (0.006)(0.006)$
$= 4 \times 10^{-5}$

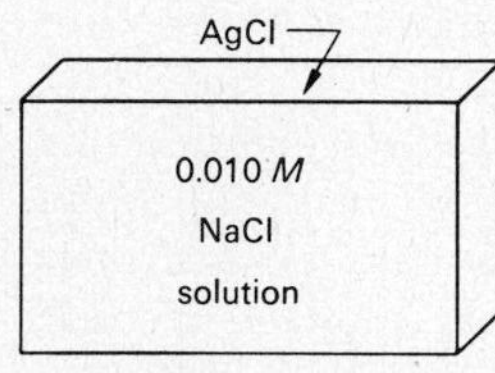

ANSWER 22 Let x = number of moles per liter of $BaSO_4$ that dissolve.
$K_{sp} = [Ba^{2+}][SO_4^{2-}]$
$9.9 \times 10^{-11} = x^2$
$x = 1.0 \times 10^{-5}$ mole liter^{-1}

$[Ca^{2+}] = x$
$[F^-] = 0.010 + 2x$
CaF_2

ANSWER 32 $K_{sp} = [Ca^{2+}][F^-]^2$
$1.7 \times 10^{-10} = x(0.010 + 2x)^2$

ANSWER 42 1.5×10^{-6}
5.1×10^{-7}
5.1×10^{-7} mole liter^{-1}

ANSWER 52 Let $x = [Sr^{2+}]$ left.
$K_{sp} = [Sr^{2+}][SO_4^{2-}]$
$2.8 \times 10^{-7} = x(0.0010)$
$x = 2.8 \times 10^{-4}$ molar

ITEM 4

The value of K_{sp} for a salt can be calculated if the concentrations of its ions in a saturated solution are known. For example, the solubility of silver chloride in water at 25°C is 1.27×10^{-5} mole liter^{-1}. That is, the maximum amount of silver chloride that will dissolve in 1 liter of water is __________ mole.

ITEM 14

If K_{sp} for a salt at a given temperature is known, it is possible to reverse the calculation and to find the equilibrium concentrations of the ions and the solubility of the salt in water. To do this calculation, it is necessary to have the balanced __________ for the solution reaction and to have the expression for the __________.

ITEM 24

To find the equilibrium concentrations of Ag^+ and Cl^- in a 0.010-molar salt solution saturated with silver chloride, the solubility product expression is again used. So long as the Ag^+ and Cl^- ions are in equilibrium with the solid, no matter what their individual concentrations, the __________ of their concentrations is fixed.

ITEM 34

The solution to the last problem is summarized at the right. The equation, $x(0.010 + 2x)^2 = 1.7 \times 10^{-10}$, need not be solved in this form. The quantity $2x$ is so small compared with 0.010 that the quantity $0.010 + 2x$ can be satisfactorily approximated by 0.010. The simplified equation is __________.

ITEM 44

Let us look at another type of problem. A solution containing Cl^- ion is added drop by drop to a solution containing Ag^+. So long as the product $[Ag^+][Cl^-]$ has a value less than the solubility product of silver chloride, no __________ forms.

ITEM 54

To 0.025 mole of Cu^+, enough water and NaBr are added to bring the final solution volume to 100 ml and the final Br^- concentration to 0.0010 molar. What fraction of the original Cu^+ is in solution at equilibrium? The solubility product of CuBr is 5.3×10^{-9}.

$1.7 \times 10^{-10} = [Ca^{2+}][F^-]^2$
$1.7 \times 10^{-10} = x(0.010 + 2x)^2$
$1.7 \times 10^{-6} = x$

ANSWER 3 saturated

ANSWER 13

$K_{sp} = [Sr^{2+}][F^-]^2$

$$\frac{7.3 \times 10^{-2} \text{ g liter}^{-1}}{126 \text{ g mole}^{-1}} = 5.8 \times 10^{-4} \text{ molar}$$

$$K_{sp} = (5.8 \times 10^{-4})(2 \times 5.8 \times 10^{-4})^2 = 7.8 \times 10^{-10}$$

ANSWER 23 0.010

ANSWER 33 1.7×10^{-6}
0.010
1.7×10^{-6} mole liter^{-1}

ANSWER 43 Let x = number of moles per liter of $Pb(IO_3)_2$ that dissolve.
$K_{sp} = [Pb^{2+}][IO_3^-]^2$
$3.2 \times 10^{-13} = (x + 0.0030)(2x)^2$
Assume that x is much smaller than 0.0030.
$3.2 \times 10^{-13} = 0.0030(2x)^2$
$x = 5.2 \times 10^{-6}$
$[IO_3^-] = 1.0 \times 10^{-5}$

ANSWER 53 Let $x = [Ba^{2+}]$.
$K_{sp} = [Ba^{2+}][CrO_4^{2-}]$
$2.4 \times 10^{-10} = x(7.5 \times 10^{-4})$
$x = 3.2 \times 10^{-7}$
$Ba(NO_3)_2$ concentration:
3.2×10^{-7} molar

ITEM 5

Accordingly, in a saturated solution of silver chloride at 25°C, there is ________ mole of Ag^+ in a liter of solution: $[Ag^+] =$ ________.

ITEM 15

The next step is to express the ion concentrations in terms of a convenient unknown. For a saturated solution at 25°C, let x be the moles per liter of silver bromide that are dissolved in water; then $[Ag^+] =$ ________ and $[Br^-] =$ ________.

ITEM 25

If x is the number of moles of silver chloride dissolved per liter of 0.010-molar NaCl solution, then $[Ag^+]$ is also equal to x. The $[Cl^-]$ resulting from the dissolved silver chloride is ________, and the total $[Cl^-]$ is ________.

ITEM 35

A summary of the solution to the NaCl–AgCl problem (Items 23 through 27) is given at the right. Again, x is so small that it makes no significant contribution to the quantity $(0.010 + x)$. The equation, $1.6 \times 10^{-10} = x(0.010 + x)$, may be simplified to ________.

ITEM 45

However, once the product $[Ag^+][Cl^-]$ has reached the value of the solubility product for silver chloride, any attempt to increase the concentration of Cl^- or Ag^+ will lead to the formation, or *precipitation*, of AgCl. As Cl^- or Ag^+ is added, the product $[Ag^+][Cl^-]$ remains ________, but the amount of AgCl(*s*) ______creases.

ITEM 55

K_{sp} for $BaSO_4$ is 9.9×10^{-11}, and K_{sp} for $CaSO_4$ is 2.4×10^{-5}. The maximum concentration of SO_4^{2-} possible in a solution that is 0.010 molar in Ca^{2+} is ________. The maximum concentration of SO_4^{2-} possible in a solution that is 0.00010 molar in Ba^{2+} is ________.

Solubility of AgCl is 1.27×10^{-5} mole liter^{-1} at 25°C

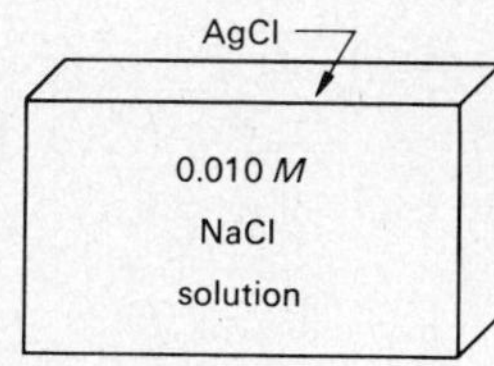

$1.6 \times 10^{-10} = [Ag^+][Cl^-]$
$1.6 \times 10^{-10} = x(0.010 + x)$
$x = 1.6 \times 10^{-8}$

ANSWER 4 1.27×10^{-5}

ANSWER 14 equation
solubility product constant (K_{sp})

ANSWER 24 product

ANSWER 34 $1.7 \times 10^{-10} = x(0.010)^2$

ANSWER 44 solid AgCl

ANSWER 54 Let x = final $[Cu^+]$.
$K_{sp} = [Cu^+][Br^-]$
$5.3 \times 10^{-9} = x(1.0 \times 10^{-3})$
$x = 5.3 \times 10^{-6}$
fraction Cu^+ in 100 ml

$$= \frac{5.3 \times 10^{-7} \text{ mole}}{0.025 \text{ mole}}$$

$$= 2.1 \times 10^{-5}$$

ITEM 6

What is $[Cl^-]$ in the saturated solution at the right?

ITEM 16

Substitution of x, and the known value of 7.7×10^{-13} for the K_{sp} at 25°C, into the solubility product constant expression gives ________. Therefore, $x =$ ________.

ITEM 26

Substitution of the concentration expressions in x and the value 1.6×10^{-10} for the K_{sp} of AgCl at 25°C into the solubility product expression gives the equation ________.

ITEM 36

In equilibrium calculations, such a simplification is often, though by no means always, possible. If solution of the equation shows that x is less than about 10% of the number to which it is added or from which it is ________, the simplification is all right.

ITEM 46

The maximum concentration of sodium chloride that can be present in a solution that is 1.0×10^{-4} molar in Ag^+ (from $AgNO_3$) is ________. The solubility product for AgCl is 1.6×10^{-10}.

ITEM 56

From the results of Item 55, it is apparent that when H_2SO_4 is added to a solution containing 0.010-molar Ca^{2+} and 0.00010-molar Ba^{2+}, ________ will precipitate before ________.

Items 57 through 60 provide a review and test of your skill in solving solubility product problems. Set up the solution to each problem in an organized way and solve it completely. Use the periodic table in the inside front cover to obtain atomic weights.

$[Ag^+] = 1.27 \times 10^{-5}$
$[Cl^-] = ?$
AgCl
25°C

$[Ag^+] = x$
$[Br^-] = x$
$K_{sp} = [Ag^+][Br^-]$
AgBr

$[Ag^+] = x$
$[Cl^-] = 0.010 + x$
AgCl
$AgCl \rightleftarrows Ag^+ + Cl^-$

ANSWER 5 1.27×10^{-5}
1.27×10^{-5}

ANSWER 15 x
x

ANSWER 25 x
$0.010 + x$

ANSWER 35 $1.6 \times 10^{-10} = x(0.010)$

ANSWER 45 constant
in

ANSWER 55 Let $x = [SO_4^{2-}]$ for Ca^{2+}.
$K_{sp} = [Ca^{2+}][SO_4^{2-}]$
$2.4 \times 10^{-5} = (1.0 \times 10^{-2})x$
$x = 2.4 \times 10^{-3}$ molar
Let $x' = [SO_4^{2-}]$ for Ba^{2+}.
$K_{sp} = [Ba^{2+}][SO_4^{2-}]$
$9.9 \times 10^{-11} = (1.0 \times 10^{-4})x'$
$x' = 9.9 \times 10^{-7}$ molar

ITEM 7

The saturated solution at the right is at equilibrium; hence, we can calculate $K_{sp} = [Ag^+][Cl^-]$ to be __________.

ITEM 17

From the result of Item 16, it is apparent that $[Ag^+] = [Br^-] =$ __________, and the solubility of AgBr is __________ mole liter^{-1} at 25°C.

ITEM 27

Solution of the equation in Answer 26 yields $x = 1.6 \times 10^{-8}$; therefore, $[Ag^+] =$ __________, $[Cl^-] =$ __________, and the solubility of silver chloride in a 0.010-molar solution of salt is __________.

ITEM 37

Consider the calculation of the solubility of silver chloride in 1.0×10^{-7}-molar sodium chloride. If $[Ag^+]$ is x, the total $[Cl^-]$ is __________. The K_{sp} is 1.6×10^{-10}, so the equation to be solved for x is __________.

ITEM 47

For compounds that dissociate when dissolved in water, the maximum value that the product of the ionic molarities, each raised to a power equal to the number of such ions in the solution equation, can have is the value of __________.

ITEM 57

The solubility of silver carbonate, Ag_2CO_3, is approximately 0.032 g liter^{-1} of solution at 20°C. Calculate the solubility product constant for Ag_2CO_3 at 20°C.

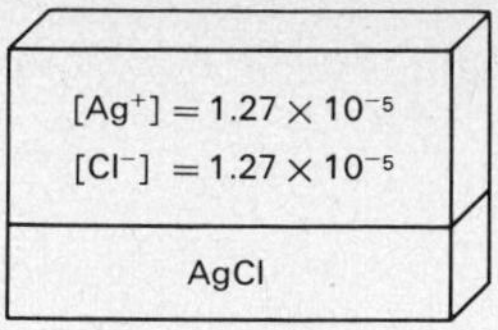

$[Ag^+] = 1.27 \times 10^{-5}$
$[Cl^-] = 1.27 \times 10^{-5}$
AgCl

$AgCl(s) \rightleftarrows Ag^+ + Cl^-$

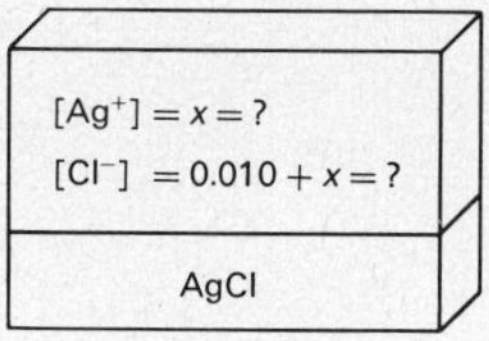

ANSWER 6 1.27×10^{-5}

ANSWER 16 $7.7 \times 10^{-13} = x^2$
8.8×10^{-7}

$[Ag^+] = x = ?$
$[Cl^-] = 0.010 + x = ?$
AgCl

ANSWER 26 $K_{sp} = [Ag^+][Cl^-]$
$1.6 \times 10^{-10} = x(0.010 + x)$

ANSWER 36 subtracted

ANSWER 46 $K_{sp} = [Ag^+][Cl^-]$
Let $x = [Cl^-]$ at precipitation point.
$1.6 \times 10^{-10} = (1.0 \times 10^{-4})x$
$x = 1.6 \times 10^{-6}$ mole liter^{-1}

$Ag_2CO_3(s) \rightleftharpoons 2Ag^+ + CO_3^{2-}$

ANSWER 56 $BaSO_4$
$CaSO_4$

ITEM 8

The equation for the solution of PbF_2 in water is $PbF_2 \rightleftharpoons Pb^{2+} + 2F^-$. Therefore, $K_{sp} =$ ________.

ITEM 18

For every mole of calcium fluoride, CaF_2, dissolved in water, there are produced ________ mole(s) of Ca^{2+} and ________ mole(s) of F^-. The expression for the K_{sp} is ________.

ITEM 28

Calculate the solubility of silver chloride ($K_{sp} = 1.6 \times 10^{-10}$) in pure water.

ITEM 38

Solve the equation with the assumption that x is much smaller than 10^{-7}. Is this simplification justified?

ITEM 48

Enough water is added to 0.0010 mole of Ba^{2+} [as $Ba(NO_3)_2$] and 0.00010 mole of CrO_4^{2-} (as K_2CrO_4) to make 1 liter of solution. K_{sp} for $BaCrO_4$ is 2.4×10^{-10}. A precipitate of $BaCrO_4$ (will, will not) form.

ITEM 58

Compare the solubility of CuBr ($K_{sp} = 5.3 \times 10^{-9}$) (a) in pure water and (b) in 0.0010-molar $CuNO_3$.

ANSWER 7 $(1.27 \times 10^{-5})(1.27 \times 10^{-5})$
$= 1.61 \times 10^{-10}$

ANSWER 17 8.8×10^{-7}
8.8×10^{-7}

ANSWER 27 1.6×10^{-8}
0.010
(to two significant figures)
1.6×10^{-8} mole liter^{-1}

ANSWER 37 $x + 1.0 \times 10^{-7}$
$x(x + 1.0 \times 10^{-7})$
$= 1.6 \times 10^{-10}$

ANSWER 47 the solubility product constant

ANSWER 57

molarity of dissolved Ag_2CO_3

$$= \frac{3.2 \times 10^{-2} \text{ g liter}^{-1}}{276 \text{ g mole}^{-1}}$$

$= 1.16 \times 10^{-4}$ mole liter^{-1}

$$K_{sp} = [Ag^+]^2[CO_3^{2-}]$$
$$= (2.32 \times 10^{-4})^2(1.16 \times 10^{-4})$$
$$= 6.2 \times 10^{-12}$$

(If your method of solution was wrong, review Items 4 through 13.)

ITEM 9

The maximum amount of PbF_2 that will dissolve in water at 20°C is 1.9×10^{-3} mole liter^{-1} of solution. In a saturated solution of PbF_2, then, $[Pb^{2+}] =$ __________ and $[F^-] =$ __________.

ITEM 19

If x represents the number of moles of calcium fluoride that are dissolved per liter of solution, the concentrations are $[Ca^{2+}] =$ __________ and $[F^-] =$ __________.

ITEM 29

In 0.010-molar NaCl the solubility of AgCl is 1.6×10^{-8} mole liter^{-1}, and in pure water the solubility of AgCl is 1.3×10^{-5} mole liter^{-1}. It is apparent that the presence of Cl^-, a common ion, ______creases the solubility of the silver chloride.

ITEM 39

Let x be the number of moles per liter of silver bromide ($K_{sp} = 7.7 \times 10^{-13}$) that are dissolved in 1.0×10^{-6}-molar silver nitrate solution saturated with silver bromide. Substitution of numerical values in the general expression for $K_{sp} =$ __________ yields the equation __________.

ITEM 49

To 0.010 mole of lead nitrate and 0.010 mole of potassium chloride, enough water is added to make the volume of the solution 1.0 liter. Since K_{sp} for $PbCl_2$ is 1.7×10^{-5}, a precipitate of $PbCl_2$ (will, will not) form.

ITEM 59

K_{sp} for BaF_2 is 1.7×10^{-6}. Enough water is added to 0.0010 mole of Ba^{2+} [as $Ba(NO_3)_2$] and 0.0050 mole of F^- (as NaF) to make the final volume 1.00 liter. Is a precipitate of BaF_2 present at equilibrium?

$CaF_2 \rightleftharpoons Ca^{2+} + 2F^-$

ANSWER 8 $[Pb^{2+}][F^-]^2$

ANSWER 18 1
2
$K_{sp} = [Ca^{2+}][F^-]^2$

ANSWER 28 Let x = number of moles per liter of AgCl that dissolve.
$K_{sp} = [Ag^+][Cl^-]$
$1.6 \times 10^{-10} = x^2$
$x = 1.3 \times 10^{-5}$
solubility
$= 1.3 \times 10^{-5}$ mole liter^{-1}

ANSWER 38 1.6×10^{-10}
$= x(x + 1.0 \times 10^{-7})$
$1.6 \times 10^{-10} = x(1.0 \times 10^{-7})$
$x = 1.6 \times 10^{-3}$
Simplification is *not* justified: 1.6×10^{-3} is not small compared with 1.0×10^{-7}.

ANSWER 48

$[Ba^{2+}][CrO_4^{2-}] = (1.0 \times 10^{-3})(1.0 \times 10^{-4})$
$= 1.0 \times 10^{-7}$
1.0×10^{-7} is greater than 2.4×10^{-10}, the K_{sp}. Therefore, a precipitate will be present.

ANSWER 58

$CuBr(s) \rightleftharpoons Cu^+ + Br^-$
$K_{sp} = [Cu^+][Br^-]$
Let x = number of moles per liter of CuBr that dissolve.
(a) $5.3 \times 10^{-9} = (x)(x)$
$x = 7.3 \times 10^{-5}$ mole liter^{-1}
(b) $5.3 \times 10^{-9} = (x + 0.0010)x$
Assume that x is much smaller than 0.0010.
Then $5.3 \times 10^{-9} = 0.0010\ x$.
$x = 5.3 \times 10^{-6}$ mole liter^{-1}
(If your method of solution was wrong, review Items 14 through 43.)

ITEM 10

From the concentrations of Pb^{2+} and F^- in a saturated solution, the numerical value of K_{sp} for lead difluoride at 20°C is ________.

ITEM 20

The solubility product for calcium fluoride is 1.7×10^{-10} at 20° C. Substitution of this and the concentration expressions into the equilibrium expression gives the equation ________.

ITEM 30

In 0.010-molar sodium fluoride solution saturated with calcium fluoride, the concentration of the dissolved calcium fluoride can be represented as x mole liter^{-1}. In this saturated solution $[Ca^{2+}]$ is ________.

ITEM 40

Solution of the equation with the assumption that x is much smaller than 1.0×10^{-6} yields $x=$________. Is the assumption justified?

ITEM 50

In Item 49, $[Pb^{2+}]$ is 0.010. The maximum possible concentration of Cl^- in this solution is ________.

ITEM 60

Sulfide ions are added to a solution containing Pb^{2+} [as $Pb(NO_3)_2$] until only 5×10^{-10} mole of Pb^{2+} remains. The final volume of the solution is 1.0 liter. What is the concentration of S^{2-}? The solubility product for PbS is 1.0×10^{-29}.

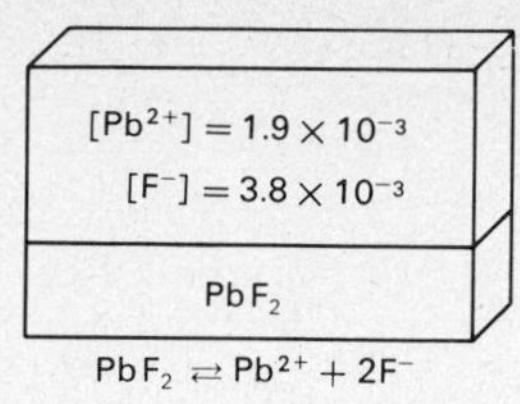

ANSWER 9 1.9×10^{-3}
3.8×10^{-3}

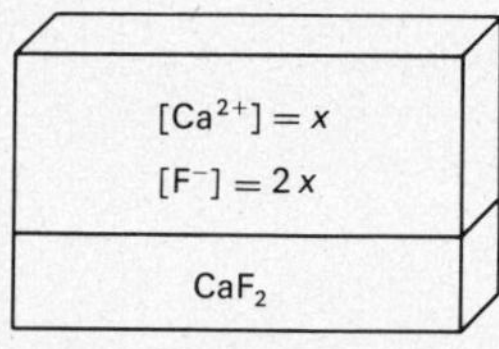

ANSWER 19 x
$2x$

ANSWER 29 de

ANSWER 39 $K_{sp} = [Ag^+][Br^-]$
7.7×10^{-13}
$= (1.0 \times 10^{-6} + x)(x)$

K_{sp} for $PbCl_2$ is 1.7×10^{-5}

ANSWER 49

$[Pb^{2+}][Cl^-]^2 = (0.010)(0.010)^2 = 1.0 \times 10^{-6}$
1.0×10^{-6} is less than 1.7×10^{-5}
Therefore, $PbCl_2(s)$ will not form.

ANSWER 59

$[Ba^{2+}][F^-]^2 = (0.0010)(0.0050)^2$
$= 2.5 \times 10^{-8}$
$K_{sp} = 1.7 \times 10^{-6}$
Since the product of $[Ba^{2+}][F^-]^2$ is less than K_{sp}, no precipitation will occur. (If your method of solution was wrong, review Items 44 through 49.)

ANSWER 10 $(1.9 \times 10^{-3})(3.8 \times 10^{-3})^2$
$= 2.7 \times 10^{-8}$

(*Item 11 is on page 292.*)

ANSWER 20 $1.7 \times 10^{-10} = x(2x)^2$
(or $1.7 \times 10^{-10} = 4x^3$)

(*Item 21 is on page 292.*)

ANSWER 30 x

(*Item 31 is on page 292.*)

ANSWER 40 $x(1.0 \times 10^{-6}) = 7.7 \times 10^{-13}$
$x = 7.7 \times 10^{-7} - 0.77 \times 10^{-6}$
Not justified: 0.77×10^{-6} is not small compared with 1.0×10^{-6}.

(*Item 41 is on page 292.*)

ANSWER 50 Let $x = [Cl^-]$.
$K_{sp} = [Pb^{2+}][Cl^-]^2$
$1.7 \times 10^{-5} = (0.010)(x)^2$
$x = 4.1 \times 10^{-2} = 0.041$ molar

(*Item 51 is on page 292.*)

ANSWER 60 Let $x =$ final $[S^{2-}]$.
$K_{sp} = [Pb^{2+}][S^{2-}]$
$1.0 \times 10^{-29} = (5 \times 10^{-10})x$
$x = 2 \times 10^{-20}$ molar
(If your method of solution was wrong, review Items 44 through 55.)

(*Turn to page 314.*)

In this last section you have added to your understanding of chemical systems by learning how to do calculations pertaining to the equilibria between slightly soluble salts and their ions in aqueous solution. You know the mathematical relationship between the solubility and the solubility product constant, K_{sp}, of a salt. Knowing the K_{sp}, you can calculate the solubility of a salt in water or in a solution containing an ion common to the salt. You have seen that equilibrium problems frequently may be simplified by using approximations, which then must be justified at the end of the calculation.

Once the basic concepts of chemical equilibrium presented in this chapter are mastered, you are ready to study the underlying energy relationships and the application of the concepts to all sorts of chemical systems, including those that are part of your own physiology.

REVIEW 6 THE REPRESENTATIVE ELEMENTS

The building of a chemical vocabulary requires not only the memorization of names but also a knowledge of charges, oxidation numbers, and formulas. Using the arrangement of the representative elements in the periodic table as a basis, the first section of this review will help you to become familiar with these elements, their names, their common ions and oxidation numbers, and their simplest compounds. The second section is concerned with the most common anions formed from the representative elements and oxygen.

6-1 THE PERIODIC TABLE AND SIMPLEST COMPOUNDS

The periodic table is of real use to a chemist: It arranges the elements by atomic number, an arrangement that correlates the chemical and physical properties of the elements. As in Section 6–3 of *Chemical Principles,* you will build part of the periodic table and, in doing so, will learn how to name and write formulas for the simplest compounds of the representative elements.

In this chapter you will need to know that electrons of atoms are arranged in energy levels or shells, as mentioned in Section 6–4 of *Chemical Principles.* Energy levels will be discussed in detail in Review 8. For the purposes of this review it is sufficient that you realize that the first shell can contain a maximum of two electrons and the second shell a maximum of eight electrons. Although the higher energy shells can contain more electrons, they reach stability when they are occupied by eight electrons. The properties of atoms are usually determined by the number of electrons in the outermost shell, the highest energy level that is populated by electrons.

If you can write the formulas and names of the metallic oxides of the representative elements and if you can place the representative elements in their proper places in the periodic table, you are already finished with this chapter. The transition metals will be discussed in Review 9.

THE ELEMENTS

1	H	Hydrogen
2	He	Helium
3	Li	Lithium
4	Be	Beryllium
5	B	Boron
6	C	Carbon
7	N	Nitrogen
8	O	Oxygen
9	F	Fluorine
10	Ne	Neon
11	Na	Sodium
12	Mg	Magnesium
13	Al	Aluminum
14	Si	Silicon
15	P	Phosphorus
16	S	Sulfur
17	Cl	Chlorine
18	Ar	Argon
19	K	Potassium
20	Ca	Calcium
21	Sc	Scandium
22	Ti	Titanium
23	V	Vanadium
24	Cr	Chromium
25	Mn	Manganese
26	Fe	Iron
27	Co	Cobalt
28	Ni	Nickel
29	Cu	Copper
30	Zn	Zinc
31	Ga	Gallium
32	Ge	Germanium
33	As	Arsenic
34	Se	Selenium
35	Br	Bromine
36	Kr	Krypton
37	Rb	Rubidium
38	Sr	Strontium
39	Y	Yttrium
40	Zr	Zirconium
41	Nb	Niobium
42	Mo	Molybdenum
43	Tc	Technetium
44	Ru	Ruthenium
45	Rh	Rhodium
46	Pd	Palladium
47	Ag	Silver
48	Cd	Cadmium
49	In	Indium
50	Sn	Tin
51	Sb	Antimony
52	Te	Tellurium
53	I	Iodine
54	Xe	Xenon
55	Cs	Cesium
56	Ba	Barium
57	La	Lanthanum
58	Ce	Cerium
59	Pr	Praseodymium
60	Nd	Neodymium
61	Pm	Promethium
62	Sm	Samarium
63	Eu	Europium
64	Gd	Gadolinium
65	Tb	Terbium
66	Dy	Dysprosium
67	Ho	Holmium
68	Er	Erbium
69	Tm	Thulium
70	Yb	Ytterbium
71	Lu	Lutetium
72	Hf	Hafnium
73	Ta	Tantalum
74	W	Tungsten
75	Re	Rhenium
76	Os	Osmium
77	Ir	Iridium
78	Pt	Platinum
79	Au	Gold
80	Hg	Mercury
81	Tl	Thallium
82	Pb	Lead
83	Bi	Bismuth
84	Po	Polonium
85	At	Astatine
86	Rn	Radon
87	Fr	Francium
88	Ra	Radium
89	Ac	Actinium
90	Th	Thorium
91	Pa	Protoactinium
92	U	Uranium
93	Np	Neptunium
94	Pu	Plutonium
95	Am	Americium
96	Cm	Curium
97	Bk	Berkelium
98	Cf	Californium
99	Es	Einsteinium
100	Fm	Fermium
101	Md	Mendelevium
102	No	Nobelium
103	Lr	Lawrencium
104		

ITEM 1

The atomic number of an element is the same as the number of __________ in the nucleus of each atom of the element and the same as the number of __________ around the nucleus.

ITEM 18

An electron can be added to a halogen atom in two ways. One way is by complete transfer of an electron from an atom of another element. This produces a *halide* ion that has __________ valence electron(s).

ITEM 35

The symbol for the lithium ion is Li^+. Similarly, the sodium ion is written __________. It has the electron arrangement of atoms of the element __________.

ITEM 52

The charge on a monatomic ion is its *oxidation state* or *oxidation number,* which is defined as the number of electrons that must be added or removed to convert the ion to a neutral atom. The oxidation number of Ca^{2+} is +2, of Mg^{2+} is __________, and of Cl^- is __________.

ITEM 69

Five elements are characterized by four valence electrons. The atomic numbers and names of these elements are 6, __________; 14, __________; 32, __________; 50, __________; and 82, __________.

ITEM 86

"Arsenious" is the name given to the __________ oxidation state of arsenic. "Arsenic" can refer not only to the name of the element but also to the __________ oxidation state.

ITEM 103

How much of this partial periodic table can you now complete?

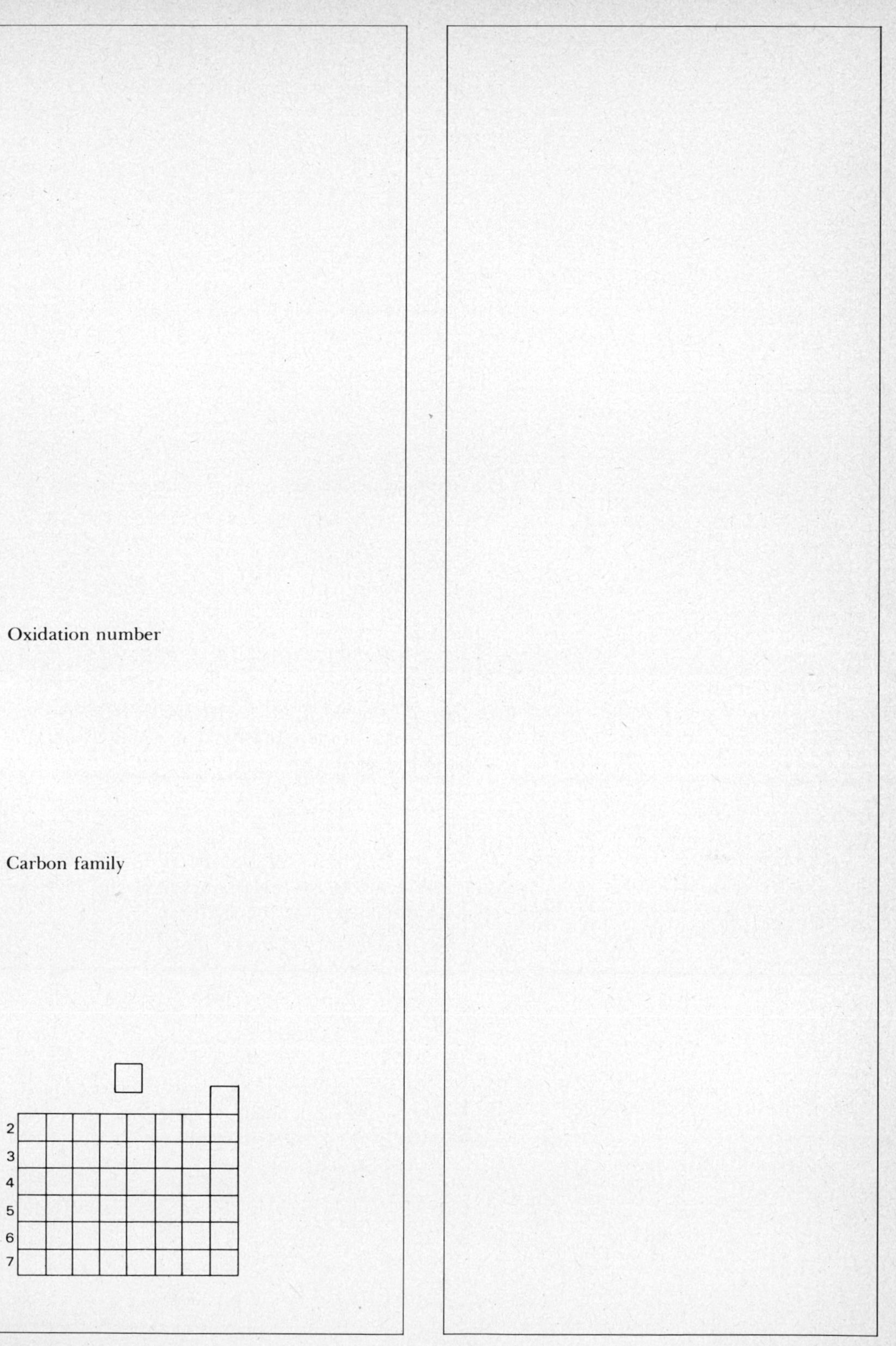
Oxidation number
Carbon family
2
3
4
5
6
7

ITEM 2

The chemical properties of atoms are dependent on the number of ________ and the energy levels that they occupy.

ITEM 19

A halide ion has ________ more electron(s) than it has protons so each ion has a charge of ________.

ITEM 36

The symbols of the remaining alkali metal cations are ________, ________, ________, and ________. They have the electron arrangements of atoms of the elements ________, ________, ________, and ________, respectively.

ITEM 53

Just as the sum of the ionic charges of an electrically neutral salt must add algebraically to zero, the sum of the oxidation numbers of the ions of a neutral compound must add to ________.

ITEM 70

Enter the symbols for the elements of the carbon family in the diagram at the right.

ITEM 87

Usually the arsenic and antimony halides are named as trihalides or pentahalides. Thus, the chloride compounds of these elements can be named ________, ________, ________, and ________.

				H				
								He
2	Li	Be	B				F	Ne
3	Na	Mg	Al				Cl	Ar
4	K	Ca	Ga				Br	Kr
5	Rb	Sr	In				I	Xe
6	Cs	Ba	Tl				At	Rn
7	Fr	Ra						

ANSWER 1 protons
electrons

ANSWER 18 eight

ANSWER 35 Na^+
neon

ANSWER 52 +2
−1

ANSWER 69 carbon
silicon
germanium
tin
lead

ANSWER 86 +3 (lower)
+5 (higher)

ANSWER 103

					H			
	IA	IIA	IIIA	IVA	VA	VIA	VIIA	He
2	Li	Be	B	C	N	O	F	Ne
3	Na	Mg	Al	Si	P	S	Cl	Ar
4	K	Ca	Ga	Ge	As	Se	Br	Kr
5	Rb	Sr	In	Sn	Sb	Te	I	Xe
6	Cs	Ba	Tl	Pb	Bi	Po	At	Rn
7	Fr	Ra						

(*Turn to page 354.*)

ITEM 3

The first energy level of an atom can hold a maximum of __________ electrons. The element of atomic number 2, named __________, is a gas that is totally inert to chemical reaction. (Use the table of elements at the beginning of the section as needed.)

ITEM 20

Negative ions are named by replacing "ine" in the element name by "ide"; thus, fluorine becomes __________, chlorine becomes __________, bromine becomes __________, iodine becomes __________, and astatine becomes __________.

ITEM 37

In chemical reactions the valence electrons of alkali metals often are donated to elements of high electron affinity. Each of the ions so formed has __________ electrons in the outermost occupied energy level.

ITEM 54

Since the oxidation number of the barium ion is +2, an ionic compound of barium and chlorine would have the formula __________.

ITEM 71

From its position in the diagram, you can predict that an atom of a carbon family element must share, donate, or accept __________ electrons. An oxidation number of either __________ or __________ might be expected.

ITEM 88

Roman numerals also can be used to indicate the oxidation numbers. In this system, known as *Stock nomenclature,* the arsenic oxidation states are written as __________ and __________.

ANSWER 2 electrons

ANSWER 19 one
1−

ANSWER 36 K^+, Rb^+, Cs^+, Fr^+
argon
krypton
xenon
radon

ANSWER 53 zero

ANSWER 70

H

								He
2	Li	Be	B	C			F	Ne
3	Na	Mg	Al	Si			Cl	Ar
4	K	Ca	Ga	Ge			Br	Kr
5	Rb	Sr	In	Sn			I	Xe
6	Cs	Ba	Tl	Pb			At	Rn
7	Fr	Ra						

ANSWER 87 arsenic trichloride
arsenic pentachloride
antimony trichloride
antimony pentachloride

ITEM 4

The second energy level of an atom can hold a maximum of ________ electrons. The element of atomic number 10, named ________, is also a chemically inert gas.

ITEM 21

The other way in which an electron can be added to the valence shell of a halogen is by sharing. Two electrons (one from each atom) in a molecule of F—F are shared so both atoms have, at least part of the time, ________ valence electron(s).

ITEM 38

The attraction of oppositely charged ions formed by electron transfer between atoms can be called an *ionic* (or *electrovalent*) *bond.* This is in contrast to the bond formed by sharing of electrons, which is called a ________ bond.

ITEM 55

Oxidation numbers are defined also for atoms that form covalent bonds: Two shared electrons are assigned to the atom that has a greater attraction for them. In compounds of a metal and a nonmetal, the ________ has the greater attraction for electrons.

ITEM 72

In fact, the oxidation number of carbon (except in some organic compounds), silicon, and germanium is usually +4. But germanium, tin, and lead have oxidation numbers of both +2 and +4. Write the formulas of the chlorides of the carbon family elements by using the oxidation numbers given.

ITEM 89

With hydrogen the elements of the nitrogen family form compounds in which they have oxidation states of −3 or +3 (depending on whether the nitrogen family element has greater or lesser attraction for electrons than does hydrogen). The formulas of these compounds are ________, ________, ________, ________, and ________.

Ionic bond

ANSWER 3 two
helium

ANSWER 20 fluoride
chloride
bromide
iodide
astatide

ANSWER 37 eight, except two in Li^+

ANSWER 54 $BaCl_2$

ANSWER 71 4
+4
−4

ANSWER 88 arsenic(III)
arsenic(V)

ITEM 5

The elements of atomic numbers 18, 36, 54, and 86 have eight electrons in the outermost occupied energy level. These four elements are also extremely unreactive gases. The names of these elements are __________, __________, __________, and __________.

ITEM 22

The sharing of electrons between two atoms produces a *covalent bond.* Just as the bond between atoms in F_2 is covalent, so are the bonds in the other diatomic halogen molecules __________, __________, __________, and __________.

ITEM 39

Crystals of ionic compounds are electrically neutral; thus, the total number of charges on the negative ions must be equal to the total number of charges on the __________ ions.

ITEM 56

For example, electrons can be transferred from beryllium to fluorine, oxygen, and sometimes chlorine so that an ionic bond is formed. With other elements beryllium forms __________ bonds.

ITEM 73

The names of these compounds must reflect the oxidation number of the metal. The easiest way is by addition of a Roman numeral to the metal name; for example, $GeBr_2$ is germanium(II) bromide. Name the tin and lead bromides in this way.

ITEM 90

The diagram (see Item 91) has only one unfilled column. This is the one for the elements containing atoms with six valence electrons. The atomic numbers of these elements are 8, 16, 34, 52, and 84. Their names are __________, __________, __________, __________, and __________.

Covalent bond

Oxygen family

ANSWER 4 eight
neon

ANSWER 21 eight

ANSWER 38 covalent

ANSWER 55 nonmetal

ANSWER 72 CCl_4
$SiCl_4$
$GeCl_4$ $GeCl_2$
$SnCl_4$ $SnCl_2$
$PbCl_4$ $PbCl_2$

ANSWER 89 NH_3, PH_3, AsH_3, SbH_3, BiH_3 (These compounds have been given the names ammonia, phosphine, arsine, stibine, and bismuthine, respectively.)

ITEM 6

Thus, there is a special stability associated with atoms in which the outermost energy level populated with electrons, the so-called *valence shell,* contains either the maximum possible number of electrons or __________ of them.

ITEM 23

Place the symbols for the halogen family (atomic numbers 9, 17, 35, 53, 85) and the rare gases (2, 10, 18, 36, 54, 86) in the energy level diagram at the right.

ITEM 40

Since a halide ion has a charge of __________ and an alkali metal ion has a charge of __________, the alkali halide salts have a ratio of metal ions to halide ions of __________.

ITEM 57

In $BeBr_2$ the electrons are attracted more to the bromine atom of a Be—Br bond than to the beryllium atom. Each Br has an oxidation number of −1; Be has an oxidation number of __________.

ITEM 74

Alternatively, the number of halide ions can be specified by prefixes, as in germanium difluoride and germanium tetrafluoride. Name the tin and lead fluorides in this way.

ITEM 91

Complete the diagram at the right with the addition of the symbols of the oxygen family of elements.

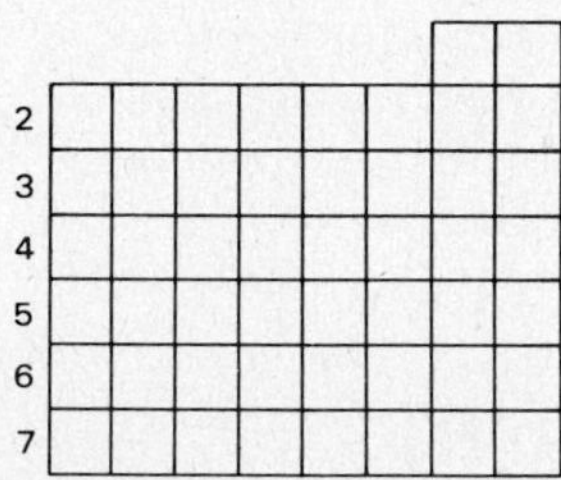

				H				
								He
2	Li	Be	B	C	N		F	Ne
3	Na	Mg	Al	Si	P		Cl	Ar
4	K	Ca	Ga	Ge	As		Br	Kr
5	Rb	Sr	In	Sn	Sb		I	Xe
6	Cs	Ba	Tl	Pb	Bi		At	Rn
7	Fr	Ra						

ANSWER 5 argon
krypton
xenon
radon

ANSWER 22 Cl_2
Br_2
I_2
At_2

ANSWER 39 positive

ANSWER 56 covalent

ANSWER 73 tin(II) bromide
tin(IV) bromide
lead(II) bromide
lead(IV) bromide

ANSWER 90 oxygen
sulfur
selenium
tellurium
polonium

ITEM 7

In other words, atoms containing eight electrons in the valence shell (two electrons for helium) show little tendency to add an electron or to ________ an electron.

ITEM 24

At the top of the halogen column is the empty space for element number 1. For a good reason, a symbol has not yet been put in. Element of atomic number 1, named ________, is unique. Its properties are not similar to those of the halogens or any of the other elements.

ITEM 41

Rare (noble) gas atoms have so little tendency to donate, accept, or share electrons that the elements are composed of single ________.

ITEM 58

All the alkaline earth metals ________, ________, ________, ________, ________, and ________ have ________ electrons in the valence shell and an oxidation number of ________ in compounds.

ITEM 75

Still another system is the use of special names, in which the suffix "ic" indicates the higher of two oxidation states and "ous" indicates the lower. Thus, "stannic" indicates that tin has an oxidation number of ________ and "stannous" indicates that ________.

ITEM 92

Although not as high as those of the halogens, the electron affinities of oxygen and sulfur are high enough that they can accept electrons from metals to form ions. Oxide and sulfide ions have a charge of ________ because ________.

H

								He
2	Li	Be					F	Ne
3	Na	Mg					Cl	Ar
4	K	Ca					Br	Kr
5	Rb	Sr					I	Xe
6	Cs	Ba					At	Rn
7	Fr	Ra						

ANSWER 6 eight

ANSWER 23

								He
2							F	Ne
3							Cl	Ar
4							Br	Kr
5							I	Xe
6							At	Rn
7								

ANSWER 40 1−
1+
1/1

ANSWER 57 +2

ANSWER 74 tin difluoride
tin tetrafluoride
lead difluoride
lead tetrafluoride

ANSWER 91

H

								He
2	Li	Be	B	C	N	O	F	Ne
3	Na	Mg	Al	Si	P	S	Cl	Ar
4	K	Ca	Ga	Ge	As	Se	Br	Kr
5	Rb	Sr	In	Sn	Sb	Te	I	Xe
6	Cs	Ba	Tl	Pb	Bi	Po	At	Rn
7	Fr	Ra						

ITEM 8

At the right is a crude representation of some of the energy levels in an atom. The first energy level can contain a maximum of ________ electrons. In the top right-hand space, write in the symbol for element number 2.

ITEM 25

Therefore, we have moved the box away from the rest of the chart; now put in the symbol for the first element.

ITEM 42

Because halogen atoms have strong tendencies to accept electrons, the pure halogen elements are comprised of ________ in which atoms share electrons.

ITEM 59

Compounds are usually named as ionic salts, regardless of whether the bonding is ionic or covalent. The "ide" ending and the negative oxidation number are given to the atom that has the greater ________.

ITEM 76

Similarly, "plumbic" indicates that lead has an oxidation number of ________ and "plumbous" indicates that ________.

ITEM 93

All elements except helium, neon, and argon form compounds with bonds to oxygen. In compounds of oxygen with nonmetals, and with some metals, electrons are shared to form ________ bonds.

2								
3								
4								
5								
6								
7								

								He
2							F	Ne
3							Cl	Ar
4							Br	Kr
5							I	Xe
6							At	Rn
7								

ANSWER 7 lose

ANSWER 24 hydrogen

ANSWER 41 atoms

ANSWER 58 beryllium
magnesium
calcium
strontium
barium
radium
two
+2

ANSWER 75 +4
tin has an oxidation number of +2

ANSWER 92 2−
addition of two electrons to oxygen or sulfur atoms gives an ion with the electron arrangement of a rare gas

ITEM 9

Helium is the first of what are called the "rare," "inert," or "noble" gases. Put the symbols of the remaining rare gases in the column at the right of the chart in order of increasing atomic number: 10, 18, 36, 54, 86.

ITEM 26

In compounds of metals with hydrogen, an electron from the metal becomes bound to a hydrogen atom. The resulting *hydride* ion has a charge of __________ and has the electron arrangement of the element __________.

ITEM 43

In contrast, a metal crystal is one large molecule composed of positive ions in a cloud of electrons that move freely. This metal structure is possible because the metallic elements have relatively __________ ionization energies.

ITEM 60

Write the names and formulas (in the form $CaCl_2$) for the iodide salts of the alkaline earth metals. (Note that because alkaline earth metals have only the oxidation number of +2 in compounds, there is no need to indicate in the name that there are two iodide ions per calcium ion.)

ITEM 77

Give the formula for each of the following: tin(IV) fluoride, __________; germanium(II) bromide, __________; carbon tetraiodide, __________; plumbous chloride, __________; stannic fluoride __________; silicon tetrachloride, __________.

ITEM 94

The compounds of oxygen are important, and it is useful to know some formulas and names. In nearly all compounds oxygen has the oxidation number of the oxide ion, which is __________. For a neutral compound, the sum of the oxidation numbers on the atoms must be __________.

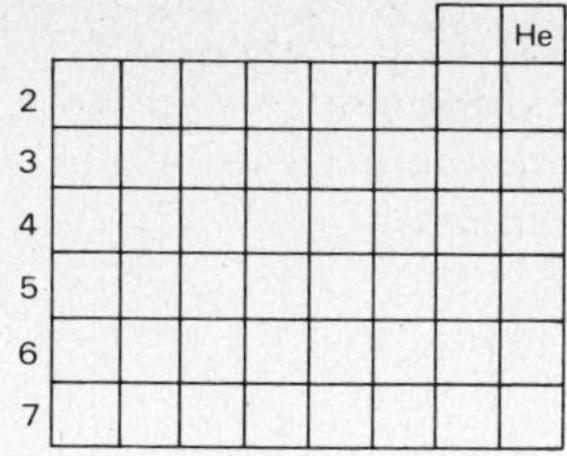

ANSWER 8 two

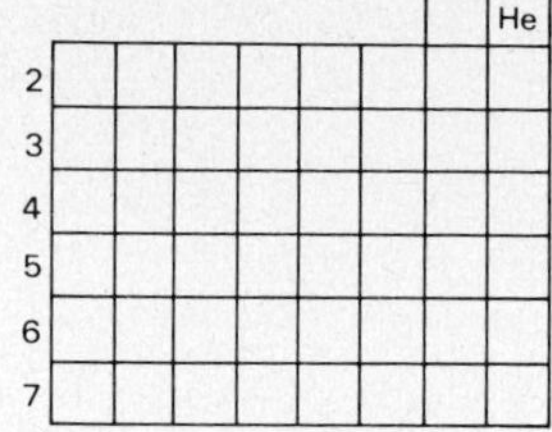

ANSWER 25

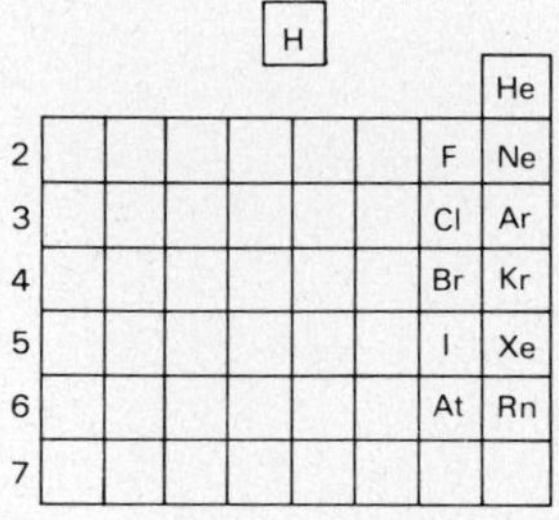

ANSWER 42 diatomic molecules

ANSWER 59 attraction for electrons

ANSWER 76 +4
lead has an oxidation number of +2

ANSWER 93 covalent

ITEM 10

Electron affinity is the energy released when an electron is added to a gaseous neutral atom. Would you expect that an electron could be bound to atoms of the rare gases in a chemical reaction?

ITEM 27

H^+, which would be formed by removal of an electron from a hydrogen atom, is never found as the free ion in chemical systems. In H_2 and other hydrogen compounds (except some metal hydrides), electrons are shared so a ________ bond is formed.

ITEM 44

So far, our diagram is filled for the atomic numbers of the rare gases (2, 10, 18, 36, 54, 86) and the elements with atomic numbers that are one more or one less than those of the rare gases. Show these elements in their proper places in the diagram at the right.

ITEM 61

Name the following salts: KF, $SrCl_2$, BaF_2, LiH, NaBr, and MgI_2.

ITEM 78

Name the following compounds: SrH_2, $BeBr_2$, $AlBr_3$, SnF_2, $MgCl_2$, PbI_2.

ITEM 95

Write the formulas and names of the oxides of the alkali metals.

Electron affinity

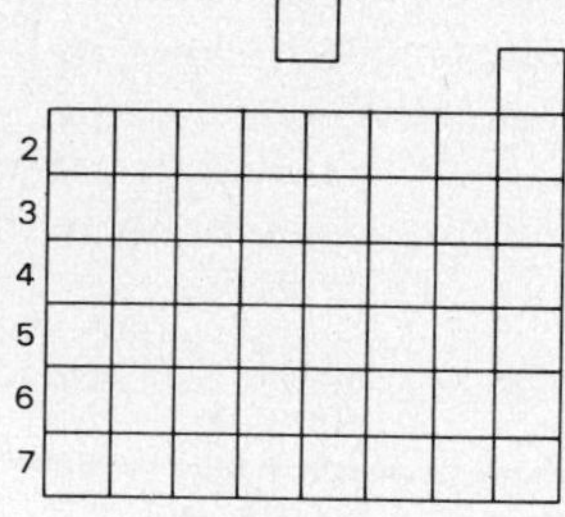

ANSWER 9

								He
2								Ne
3								Ar
4								Kr
5								Xe
6								Rn
7								

ANSWER 26 1−
helium

ANSWER 43 low

ANSWER 60

beryllium iodide	BeI_2
magnesium iodide	MgI_2
calcium iodide	CaI_2
strontium iodide	SrI_2
barium iodide	BaI_2
radium iodide	RaI_2

ANSWER 77 SnF_4
$GeBr_2$
CI_4
$PbCl_2$
SnF_4
$SiCl_4$

ANSWER 94 −2
zero

ITEM 11

Conversely, there are five known elements composed of atoms that contain only seven electrons in the outermost occupied energy level. Would you expect these elements to have a high affinity for electrons?

ITEM 28

Electron affinity is the energy __________.

ITEM 45

Using the diagram in Answer 44 as a guide, write the names of the ionic compounds, or salts, formed by donation of electrons from each alkali metal to the halogen at the end of its row. (Omit francium: It has an unstable nucleus and has been prepared only in trace amounts.)

ITEM 62

Write the formula for each of the following: cesium iodide __________, calcium hydride __________, and radium bromide __________.

ITEM 79

In the diagram at the right enter the symbols of the following: the rare gases, the halogens, the alkali metals, the alkaline earth metals, boron, aluminum, carbon, silicon, germanium, tin, lead.

ITEM 96

Give the formulas and names of the oxides of the alkaline earth metals.

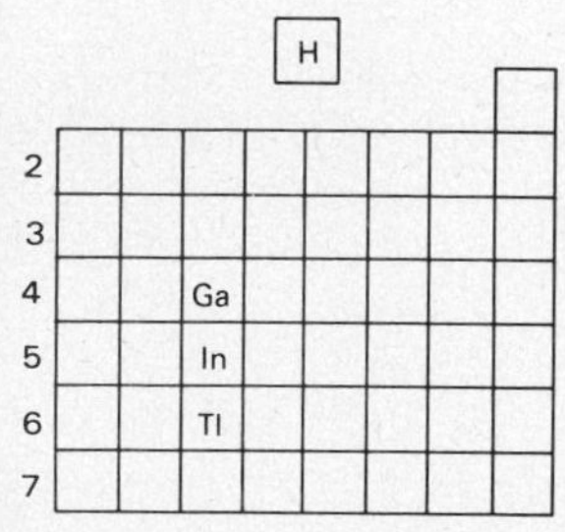

ANSWER 10 No. Atoms of the rare gases have a very stable electron arrangement, so it would require more energy to add an electron than is normally available in chemical reactions.

ANSWER 27 covalent

ANSWER 44

H

								He
2	Li						F	Ne
3	Na						Cl	Ar
4	K						Br	Kr
5	Rb						I	Xe
6	Cs						At	Rn
7	Fr							

ANSWER 61 potassium fluoride
strontium chloride
barium fluoride
lithium hydride
sodium bromide
magnesium iodide

ANSWER 78 strontium hydride
beryllium bromide
aluminum tribromide
tin(II) fluoride, tin difluoride, or stannous fluoride
magnesium chloride
lead(II) iodide, lead diiodide, or plumbous iodide

ANSWER 95

Li_2O	lithium oxide
Na_2O	sodium oxide
K_2O	potassium oxide
Rb_2O	rubidium oxide
Cs_2O	cesium oxide
Fr_2O	francium oxide

ITEM 12

When electrons are added to atoms that have only seven electrons in the valence shell, energy is released. The resulting anions are stable because there are __________ electrons in the valence shell of the ions.

ITEM 29

Whereas energy is released when electrons are added to atoms of some elements, removing an electron from the influence of the positively charged nucleus always requires __________.

ITEM 46

The formulas of these salts can be written as LiF, Li^+F^-, or $(LiF)_x$, in which x is an indeterminate number. Write formulas of the type LiF for the rest of the salts in Item 45.

ITEM 63

Atoms of five of the elements have three valence electrons. These elements have the atomic numbers 5, 13, 31, 49, and 81. Their names are __________, __________, __________, __________, and __________.

ITEM 80

Atoms of five elements have five valence electrons. The atomic numbers of these elements are 7, 15, 33, 51, and 83. Their names are __________, __________, __________, __________, and __________.

ITEM 97

The formula for aluminum trioxide is __________.

Boron family

Nitrogen family

ANSWER 11 Yes. If an electron is added, there will be eight electrons in the valence shell; this will yield the stable electron arrangement of a rare gas.

ANSWER 28 released when an electron is added to a gaseous, neutral atom

ANSWER 45 lithium fluoride
sodium chloride
potassium bromide
rubidium iodide
cesium astatide

ANSWER 62 CsI
CaH_2
$RaBr_2$

ANSWER 79

H

								He
2	Li	Be	B	C			F	Ne
3	Na	Mg	Al	Si			Cl	Ar
4	K	Ca	Ga	Ge			Br	Kr
5	Rb	Sr	In	Sn			I	Xe
6	Cs	Ba	Tl	Pb			At	Rn
7	Fr	Ra						

ANSWER 96

BeO	beryllium oxide
MgO	magnesium oxide
CaO	calcium oxide
SrO	strontium oxide
BaO	barium oxide
RaO	radium oxide

ITEM 13

The atomic numbers of the rare gas elements are 2, 10, 18, 36, 54, and 86. The atomic numbers of the five elements that have seven electrons in the valence shell are __________.

ITEM 30

The energy required to remove an electron from a neutral, gaseous atom is called the *ionization energy.* Would you expect an element of the helium family to have a high or a low ionization energy?

ITEM 47

There are six elements that have two electrons in the valence shell and the rest of the electrons in an inert gas arrangement. It is not difficult to predict that ions formed from these elements would have a charge of __________.

ITEM 64

Write the symbols of the boron family of elements in the diagram at the right.

ITEM 81

Place the symbols of the nitrogen family of elements in the diagram at the right.

ITEM 98

Give the formula for each of the following: silicon dioxide, __________; tin monoxide, __________; stannic oxide, __________; stannous sulfide, __________; plumbous oxide, __________; and carbon disulfide, __________.

Ionization energy

				H				
								He
2	Li	Be					F	Ne
3	Na	Mg					Cl	Ar
4	K	Ca					Br	Kr
5	Rb	Sr					I	Xe
6	Cs	Ba					At	Rn
7	Fr	Ra						

				H				
								He
2	Li	Be	B	C			F	Ne
3	Na	Mg	Al	Si			Cl	Ar
4	K	Ca	Ga	Ge			Br	Kr
5	Rb	Sr	In	Sn			I	Xe
6	Cs	Ba	Tl	Pb			At	Rn
7	Fr	Ra						

ANSWER 12 eight

ANSWER 29 energy

ANSWER 46 NaCl
KBr
RbI
CsAt

ANSWER 63 boron
aluminum
gallium
indium
thallium

ANSWER 80 nitrogen
phosphorus
arsenic
antimony
bismuth

ANSWER 97 Al_2O_3

ITEM 14

Write the names of the elements containing seven electrons in the outermost occupied energy level.

ITEM 31

However, six of the known elements have an atomic number that is one greater than those of the rare gases. These elements have relatively low ionization energies because __________.

ITEM 48

The atomic numbers of these elements, the *alkaline earth metals,* are 4, 12, 20, 38, 56, and 88. The names of the alkaline earth metals are __________, __________, __________, __________, __________, and __________.

ITEM 65

The elements of the boron family, except boron itself, are metals. With nonmetals they form ionic or covalent compounds in which the oxidation number of the metal can be predicted to be __________.

ITEM 82

From the diagram in Answer 81 it is clear that if atoms of these elements gain three electrons, they will have eight valence electrons. You can predict that elements of the nitrogen family could have an oxidation number of __________.

ITEM 99

The formulas and names of the arsenic oxides are As_2O_3: arsenic trioxide, arsenic(III) oxide, or arsenious oxide; and __________.

Alkaline earth metals

ANSWER 13 9, 17, 35, 53, 85

ANSWER 30 High, because removal of an electron would result in a less stable electron arrangement. A large amount of energy is required to remove an electron from a rare gas atom.

ANSWER 47 2+

ANSWER 64

				H				
								He
2	Li	Be	B				F	Ne
3	Na	Mg	Al				Cl	Ar
4	K	Ca	Ga				Br	Kr
5	Rb	Sr	In				I	Xe
6	Cs	Ba	Tl				At	Rn
7	Fr	Ra						

ANSWER 81

				H				
								He
2	Li	Be	B	C	N		F	Ne
3	Na	Mg	Al	Si	P		Cl	Ar
4	K	Ca	Ga	Ge	As		Br	Kr
5	Rb	Sr	In	Sn	Sb		I	Xe
6	Cs	Ba	Tl	Pb	Bi		At	Rn
7	Fr	Ra						

ANSWER 98 SiO_2
SnO
SnO_2
SnS
PbO
CS_2

ITEM 15

These elements are known collectively as the *halogen* family. Place the symbols for the halogens in rows 2 to 6 of column 7 in the diagram, in order of increasing atomic number: 9, 17, 35, 53, 85.

ITEM 32

Atoms of elements numbered 3, 11, 19, 37, 55, and 87 have one valence electron. These elements, known as the *alkali metals,* are named ________, ________, ________, ________, ________, and ________.

ITEM 49

Enter the symbols for the alkaline earth metals in column 2 of the diagram at the right.

ITEM 66

Aluminum is by far the most important of the boron family metals. Write the formulas for the compounds of aluminum and the halogens. (Omit astatine: It has an unstable nucleus and is extremely rare.)

ITEM 83

Nitrogen and phosphorus have an oxidation number of -3 in some compounds. The symbol for the nitride ion is ________ and for the phosphide ion is ________.

ITEM 100

The formulas and names of the antimony oxides are ________.

								He
2								Ne
3								Ar
4								Kr
5								Xe
6								Rn
7								

Alkali metals

				H				
								He
2	Li						F	Ne
3	Na						Cl	Ar
4	K						Br	Kr
5	Rb						I	Xe
6	Cs						At	Rn
7	Fr							

ANSWER 14 fluorine
chlorine
bromine
iodine
astatine

ANSWER 31 removal of one electron leaves an ion with the electron arrangement of a rare gas atom

ANSWER 48 beryllium
magnesium
calcium
strontium
barium
radium

ANSWER 65 +3

ANSWER 82 −3

ANSWER 99

As_2O_5: arsenic pentoxide, arsenic(V) oxide, or arsenic oxide (least preferable)

ITEM 16

The electrons in the outermost occupied energy level are called *valence electrons.* Halogen atoms have __________ valence electrons, and atoms of the inert gases have __________ valence electrons.

ITEM 33

Enter the symbols for the alkali metals in column 1 of the chart at the right in order of increasing atomic number: 3, 11, 19, 37, 55, 87.

ITEM 50

Two electrons can be donated from calcium atoms to atoms of high electron affinity. A halogen atom can accept only __________ electron(s).

ITEM 67

The names of aluminum compounds should include a reference to the number of halogen atoms. (It can form polyatomic ions containing four halide ions.) Thus, the prefix "tri" is used, as in aluminum trifluoride. Name the other three important halides of aluminum.

ITEM 84

The formula for sodium nitride is __________, and the formula for calcium phosphide is __________.

ITEM 101

The diagram at the right of the representative elements will be a partial periodic table if the group numbers are added. The group number is the same as the number of valence electrons. It is written in Roman numerals I to VII, followed by the letter "A," at the top of each column. Add group numbers.

H

								He
2							F	Ne
3							Cl	Ar
4							Br	Kr
5							I	Xe
6							At	Rn
7								

H

								He
2	Li	Be	B	C	N	O	F	Ne
3	Na	Mg	Al	Si	P	S	Cl	Ar
4	K	Ca	Ga	Ge	As	Se	Br	Kr
5	Rb	Sr	In	Sn	Sb	Te	I	Xe
6	Cs	Ba	Tl	Pb	Bi	Po	At	Rn
7	Fr	Ra						

ANSWER 15

								He
2							F	Ne
3							Cl	Ar
4							Br	Kr
5							I	Xe
6							At	Rn
7								

ANSWER 32 lithium
sodium
potassium
rubidium
cesium
francium

ANSWER 49

H

								He
2	Li	Be					F	Ne
3	Na	Mg					Cl	Ar
4	K	Ca					Br	Kr
5	Rb	Sr					I	Xe
6	Cs	Ba					At	Rn
7	Fr	Ra						

ANSWER 66 AlF_3
$AlCl_3$
$AlBr_3$
AlI_3

ANSWER 83 N^{3-}
P^{3-}

ANSWER 100 Sb_2O_3: antimony trioxide
or antimony(III) oxide
Sb_2O_5: antimony pentoxide
or antimony(V) oxide

ITEM 17

The chemical properties of all the halogens are very similar: They often react with other substances in such a way that __________ is added to the valence shell.

ITEM 34

When a lithium atom loses one electron, the resulting ion has the electron arrangement of the element __________. The lithium ion has a charge of ________.

ITEM 51

Thus, in an ionic compound of calcium and chlorine, the ratio of chloride ions to calcium ions must be __________ and the formula would be __________.

ITEM 68

Boron is not a metal, a fact from which one can deduce that the ionization energies for the three valence electrons are relatively __________. A boron atom does not donate electrons to other atoms but shares electrons to form __________ bonds in strange and interesting molecules.

ITEM 85

But nitrogen and phosphorus can have any oxidation number from −3 to +5; the oxidation number of arsenic, antimony, and bismuth can be +3 or +5. (The reasons must be bypassed here.) Write the formulas of the bromides of arsenic and antimony.

ITEM 102

In the diagram at the right shade with pencil the spaces filled with symbols of metallic elements. Half-fill the spaces of the "borderline" semimetallic elements germanium, arsenic, antimony, selenium, and tellurium.

				H				
	IA	IIA	IIIA	IVA	VA	VIA	VIIA	He
2	Li	Be	B	C	N	O	F	Ne
3	Na	Mg	Al	Si	P	S	Cl	Ar
4	K	Ca	Ga	Ge	As	Se	Br	Kr
5	Rb	Sr	In	Sn	Sb	Te	I	Xe
6	Cs	Ba	Tl	Pb	Bi	Po	At	Rn
7	Fr	Ra						

ANSWER 16 seven
eight

ANSWER 33

				H				
								He
2	Li						F	Ne
3	Na						Cl	Ar
4	K						Br	Kr
5	Rb						I	Xe
6	Cs						At	Rn
7	Fr							

ANSWER 50 one

ANSWER 67 aluminum trichloride
aluminum tribromide
aluminum triiodide

ANSWER 84 Na_3N
Ca_3P_2

ANSWER 101

				H				
	IA	IIA	IIIA	IVA	VA	VIA	VIIA	He
2	Li	Be	B	C	N	O	F	Ne
3	Na	Mg	Al	Si	P	S	Cl	Ar
4	K	Ca	Ga	Ge	As	Se	Br	Kr
5	Rb	Sr	In	Sn	Sb	Te	I	Xe
6	Cs	Ba	Tl	Pb	Bi	Po	At	Rn
7	Fr	Ra						

ANSWER 17 one electron

(*Item 18 is on page 318.*)

ANSWER 34 helium
1+

(*Item 35 is on page 318.*)

ANSWER 51 2/1
$CaCl_2$ or $Ca^{2+}(Cl^-)_2$
or $(CaCl_2)_x$

(*Item 52 is on page 318.*)

ANSWER 68 high
covalent

(*Item 69 is on page 318.*)

ANSWER 85 $AsBr_3$
$AsBr_5$
$SbBr_3$
$SbBr_5$

(*Item 86 is on page 318.*)

ANSWER 102

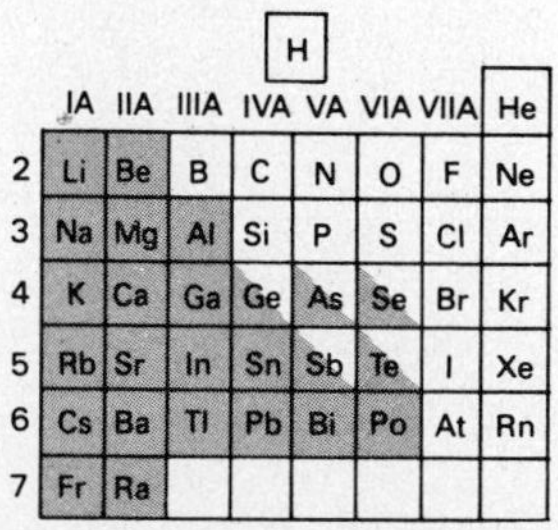

H

	IA	IIA	IIIA	IVA	VA	VIA	VIIA	He
2	Li	Be	B	C	N	O	F	Ne
3	Na	Mg	Al	Si	P	S	Cl	Ar
4	K	Ca	Ga	Ge	As	Se	Br	Kr
5	Rb	Sr	In	Sn	Sb	Te	I	Xe
6	Cs	Ba	Tl	Pb	Bi	Po	At	Rn
7	Fr	Ra						

(*Item 103 is on page 318.*)

By now you need not have memorized the periodic table of the representative elements, but you should know the positions of the more common elements and the relationship between the position of an element and its usual oxidation number.

6–2 NAMES AND FORMULAS OF THE POLYATOMIC ANIONS

The nonmetals and semimetals of the representative elements can be bonded covalently to oxygen in molecules and polyatomic ions. Some of the possible anions are found so often in minerals and common chemicals that it is important to know their names and to be able to write formulas for compounds in which they occur. This section provides practice in writing the names and formulas of the important representative element anions. More of these anions are given in Table 3–4 of *Chemical Principles*.

The name of a polyatomic ion depends on the oxidation state of the element that is combined with oxygen. For this reason it is necessary for you to know that the oxidation number of oxygen in all the anions is -2. Further, the sum of the oxidation numbers of the atoms in an ion must be equal to the charge on the ion.

If you can answer correctly Items 37–40 at the end of the section, there is no reason for you to do the rest of the items.

ITEM 1

The anion formed by the combination of an oxide ion with hydrogen is the hydroxide ion for which the symbol is ________. Remembering that NH_4^+ is the ammonium ion, name the following compounds: LiOH, $Al(OH)_3$, $Mg(OH)_2$, NH_4OH.

ITEM 8

Name the following ions: ClO_4^-, ClO_2^-, ClO^-, ClO_3^-.

ITEM 15

The acids corresponding to the chlorine anions are *hypochlorous*, HClO (or HOCl, which indicates the structure); *chlorous*, $HClO_2$; *chloric*, $HClO_3$; and *perchloric*, ________.

ITEM 22

Write formulas for the compounds sodium thiosulfate ________, calcium sulfite ________, magnesium sulfate ________, cesium sulfide ________, lithium iodate ________, and potassium hydrogen sulfite ________.

ITEM 29

Give the names corresponding to K_3PO_4, NaH_2PO_4, $Mg_3(PO_4)_2$, $CaHPO_4$, $Ca(H_2PO_4)_2$.

ITEM 36

Write the formulas of the following: magnesium borate, ammonium carbonate, potassium hydrogen sulfate, sodium chlorate.

ITEM 2

As a monatomic ion, a halogen has the oxidation number __________. When a halogen (except fluorine) is bonded to one or more oxygen atoms, it has other oxidation numbers.

ITEM 9

Write formulas for the following compounds: lithium hypochlorite, __________; potassium perchlorate, __________; magnesium chlorate, __________; ammonium chlorite, __________; beryllium chloride, __________.

ITEM 16

It is apparent that "ous" in the acid name is converted to __________ in the anion name; and "ic" in the acid name is converted to __________ in the anion name.

ITEM 23

Give the names corresponding to the formulas $(NH_4)_2SO_4$, BaS_2O_3, $CsHSO_4$, $Mg(ClO)_2$.

ITEM 30

Write the formulas for the sodium salts containing the anions nitrite, sulfite, phosphate, perchlorate, and hypobromite.

ITEM 37

Write the formulas of the lithium compounds containing the anions thiosulfate, iodate, borate, hydrogen sulfate, and bicarbonate.

ANSWER 1 OH^-
lithium hydroxide
aluminum trihydroxide
(or aluminum hydroxide)
magnesium hydroxide
ammonium hydroxide

ANSWER 8 perchlorate
chlorite
hypochlorite
chlorate

ANSWER 15 $HClO_4$

ANSWER 22 $Na_2S_2O_3$
$CaSO_3$
$MgSO_4$
Cs_2S
$LiIO_3$
$KHSO_3$

ANSWER 29 potassium phosphate
(or tripotassium phosphate)
sodium dihydrogen phosphate
magnesium phosphate
calcium hydrogen phosphate
calcium dihydrogen phosphate

ANSWER 36 $Mg_3(BO_3)_2$
$(NH_4)_2CO_3$
$KHSO_4$
$NaClO_3$

ITEM 3

For example, in the *hypochlorite* ion, ClO^-, oxygen has an oxidation number of −2, so Cl must have an oxidation number of __________.

ITEM 10

Name the following compounds: $NaClO$, $Ca(ClO_3)_2$, NH_4ClO_4, $KClO_2$.

ITEM 17

The names and formulas of the iodine acids are *hypoiodous*, __________; __________, __________; __________, __________; and __________, __________.

ITEM 24

The two important nitrogen–oxygen anions are derived from *nitric acid*, HNO_3, and *nitrous acid*, HNO_2. The ion NO_3^- is named __________, and NO_2^- is named __________.

ITEM 31

Phosphorous acid is H_2HPO_3. As you might expect, all the anions derived from it contain the word __________.

ITEM 38

Write the formulas of the magnesium compounds containing the ions nitrate, hydrogen phosphate, bromide, bromite, and sulfite.

ANSWER 2 -1

ANSWER 9 LiClO (or LiOCl)
$KClO_4$
$Mg(ClO_3)_2$
NH_4ClO_2
$BeCl_2$

ANSWER 16 "ite"
"ate"

ANSWER 23

ammonium sulfate (or diammonium sulfate)
barium thiosulfate
cesium hydrogen sulfate (cesium acid sulfate or cesium bisulfate)
magnesium hypochlorite

ANSWER 30 $NaNO_2$
Na_2SO_3
Na_3PO_4
$NaClO_4$
NaOBr (or NaBrO)

ANSWER 37 $Li_2S_2O_3$
$LiIO_3$
Li_3BO_3
$LiHSO_4$
$LiHCO_3$

ITEM 4

The oxidation number of Cl in the *chlorite* ion, ClO_2^-, is ________; in the *chlorate* ion, ClO_3^-, it is ________; and in the *perchlorate* ion, ClO_4^-, it is ________.

ITEM 11

By analogy with the chlorine–oxygen anions, you realize that the name of BrO^- is *hypobromite,* that of BrO_2^- is ________, that of BrO_3^- is ________, and that of BrO_4^- is ________.

ITEM 18

When sulfuric acid, H_2SO_4, reacts with bases, either one or two hydrogen ions can be donated to bases, so two anions are possible.

$$H_2SO_4 + OH^- \rightarrow H_2O + HSO_4^-$$
$$HSO_4^- + OH^- \rightarrow \_\_\_\_\_\_\_\_.$$

ITEM 25

Give the names of the following compounds: $Ca(NO_2)_2$, $Ba(NO_3)_2$, K_2SO_3, MgS_2O_3, KNO_2.

ITEM 32

One hydrogen in H_2HPO_3 is bonded to phosphorus and cannot be donated to a base. Thus, there are only two anions possible: *phosphite,* HPO_3^{2-}, and *hydrogen phosphite,* ________.

ITEM 39

Write the formulas of the aluminum compounds containing the ions sulfate, sulfide, phosphate, hydroxide, and perchlorate.

ANSWER 3 +1

ANSWER 10 sodium hypochlorite
calcium chlorate
ammonium perchlorate
potassium chlorite

ANSWER 17 HIO (or HOI)
iodous, HIO_2
iodic, HIO_3
periodic, HIO_4

ANSWER 24 nitrate
nitrite

ANSWER 31 phosphite

ANSWER 38 $Mg(NO_3)_2$
$MgHPO_4$
$MgBr_2$
$Mg(BrO_2)_2$
$MgSO_3$

ITEM 5

You notice that the names of the anions with the two lower oxidation states of Cl, hypochlorite and chlorite, have the suffix "ite" and the two higher oxidation state names, chlorate and __________, have the suffix __________.

ITEM 12

Write the names of the following compounds: $Mg(BrO_2)_2$, $LiBrO$, $KBrO_4$, $Ca(BrO_3)_2$.

ITEM 19

The anion SO_4^{2-} is named *sulfate,* and HSO_4^- is named *hydrogen sulfate* (sometimes *acid sulfate* or *bisulfate*). The formula for calcium sulfate is __________, and the one for calcium hydrogen sulfate is __________.

ITEM 26

Write formulas for the following: magnesium perchlorate, sodium nitrate, sodium nitride, sodium nitrite, lithium sulfite.

ITEM 33

Noting this exception to the usual nomenclature, write the formula for sodium phosphite and the formula for sodium hydrogen phosphite.

ITEM 40

Write the formulas of each of the following anions: borate, carbonate, dihydrogen phosphate, thiosulfate, bicarbonate, hypochlorite.

ANSWER 4 +3
+5
+7

ANSWER 11 bromite
bromate
perbromate

ANSWER 18 $H_2O + SO_4^{2-}$

ANSWER 25 calcium nitrite
barium nitrate
potassium sulfite
magnesium thiosulfate
potassium nitrite

ANSWER 32 $HHPO_3^-$ (or $H_2PO_3^-$)

ANSWER 39 $Al_2(SO_4)_3$
Al_2S_3
$AlPO_4$
$Al(OH)_3$
$Al(ClO_4)_3$

ITEM 6

The prefix "hypo" indicates a lower oxidation state than the "ite" anion. For chlorine–oxygen anions, *hypochlorite* indicates the Cl oxidation number __________ and the formula __________; *chlorite* indicates the oxidation number __________ and the formula __________.

ITEM 13

Similarly, IO_3^- is named *iodate,* and IO_4^- is named __________. IO^- is named __________, and IO_2^- is named __________.

ITEM 20

The names and formulas for the potassium salts containing the two sulfate ions are __________, __________; and __________, __________.

ITEM 27

Phosphoric acid, H_3PO_4, can donate three hydrogen ions to bases. Therefore, there are three phosphate ions: the *normal phosphate,* PO_4^{3-}; *hydrogen phosphate,* HPO_4^{2-}; and *dihydrogen phosphate,* __________.

ITEM 34

Three important anions remaining are *carbonate,* CO_3^{2-}; *bicarbonate* (or *hydrogen carbonate*), HCO_3^-; and *borate,* BO_3^{3-}. The two acids from which they are derived are named __________ and __________ and have the formulas __________ and __________, respectively.

ANSWER 5 perchlorate
"ate"

ANSWER 12 magnesium bromite
lithium hypobromite
potassium perbromate
calcium bromate

ANSWER 19 $CaSO_4$
$Ca(HSO_4)_2$

ANSWER 26 $Mg(ClO_4)_2$
$NaNO_3$
Na_3N
$NaNO_2$
Li_2SO_3

ANSWER 33 Na_2HPO_3
$NaHHPO_3$ (or NaH_2PO_3)

ANSWER 40 BO_3^{3-}
CO_3^{2-}
$H_2PO_4^-$
$S_2O_3^{2-}$
HCO_3^-
ClO^-

ITEM 7

The prefix "per" indicates a higher oxidation state than the "ate" anion. For chlorine anions, *chlorate* indicates the oxidation number ________ and the formula ________; *perchlorate* indicates the oxidation number ________ and the formula ________.

ITEM 14

Name the following: NH_4BrO_3, $CsIO_4$, $NaBrO$, KIO_3, $Ca(IO)_2$.

ITEM 21

Other important sulfur anions are *sulfite*, SO_3^{2-}; *hydrogen sulfite* (*acid sulfite* or *bisulfite*), HSO_3^-; and *thiosulfate*, $S_2O_3^{2-}$. Give the formulas and names of the lithium salts containing these anions.

ITEM 28

Write formulas for the following: disodium hydrogen phosphate ________, potassium dihydrogen phosphate ________, magnesium dihydrogen phosphate ________, sodium phosphate (or trisodium phosphate) ________, calcium phosphate ________.

ITEM 35

Write the names of the compounds $CaCO_3$, K_3BO_3, $Mg(HCO_3)_2$.

ANSWER 6

+1

ClO^-

+3

ClO_2^-

ANSWER 13

periodate

hypoiodite

iodite

ANSWER 20

potassium sulfate (or dipotassium sulfate), K_2SO_4

potassium hydrogen sulfate (or potassium acid sulfate or potassium bisulfate), $KHSO_4$

ANSWER 27 $H_2PO_4^-$

ANSWER 34

carbonic

boric

H_2CO_3

H_3BO_3

ANSWER 7 +5
ClO_3^-
+7
ClO_4^-

(*Item 8 is on page 356.*)

ANSWER 14 ammonium bromate
cesium periodate
sodium hypobromite
potassium iodate
calcium hypoiodite

(*Item 15 is on page 356.*)

ANSWER 21

Li_2SO_3, lithium sulfite (or dilithium sulfite)
$LiHSO_3$, lithium hydrogen sulfite, (lithium acid sulfite or lithium bisulfite)
$Li_2S_2O_3$, lithium thiosulfate

(*Item 22 is on page 356.*)

ANSWER 28 Na_2HPO_4
KH_2PO_4
$Mg(H_2PO_4)_2$
Na_3PO_4
$Ca_3(PO_4)_2$

(*Item 29 is on page 356.*)

ANSWER 35 calcium carbonate
potassium borate
(or tripotassium borate)
magnesium bicarbonate
(or magnesium hydrogen carbonate)

(*Item 36 is on page 356.*)

REVIEW 7 BALANCING REDOX EQUATIONS

The ability to balance chemical equations is essential to understanding chemical reactions. Balanced equations show the relative numbers of ions and molecules that are involved in a chemical reaction and that are therefore necessary for calculations of relative weights of reagents and products. Balanced equations also provide the first step for quantitative treatments of chemical equilibria and electrochemistry.

Many simple equations often can be balanced by inspection, but equations involving oxidation and reduction may be difficult to balance so directly. An easy system for balancing such equations is the ion–electron method discussed in *Chemical Principles,* Section 7–4.

This review provides a step-by-step development of the method and practice in balancing equations. When you have completed the items, you will be able to balance any reaction involving oxidation and reduction by this method, once you know what the reactants and products are.

It is assumed in this review that you already have learned how to assign oxidation numbers to the atoms of an ion or a molecule. This procedure is discussed in detail in Section 7–2 of *Chemical Principles,* and a short review is included here.

ASSIGNMENT OF OXIDATION NUMBERS

Oxidation numbers provide a "bookkeeping" procedure for counting the electrons that can be associated with, or assigned to, each atom of a molecule or ion. An oxidation number of zero implies that the atom in a molecule or ion is assigned the same number of electrons as it has in the free atom. Negative values of oxidation numbers correspond to a greater number of electrons than in the neutral atom, whereas positive values correspond to a smaller number of electrons.

The rules for assigning oxidation numbers to atoms in molecules or ions can be summarized as follows.

(1) The atoms in an element, such as He, H_2, O_2, S_8, C, Zn, Li, have an oxidation number of zero.

(2) Simple, one-atom ions, such as Cl^-, Zn^{2+}, Li^+, have oxidation numbers equal to the charge on the ion. (This means that the alkali metals, except as the element or when combined with each other, have an oxidation number of +1. The alkaline earth metals form only 2+ ions and have oxidation numbers of +2.)

(3) Except for H_2 and the metal hydrides, the oxidation number of the hydrogen atom in any molecule or ion is +1. In the metal hydrides, such as NaH and CaH_2, H is assigned the value −1.

(4) Except for O_2, O_3 (ozone), OF_2, peroxides, and the superoxides (KO_2, RbO_2, CsO_2), the oxidation number of the oxygen atom in any molecule or ion is −2. In peroxides, such as HOOH, the oxygen atoms are assigned the oxidation number −1, and in the superoxides, $-\frac{1}{2}$.

(5) When two nonmetal atoms are bonded in an ion or molecule, the nonmetal considered the most negative is assigned an oxidation number of the same value as its most common negative ion. Which of the two atoms is more negative depends on the positions of the elements in the periodic table: The element either above or to the right of the other is more negative. For example, in OF_2, F is assigned the oxidation number −1, and O has the oxidation number +2.

(6) The algebraic sum of the oxidation numbers of all the atoms in a molecule must be zero. For example, for CH_4, since each H has an oxidation number of +1, the carbon atom

must have the value −4 to give a total value for the molecule of zero.

(7) The sum of the oxidation numbers of all the atoms of an ion must equal the charge on the ion. For the sulfate ion, SO_4^{2-}, since each oxygen has an oxidation number of −2, the sulfur atom in SO_4^{2-} must have an oxidation number of +6. Then the total for the ion is −2, which is equal to the charge on the ion.

To test whether you know and understand these rules, whether you can assign oxidation numbers, and whether you are ready to start balancing equations, assign oxidation numbers to the atoms in each of the following examples:

(1) N_2 (2) NH_3 (3) H_2O_2 (4) Na_2CO_3

(5) CO_3^{2-} (6) Ce_2O_3 (7) HBr (8) BrO^-

(9) $KMnO_4$ (10) MnO_2 (11) $Cr_2O_7^{2-}$ (12)

PCl_3 (13) ClF (14) $P(OH)_3$ (15) C_2H_6.

ANSWERS

(1) N has an oxidation number of zero (Rule 1).

(2) H has an oxidation number of +1 (Rule 3), and to make the total of the oxidation numbers of the molecule equal to zero, N must have an oxidation number of −3 (Rule 6).

(3) This is a peroxide. Each H has an oxidation number of +1; each O has an oxidation number of −1 (Rule 4).

(4) Each Na has an oxidation number of +1; each O has an oxidation number of −2. For the total of the molecule to be zero, the carbon atom must be assigned the value +4 (Rule 6).

(5) Each O is −2, and C must be +4 to make the total for the ion equal to the ion's charge (Rule 7).

(6) Each O is −2, and each cerium atom must be +3 to give a total for the molecule of zero.

(7) Since each H is +1, Br must be −1.

(8) Since O is −2, Br must be +1 so the charge on the ion is −1.

(9) Since each O is −2 and K is +1, Mn must be +7 to give a total of zero for the molecule.

(10) Since each O is −2, Mn must be +4.

(11) Since each O is −2, each Cr must be +6 for the total of the oxidation numbers to be equal to the charge on the ion.

(12) Cl is to the right of phosphorus in the periodic table; therefore, Cl is assigned its usual negative charge of −1 and P must be +3, so the total for the molecule is zero.

(13) F is above Cl in the periodic table, so F is assigned the value −1 and Cl is +1.

(14) Each O is −2 and each H is +1, so the total of 3 OH^- is −3. Therefore, P must be +3, so the total for the molecule will be zero.

(15) Each H is +1; each C in this compound is then −3.

ITEM 1

The word *redox* describes a reaction in which electrons are transferred from one substance to another. It is derived from the words ________duction and ________dation.

ITEM 15

The equation for a half-reaction is balanced if, on each side of the equation, there are the same number of atoms of each element and the same net charge. The net charge is determined by adding the charges on the ions and the electrons. In the preceding equation (Answer 14) the charge on each electron is, of course, ________, and on Fe^{3+} it is ________.

ITEM 29

Because the charge is not equal on the two sides of the equation, we can infer that some additional charged species is involved in the half-reaction. If the reaction occurs in basic aqueous solution, we generally assume that OH^- ions are involved. Rewrite the equation at the right and add enough OH^- ions to balance the charges.

ITEM 43

Check the number of oxygen atoms on both sides. Is the equation completely balanced?

ITEM 57

Completely balance the equation with respect to atoms and check it.

ITEM 71

In this equation, if H^+ is changed to H_3O^+, the final equation is ________.

$\overset{+5}{NO_3^-} + 3e^- \rightarrow \overset{+2}{NO}$

$5H_2O + \overset{+2}{S_2O_3^{2-}} \rightarrow$
$\overset{+6}{2SO_4^{2-}} + 8e^- + 10H^+$

$\overset{0}{I_2} \rightarrow \overset{+5}{2IO_3^-} + 10e^- + 12H^+$

$3Cu + 8H^+ + 2NO_3^- \rightarrow$
$3Cu^{2+} + 2NO + 4H_2O$

ITEM 2

Oxidation numbers help us to keep track of the number of electrons associated with a particular atom or ion. Since in redox reactions __________ are transferred from one reagent to another, the __________ of the atoms of the reagents change.

ITEM 16

Therefore, in the equation $Fe^{3+} + 3e^- \rightarrow Fe$, the net charge on each side is __________. This equation is balanced with respect to atoms, electrons, and net charge.

ITEM 30

If the half-reaction at the right occurs in acidic solution, H_3O^+ is the reagent that is generally assumed to be involved, but H_3O^+ is usually abbreviated to H^+. Rewrite the equation to show this addition.

ITEM 44

Again, the same basic procedure has been followed: first, assignment of __________ and balancing of the number of __________ of the element being reduced or __________; second, addition of __________ to balance the change in __________.

ITEM 58

Balance the half-reaction for the reduction $H_3AsO_4 \rightarrow HAsO_2$ (acidic solution) with respect to electrons, charge, and atoms.

ITEM 72

Check the total equation. Is it balanced with respect to both charge and atoms?

$\overset{+5}{NO_3^-} + 3e^- \rightarrow \overset{+2}{NO}$

ANSWER 1 re
oxi

ANSWER 15 1−
3+

ANSWER 29 $\overset{+5}{NO_3^-} + 3e^- \rightarrow \overset{+2}{NO} + 4OH^-$

ANSWER 43 Yes
(left: 8)
(right: 8)

ANSWER 57

$6H_2O + \overset{0}{I_2} \rightarrow 2\overset{+5}{IO_3^-} + 10e^- + 12H^+$

ANSWER 71

$3Cu + 8H_3O^+ + 2NO_3^- \rightarrow 3Cu^{2+} + 2NO + 12H_2O$

ITEM 3

If a reagent loses electrons, its oxidation number is increased. In the example $Cu^+ - e^- \rightarrow Cu^{2+}$, the oxidation number of copper increases from __________ to __________; Cu^+ has been *oxidized* to Cu^{2+}.

ITEM 17

In the equation

$$\overset{0}{Cr} + 4\overset{-2+1}{OH^-} \rightarrow \overset{+3-2}{CrO_2^-} + 2\overset{+1-2}{H_2O} + 3e^-$$

the net charge on the left is __________ and on the right is __________.

ITEM 31

For reactions that occur in acidic solution, equations are balanced by adding H^+. For reactions occurring in basic solution, __________ ions are added to balance charge.

ITEM 45

The third step has been to balance the charge on both sides of the equation by adding __________ for reactions in acidic solution or __________ for reactions in basic solution. Next, the number of H atoms has been balanced by adding __________. Finally, the equation has been checked by counting __________.

ITEM 59

Completely balance the equation for the half-reaction $HSnO_2^- \rightarrow Sn(OH)_6^{2-}$ (basic solution).

The overall reaction is a combination of two half-reactions. You must recognize the two half-reactions (after assignment of oxidation numbers), balance the half-reactions separately, and finally combine them so the number of electrons produced in one half-reaction equals the number consumed in the other.

ITEM 73

Assign oxidation numbers to the atoms in the reaction $Cl_2 \rightarrow Cl^- + HClO$ (acidic solution).

ANSWER 2 electrons
oxidation numbers

ANSWER 16 left: $(3+) + 3(1-) = 0$
right: 0

ANSWER 30

$$\overset{+5}{NO_3^-} + 3e^- + 4H^+ \rightarrow \overset{+2}{NO}$$

ANSWER 44 oxidation numbers
atoms
oxidized
electrons
oxidation number

ANSWER 58

$$\overset{+1+5-2}{H_3AsO_4} \rightarrow \overset{+1+3-2}{HAsO_2}$$

$$2e^- + H_3AsO_4 \rightarrow HAsO_2$$

$$2H^+ + 2e^- + H_3AsO_4 \rightarrow HAsO_2$$

$$2H^+ + 2e^- + H_3AsO_4 \rightarrow HAsO_2 + 2H_2O$$

ANSWER 72 Yes.

ITEM 4

It is customary to write equations showing the *addition* of electrons to the appropriate side and to place the oxidation number of each atom above its symbol. Rewrite the equation at the right for the oxidation of Cu^+ in this way.

ITEM 18

A simple example will illustrate how a half-reaction is balanced: To begin balancing the half-reaction $Cl_2 \rightarrow Cl^-$, first assign oxidation numbers to the reagents.

ITEM 32

The equation at the right is balanced with respect to electrons and charge. It is not balanced with respect to ________.

ITEM 46

In basic aqueous solution $Be_2O_3^{2-}$ can be converted to Be. Write this much of the equation and assign oxidation numbers.

ITEM 60

Metallic silver dissolves in nitric acid to give Ag^+ and NO. To begin balancing the equation for this reaction, assign oxidation numbers to the atoms.

ITEM 74

In this reaction, part of the Cl_2 is reduced to ________ and part is oxidized to ________.

$Cu^+ - e^- \rightarrow Cu^{2+}$

ANSWER 3 +1
+2

ANSWER 17 left: $4(1-) = 4-$
right: $(1-) + 3(1-) = 4-$

ANSWER 31 OH^-

$\overset{+5}{NO_3^-} + 3e^- + 4H^+ \rightarrow \overset{+2}{NO}$

ANSWER 45 H^+
OH^-
H_2O
oxygen atoms

ANSWER 59

$\overset{+1+2-2}{HSnO_2^-} \rightarrow \overset{+4-2+1}{Sn(OH)_6^{2-}}$

$HSnO_2^- \rightarrow Sn(OH)_6^{2-} + 2e^-$

$3OH^- + HSnO_2^- \rightarrow Sn(OH)_6^{2-} + 2e^-$

$H_2O + 3OH^- + HSnO_2^- \rightarrow Sn(OH)_6^{2-} + 2e^-$

$Ag + NO_3^- \rightarrow Ag^+ + NO$

ANSWER 73 $\overset{0}{Cl_2} \rightarrow \overset{-1}{Cl^-} + \overset{+1}{H}\,\overset{+1}{Cl}\,\overset{-2}{O}$

ITEM 5

When an atom or ion is oxidized, its oxidation number is __________ creased.

ITEM 19

Next, balance the number of atoms of the element being oxidized or reduced.

ITEM 33

If the reaction occurs in aqueous solution, H_2O molecules may be involved as reactants or products. Balance the hydrogen atoms in the equation by adding an appropriate number of H_2O molecules.

ITEM 47

Is the half-reaction $Be_2O_3^{2-} \rightarrow Be$ a reduction or an oxidation?

ITEM 61

The element undergoing oxidation is __________. Therefore, the easily balanced oxidation half-reaction is __________.

ITEM 75

The two balanced half-reactions are

$$2e^- + Cl_2 \rightarrow 2Cl^-$$
$$2H_2O + Cl_2 \rightarrow 2HClO + 2e^- + 2H^+$$

They combine to give the total balanced equation __________.

Additional examples of redox equations follow. Balance each equation, using the method outlined in this chapter. The steps in balancing each equation appear on the page following the question. Practice on these equations until you have mastered the procedure and can balance them without help.

$\overset{+5}{NO_3^-} + 3e^- + 4H^+ \rightarrow \overset{+2}{NO}$

ANSWER 4 $\overset{+1}{Cu^+} \rightarrow \overset{+2}{Cu^{2+}} + e^-$

ANSWER 18 $\overset{0}{Cl_2} \rightarrow \overset{-1}{Cl^-}$

ANSWER 32 the number of atoms of H and O

ANSWER 46 $\overset{+2-2}{Be_2O_3^{2-}} \rightarrow \overset{0}{Be}$

ANSWER 60 $\overset{0}{Ag} + \overset{+5-2}{NO_3^-} \rightarrow \overset{+1}{Ag^+} + \overset{+2-2}{NO}$

ANSWER 74 $\overset{-1}{Cl^-}$ (Cl)

$\overset{+1}{HClO}$ (Cl)

ITEM 6

Complete the equation at the right. Show the oxidation of H^- to H_2 by adding electrons to the appropriate side and assigning oxidation numbers.

ITEM 20

The oxidation numbers are used in balancing electrons: Since each of the Cl atoms shows a decrease in oxidation number from ________ to ________, a total of ________ electrons must be picked up by Cl_2. Rewrite the equation to show the addition of these electrons.

ITEM 34

If the equation is balanced also with respect to oxygen atoms, the equation is completely balanced. Check this.
Number of O atoms on left: ________.
Number of O atoms on right: ________.

ITEM 48

After assignment of oxidation numbers, the next step is to balance ________.

ITEM 62

The element undergoing reduction is ________.
The unbalanced half-reaction is ________.

ITEM 76

$H_2S + Cr_2O_7^{2-} \rightarrow S + Cr^{3+}$ (acid solution)

(Use the summary on page 406 if you need it.)

$2H^- \rightarrow H_2$

$$\overset{0}{Ag} + \overset{+5-2}{NO_3^-} \rightarrow \overset{+1}{Ag^+} + \overset{+2-2}{NO}$$

ANSWER 5 in

ANSWER 19

$$\overset{0}{Cl_2} \rightarrow \overset{-1}{2Cl^-}$$

ANSWER 33

$$\overset{+5}{NO_3^-} + 3e^- + 4H^+ \rightarrow \overset{+2}{NO} + 2H_2O$$

ANSWER 47 reduction

ANSWER 61 Ag

$$\overset{0}{Ag} \rightarrow \overset{+1}{Ag^+} + e^-$$

ANSWER 75

$2H_2O + 2Cl_2 \rightarrow 2Cl^- + 2HClO + 2H^+$

or

$H_2O + Cl_2 \rightarrow Cl^- + HClO + H^+$

or

$2H_2O + Cl_2 \rightarrow Cl^- + HClO + H_3O^+$

ITEM 7

Reduction of a reagent corresponds to an increase in the number of electrons assigned to an atom of the reagent. Therefore, when an atom is reduced, its oxidation number is ________.

ITEM 21

This equation has been balanced with respect to atoms and electrons. What is the net charge on each side?

ITEM 35

Tabulate the steps that have been taken to balance the half-reaction treated in the preceding items: (a) Assign oxidation numbers and balance the number of atoms of the element being oxidized or reduced. (b) Add ________ to balance the change in oxidation number. (c) Add ________ or ________ to balance the charge. (d) Add ________. (e) Check by seeing that the oxygen atoms are balanced.

ITEM 49

Then the equation becomes ________.

ITEM 63

Balance this half-reaction.

ITEM 77

$ClO^- + CrO_2^- \rightarrow Cl^- + CrO_4^{2-}$ (basic solution)

$$\overset{+2}{Be_2O_3^{2-}} \rightarrow \overset{0}{Be}$$

ANSWER 6

$$2\overset{-1}{H^-} \rightarrow \overset{0}{H_2} + 2e^-$$

ANSWER 20 0

−1

two

$$\overset{0}{Cl_2} + 2e^- \rightarrow 2\overset{-1}{Cl^-}$$

ANSWER 34 left: 3

right: 3

ANSWER 48 the number of atoms of Be (the element being reduced)

ANSWER 62 nitrogen

$$\overset{+5}{NO_3^-} \rightarrow \overset{+2}{NO}$$

ANSWER 76

$H_2S + Cr_2O_7^{2-} \rightarrow S + Cr^{3+}$ (acid solution)

$$\overset{-2}{H_2S} \rightarrow \overset{0}{S} + 2H^+ + 2e^-$$

$$\overset{+6}{Cr_2O_7^{2-}} + 14H^+ + 6e^- \rightarrow 2\overset{+3}{Cr^{3+}} + 7H_2O$$

$$3H_2S \rightarrow 3S + 6H^+ + 6e^-$$

$$\underline{Cr_2O_7^{2-} + 14H^+ + 6e^- \rightarrow 2Cr^{3+} + 7H_2O}$$

$$3H_2S + Cr_2O_7^{2-} + 8H^+ \rightarrow 3S + 2Cr^{3+} + 7H_2O$$

or $3H_2S + Cr_2O_7^{2-} + 8H_3O^+ \rightarrow 3S + 2Cr^{3+} + 15H_2O$

ITEM 8

Write the equation and include oxidation numbers for the reduction of Cu^{2+} to Cu^{+}.

ITEM 22

Balance the equation for the half-reaction $S_8 \rightarrow S^{2-}$ by first assigning oxidation numbers and balancing atoms.

ITEM 36

In acidic solution, $S_2O_3^{2-}$ can be converted to SO_4^{2-}. Assign oxidation numbers in the half-reaction

$$S_2O_3^{2-} \rightarrow SO_4^{2-}$$

ITEM 50

The next step is to add ________. The equation becomes ________.

ITEM 64

To obtain the balanced total equation, the half-reactions must be added so that the electrons of each half-reaction cancel. For the electrons to cancel in the two half-reactions at the right, the oxidation equation must be multiplied by ________.

ITEM 78

$Zn + NO_3^- \rightarrow Zn^{2+} + NH_4^+$ (acid solution)

$Ag \rightarrow Ag^+ + e^-$

$NO_3^- + 3e^- + 4H^+ \rightarrow NO + 2H_2O$

ANSWER 7 decreased

ANSWER 21 left: 2−
right: 2−

ANSWER 35 electrons
H^+
OH^-
H_2O to balance number of H atoms

ANSWER 49

$$\overset{+2}{Be_2O_3^{2-}} \rightarrow \overset{0}{2Be}$$

ANSWER 63

$$\overset{+5}{NO_3^-} \rightarrow \overset{+2}{NO}$$

$$3e^- + NO_3^- \rightarrow NO$$

$$4H^+ + 3e^- + NO_3^- \rightarrow NO + 2H_2O$$

ANSWER 77

$ClO^- + CrO_2^- \rightarrow Cl^- + CrO_4^{2-}$ (basic solution)

$$\overset{+3}{CrO_2^-} + 4OH^- \rightarrow \overset{+6}{CrO_4^{2-}} + 2H_2O + 3e^-$$

$$\overset{+1}{ClO^-} + H_2O + 2e^- \rightarrow \overset{-1}{Cl^-} + 2OH^-$$

$$2CrO_2^- + 8OH^- \rightarrow 2CrO_4^{2-} + 4H_2O + 6e^-$$

$$3ClO^- + 3H_2O + 6e^- \rightarrow 3Cl^- + 6OH^-$$

$$2CrO_2^- + 3ClO^- + 2OH^- \rightarrow 2CrO_4^{2-} + 3Cl^- + H_2O$$

ITEM 9

The relation between oxidation and reduction and the loss or gain of electrons by an element is as follows: When an element loses electrons, it is __________; conversely, when it gains electrons, it is __________.

ITEM 23

Since each S atom must add __________ electrons to become S^{2-}, the total number of electrons picked up by S_8 is __________. Rewrite the equation and add electrons to the appropriate side.

ITEM 37

Is this an oxidation or a reduction half-reaction?

ITEM 51

Once the number of electrons is balanced, it is necessary to balance __________.

ITEM 65

Addition of the two half-reactions gives the total equation, which is __________.

ITEM 79

$NO_2 \rightarrow NO_3^- + NO$ (acid solution)

$3Ag \rightarrow 3Ag^{+} + 3e^{-}$
$NO_3^{-} + 3e^{-} + 4H^{+} \rightarrow NO + 2H_2O$

ANSWER 8 $\overset{+2}{Cu^{2+}} + e^{-} \rightarrow \overset{+1}{Cu^{+}}$

ANSWER 22 $\overset{0}{S_8} \rightarrow \overset{-2}{8S^{2-}}$

ANSWER 36 $\overset{+2-2}{S_2O_3^{2-}} \rightarrow \overset{+6-2}{SO_4^{2-}}$

ANSWER 50 electrons

$\overset{+2}{Be_2O_3^{2-}} + 4e^{-} \rightarrow \overset{0}{2Be}$

ANSWER 64 3

ANSWER 78

$Zn + NO_3^{-} \rightarrow Zn^{2+} + NH_4^{+}$ (acid solution)

$\overset{0}{Zn} \rightarrow \overset{+2}{Zn^{2+}} + 2e^{-} \qquad \overset{+5}{NO_3^{-}} + 10H^{+} + 8e^{-} \rightarrow \overset{-3}{NH_4^{+}} + 3H_2O$

$$\begin{array}{l} 4Zn \rightarrow 4Zn^{2+} + 8e^{-} \\ \underline{NO_3^{-} + 10H^{+} + 8e^{-} \rightarrow NH_4^{+} + 3H_2O} \\ 4Zn + NO_3^{-} + 10H^{+} \rightarrow 4Zn^{2+} + NH_4^{+} + 3H_2O \end{array}$$

or

$$4Zn + NO_3^{-} + 10H_3O^{+} \rightarrow 4Zn^{2+} + NH_4^{+} + 13H_2O$$

ITEM 10

Thus, oxidation corresponds to a(n) __________ in oxidation number; reduction corresponds to a(n) __________ in oxidation number.

ITEM 24

Is the equation balanced with respect to charge?

ITEM 38

Balance this half-reaction with respect to the number of atoms of the element being oxidized.

ITEM 52

Since this reaction is carried out in basic solution, the charge is balanced by __________. Do this.

ITEM 66

To change H^+ to H_3O^+, how many H_2O molecules must be added to each side of the equation? Then the equation becomes __________.

ITEM 80

$$MnO_4^- + H_2C{=}CH_2 \rightarrow MnO_2 + \underset{\displaystyle OH}{\underset{|}{H_2C}}{-}\underset{\displaystyle OH}{\underset{|}{CH_2}}$$

(basic solution)

This can also be written as

$$MnO_4^- + C_2H_4 \rightarrow MnO_2 + C_2H_6O_2$$

ANSWER 9 oxidized
reduced

ANSWER 23 two
16

$\overset{0}{S_8} + 16e^- \rightarrow \overset{-2}{8S^{2-}}$

$\overset{+2}{S_2O_3^{2-}} \rightarrow \overset{+6}{SO_4^{2-}}$

ANSWER 37 oxidation

$\overset{+2}{Be_2O_3^{2-}} + 4e^- \rightarrow \overset{0}{2Be}$

ANSWER 51 the charge

ANSWER 65

$3Ag + NO_3^- + 4H^+ \rightarrow 3Ag^+ + NO + 2H_2O$

ANSWER 79

$NO_2 \rightarrow NO_3^- + NO$ (acid solution)

$\overset{+4}{NO_2} + H_2O \rightarrow \overset{+5}{NO_3^-} + 2H^+ + e^-$

$\overset{+4}{NO_2} + 2H^+ + 2e^- \rightarrow \overset{+2}{NO} + H_2O$

$2NO_2 + 2H_2O \rightarrow 2NO_3^- + 4H^+ + 2e^-$
$NO_2 + 2H^+ + 2e^- \rightarrow NO + H_2O$

$3NO_2 + H_2O \rightarrow 2NO_3^- + NO + 2H^+$
or
$3NO_2 + 3H_2O \rightarrow 2NO_3^- + NO + 2H_3O^+$

ITEM 11

When Cl^- ions are converted to Cl_2, the oxidation number of chlorine increases from ________ to ________. Is this conversion a reduction or an oxidation?

ITEM 25

To begin balancing the equation for the half-reaction $NO_3^- \rightarrow NO$, we first assign ________ to each element. The equation now is ________.

ITEM 39

The half-reaction must now be balanced with respect to the number of electrons. Add electrons to balance the change in oxidation number of the two S atoms.

ITEM 53

Now the equation must be balanced with respect to ________. The equation becomes ________.

ITEM 67

Another example is the reaction of copper with nitric acid: $Cu + NO_3^- \rightarrow Cu^{2+} + NO$. Assign oxidation numbers to the atoms that you expect are involved in oxidation and reduction. The two half-reactions are
(oxidation) ________
(reduction) ________.

(*Item 81 is on page 398.*)

ANSWER 10 increase
decrease

ANSWER 24 Yes
(left: 16−)
(right: 16−)

ANSWER 38

$$\overset{+2}{S_2O_3^{2-}} \rightarrow \overset{+6}{2SO_4^{2-}}$$

ANSWER 52 adding OH^-

$$\overset{+2}{Be_2O_3^{2-}} + 4e^- \rightarrow \overset{0}{2Be} + 6OH^-$$

ANSWER 66

Four.

$$3Ag + NO_3^- + 4H_3O^+ \rightarrow 3Ag^+ + NO + 6H_2O$$

ANSWER 80

$$MnO_4^- + H_2C{=}CH_2 \rightarrow MnO_2 + \underset{\text{OH}}{H_2C}{-}\underset{\text{OH}}{CH_2} \text{ (basic solution)}$$

$$\overset{-2}{C_2H_4} + 2OH^- \rightarrow \overset{-1}{C_2H_6O_2} + 2e^-$$

$$2H_2O + \overset{+7}{MnO_4^-} + 3e^- \rightarrow \overset{+4}{MnO_2} + 4OH^-$$

$$3C_2H_4 + 6OH^- \rightarrow 3C_2H_6O_2 + 6e^-$$

$$4H_2O + 2MnO_4^- + 6e^- \rightarrow 2MnO_2 + 8OH^-$$

$$3C_2H_4 + 4H_2O + 2MnO_4^- \rightarrow 3C_2H_6O_2 + 2MnO_2 + 2OH^-$$

ITEM 12 Write an equation to represent the oxidation of two chloride ions to Cl_2.

ITEM 26 The element undergoing reduction is __________.

ITEM 40 In its present form, the equation has a net charge on the left of __________ and a net charge on the right of __________.

ITEM 54 Check this equation. Is it balanced?

ITEM 68 The balanced equations for the two half-reactions are __________.

ITEM 81 $Al + OH^- \rightarrow Al\,(OH)_4^- + H_2$ (basic solution)

ANSWER 11 −1
0
Oxidation

ANSWER 25 oxidation numbers

$\overset{+5-2}{NO_3^-} \rightarrow \overset{+2-2}{NO}$

ANSWER 39

$\overset{+2}{S_2O_3^{2-}} \rightarrow \overset{+6}{2SO_4^{2-}} + 8e^-$

ANSWER 53

H atoms

$\overset{+2}{Be_2O_3^{2-}} + 4e^- + 3H_2O \rightarrow \overset{0}{2Be} + 6OH^-$

ANSWER 67

$\overset{0}{Cu} + \overset{+5}{NO_3^-} \rightarrow \overset{+2}{Cu^{2+}} + \overset{+2}{NO}$

(oxidation) $Cu \rightarrow Cu^{2+}$

(reduction) $NO_3^- \rightarrow NO$

ITEM 13

Write an equation for the reduction of the ion Sn^{4+} to metallic Sn.

ITEM 27

Since this equation is balanced with respect to the number of atoms of the element being reduced, the next step is to add electrons to the appropriate side. The N atom shows a decrease in oxidation number from ________ to ________ and therefore must pick up ________ electrons. Rewrite the equation to show the addition of these electrons.

ITEM 41

Since this reaction occurs in acidic solution, H^+ ions are added to balance the ________. The equation becomes ________.

ITEM 55

In acid solution, I_2 can be converted to IO_3^-. Carry out the steps necessary to balance the half-reaction with respect to electrons.

ITEM 69

In order for the electrons to cancel when the two equations are added, the oxidation half-reaction must be multiplied by ________ and the reduction half-reaction by ________.

$$\overset{+2}{S_2O_3^{2-}} \rightarrow \overset{+6}{2SO_4^{2-}} + 8e^-$$

ANSWER 12

$$2\overset{-1}{Cl^-} \rightarrow \overset{0}{Cl_2} + 2e^-$$

ANSWER 26 nitrogen

ANSWER 40 left: 2−
right: 2(2−) + 8(1−) = 12−

ANSWER 54 Yes
(left: 6O)
(right: 6O)

ANSWER 68

$Cu \rightarrow Cu^{2+} + 2e^-$
$4H^+ + 3e^- + NO_3^- \rightarrow NO + 2H_2O$

ANSWER 81

$Al + OH^- \rightarrow Al(OH)_4^- + H_2$ (basic solution)

$$\overset{0}{Al} + 4OH^- \rightarrow \overset{+3}{Al(OH)_4^-} + 3e^-$$

$$2OH^- + 2e^- + 2\overset{+1}{H_2O} \rightarrow \overset{0}{H_2} + 4OH^-$$

or

$$2H_2O + 2e^- \rightarrow H_2 + 2OH^-$$

$$2Al + 8OH^- \rightarrow 2Al(OH)_4^- + 6e^-$$
$$6H_2O + 6e^- \rightarrow 3H_2 + 6OH^-$$
$$\overline{2Al + 6H_2O + 2OH^- \rightarrow 2Al(OH)_4^- + 3H_2}$$

ITEM 14

Equations such as these are known as *half-reactions.* A half-reaction represents only a reduction or only an oxidation and shows, as part of the balanced equation, the electrons lost or gained by an atom. Write the half-reaction for the reduction of Fe^{3+} to Fe.

ITEM 28

The equation is balanced now with respect to electrons associated with N, the element undergoing reduction; the next step is to balance the net charge. As the equation stands, the net charge on the left is ________ and on the right it is ________.

ITEM 42

The equation is now unbalanced with respect to ________, so ________ is added to the ________ side. The equation then becomes ________.

ITEM 56

Balance the equation with respect to charge.

ITEM 70

The addition step is ________.

$Cu \rightarrow Cu^{2+} + 2e^-$
$4H^+ + 3e^- + NO_3^- \rightarrow NO + 2H_2O$

ANSWER 13 $\overset{+4}{Sn^{4+}} + 4e^- \rightarrow \overset{0}{Sn}$

ANSWER 27 +5
+2
3
$\overset{+5}{NO_3^-} + 3e^- \rightarrow \overset{+2}{NO}$

ANSWER 41

charge
$\overset{+2}{S_2O_3^{2-}} \rightarrow \overset{+6}{2SO_4^{2-}} + 8e^- + 10H^+$

ANSWER 55 0 +5
$I_2 \rightarrow IO_3^-$
$I_2 \rightarrow 2IO_3^-$
$I_2 \rightarrow 2IO_3^- + 10e^-$

ANSWER 69 3
2

ANSWER 14 $\overset{+3}{Fe^{3+}} + 3e^- \rightarrow \overset{0}{Fe}$

(*Item 15 is on page 376.*)

ANSWER 28 $(1-) + 3(1-) = 4-$

0

(*Item 29 is on page 376.*)

ANSWER 42

H and O atoms

H_2O

left

$$5H_2O + \overset{+2}{S_2O_3^{2-}} \rightarrow 2\overset{+6}{SO_4^{2-}} + 8e^- + 10H^+$$

(*Item 43 is on page 376.*)

ANSWER 56 $\overset{0}{I_2} \rightarrow 2\overset{+5}{IO_3^-} + 10e^- + 12H^+$

(*Item 57 is on page 376.*)

ANSWER 70

$$3Cu \rightarrow 3Cu^{2+} + 6e^-$$
$$\underline{8H^+ + 6e^- + 2NO_3^- \rightarrow 2NO + 4H_2O}$$
$$3Cu + 8H^+ + 2NO_3^- \rightarrow 3Cu^{2+} + 2NO + 4H_2O$$

(*Item 71 is on page 376.*)

SUMMARY OF STEPS FOR BALANCING REDOX EQUATIONS

(1) Write down the main reactants and products.
(2) Assign oxidation numbers.
(3) Separate the oxidation and reduction half-reactions.
(4) Balance the half-reactions with respect to
    (a) the number of atoms of the element undergoing oxidation or reduction;
    (b) electrons by adding electrons to the electron-deficient side of the equation;
    (c) charge by adding H^+ ions or OH^- ions;
    (d) H and O atoms by adding H_2O.
(5) Multiply the half-reactions by appropriate numbers so that, when they are added, the electrons cancel.
(6) Add the half-reactions and check for balance.

REVIEW 8 ELECTRONS IN ATOMS

One of the great successes of physical science in the twentieth century was the development of a model for the electronic structure of the atom. The model was derived from information obtained by the study of atomic spectra and from extensive quantum mechanical calculations. Neither of these will be treated here, but atomic spectra are discussed briefly in Section 8-2 and a short introduction to the quantum mechanical calculations is presented in Sections 8-3 through 8-6 of *Chemical Principles.*

In this review you will learn some of the results of these calculations that are particularly important for understanding the chemical behavior of atoms. This will serve as a review of Section 8-7 of *Chemical Principles* and also as an introduction to Chapter 9 of that text. When you have finished the chapter, you will be able to predict the approximate arrangement in space of an atom's electrons and to indicate the relative energies of these electrons. This knowledge will allow you to predict the electronic configurations of the elements of the first two rows of the periodic table.

ITEM 1

Although an electron of an atom does not occupy a fixed position relative to the nucleus, there is a region of space in the neighborhood of the nucleus in which the probability of finding an electron of an atom is relatively high. We can say, for example, that there is some "most probable distance" between an electron and the ________ of an atom.

ITEM 14

For the hydrogen atom, the energies of the allowed electronic states are given by the expression $E = -313.5/n^2$ kcal mole^{-1} or $E = -13.58/n^2$ electron volts (eV). Calculate the energies for the $n = 2, 3, 4$, and ∞ levels for the hydrogen atom, and show these in the diagram at the right.

ITEM 27

The value of n for an electron determines the degree to which its orbital extends from the nucleus. For small values of n, the orbital is relatively small: There is a high probability that the electron will be located relatively ________ to the nucleus.

ITEM 40

The shape of an orbital depends on the ________ quantum number, represented by the symbol ________, which can have values ________. (We shall see that the energy of the electron can also depend, to a small extent, on this quantum number.)

ITEM 53

On the coordinate axes at the right, make sketches representing orbitals to show that only one s orbital exists but that three p orbitals, differing in their orientation, occur.

ITEM 66

The *Pauli exclusion principle* states that only one electron of an atom can have any given combination of all four quantum numbers. Since $s = \pm\frac{1}{2}$, an orbital identified by any given combination of n, l, and m can be occupied by no more than ________ electrons.

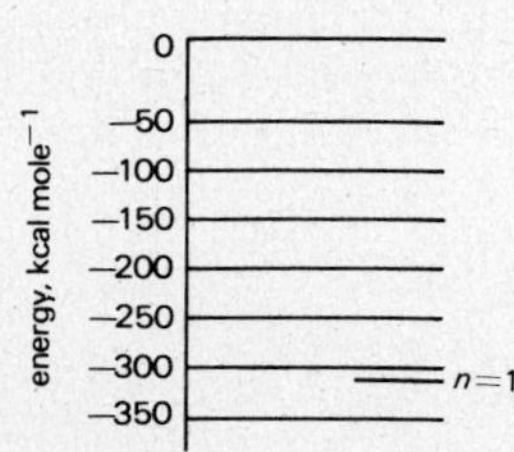
energy, kcal mole^{-1}
0
−50
−100
−150
−200
−250
−300
−350
n=1

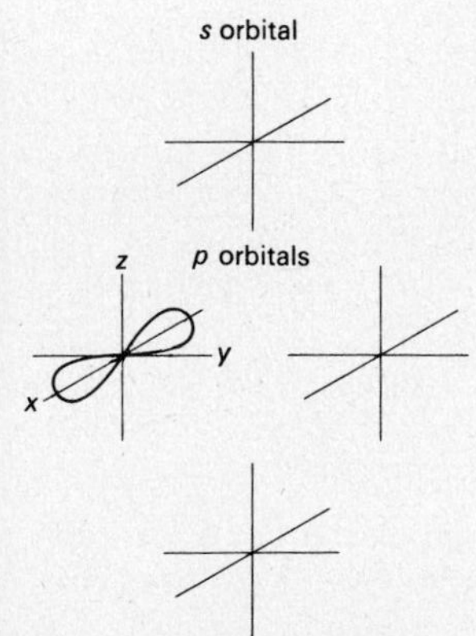
s orbital
p orbitals
z
y
x

ITEM 2

In addition, an electron has some amount of energy: the kinetic energy of its motion plus potential energy arising from the electrostatic force of attraction between it and the positively charged ___________.

ITEM 15

The difference in energy between the $n = 1$ and the $n = 2$ states of hydrogen is ___________ kcal mole^{-1}. Therefore, if an electron in the $n = 1$ state absorbed (i.e., gained) this much energy, the value of n for the new state would be ___________.

ITEM 28

The smallest orbital will be that corresponding to $n =$ ___________, whereas the largest will be that corresponding to $n =$ ___________.

ITEM 41

The energy of an electron depends primarily on its distance from the nucleus, and the distance is determined by the value of ___________, the ___________ quantum number.

ITEM 54

Since the three p orbitals can be drawn along the x, y, and z axes, they are often labeled p_x, ___________, and ___________.

ITEM 67

In an energy level diagram, an electron is often represented by an arrow indicating one of the two possible spins. Show ten electrons in their lowest energy arrangement in the energy levels at the right by adding additional arrows to represent electrons.

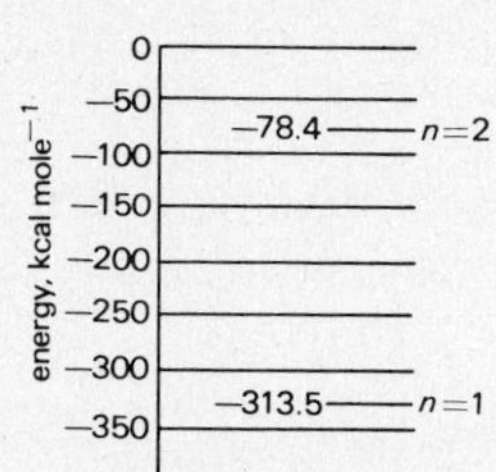

$\underline{\quad}3d_{xy}\underline{\quad}3d_{xz}\underline{\quad}3d_{yz}\underline{\quad}3d_{x^2-y^2}\underline{\quad}3d_{z^2}$

$\underline{\quad}3p_x\underline{\quad}3p_y\underline{\quad}3p_z$

$\underline{\quad}3s$

$\underline{\quad}2p_x\underline{\quad}2p_y\underline{\quad}2p_z$

$\underline{\quad}2s$

$\underline{\uparrow\downarrow}\ 1s$

ANSWER 1 nucleus

ANSWER 14

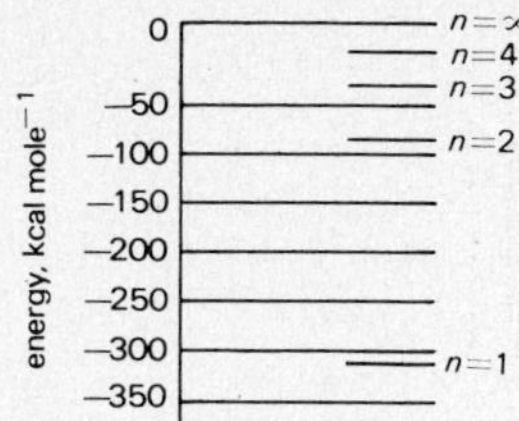

ANSWER 27 close

ANSWER 40 azimuthal

l

$0, 1, 2, \ldots, n-1$

ANSWER 53

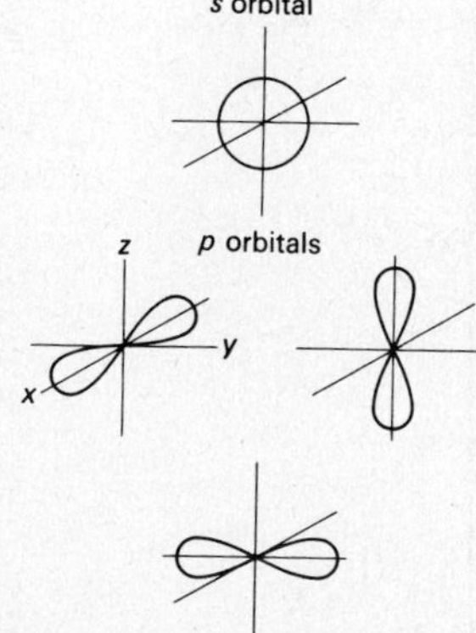

ANSWER 66 two

ITEM 3

Together, the most probable distance from the nucleus and the energy describe, in part, the state of an electron of an atom. If the state of an electron were to change, both the __________ and the __________ would change.

ITEM 16

Conversely, if a hydrogen atom having its electron in the $n = 2$ state emits 235.1 kcal mole^{-1} of energy, n will become __________.

ITEM 29

In fact, in the limit where $n = \infty$, there is no interaction between the __________ and the __________. They are completely separated.

ITEM 42

When many electrons are present in an atom, their orbitals penetrate one another. The extent to which an electron is influenced by the nucleus then depends on both the size and the shape of its orbital. For many-electron atoms, therefore, the energy of a particular electron is determined not only by the value of n, but also by the value of __________.

ITEM 55

Orbitals with $l = 2$, that is, __________ orbitals, can have any one of __________ orientations, each orientation corresponding to a different value of the quantum number __________.

ITEM 68

The atomic number of fluorine is 9. In the diagram at the right, indicate the arrangement of nine electrons in a fluorine atom in the ground state and label the occupied levels.

ANSWER 2 nucleus

ANSWER 15 235.1
2

ANSWER 28 1
∞

ANSWER 41 n
principal

Spatial representations of all d orbitals are shown at the beginning of Review 11.

ANSWER 54 p_y
p_z

___ ___ ___ ___ ___
___ ___ ___
___
___ ___ ___
___
___

ANSWER 67

___ $3d_{xy}$ ___ $3d_{xz}$ ___ $3d_{yz}$ ___ $3d_{x^2-y^2}$ ___ $3d_{z^2}$
___ $3p_x$ ___ $3p_y$ ___ $3p_z$
___ $3s$
$\uparrow\downarrow$ $2p_x$ $\uparrow\downarrow$ $2p_y$ $\uparrow\downarrow$ $2p_z$
$\uparrow\downarrow$ $2s$
$\uparrow\downarrow$ $1s$

ITEM 4

Since the states available to an electron are governed by the laws of quantum mechanics, the numbers used to identify these states are known as __________ numbers.

ITEM 17

An atom is in the *ground state* when its electrons have the lowest possible energy. The electron of a hydrogen atom in the ground state has a principal quantum number of __________.

ITEM 30

The quantum number n, which can have the values __________, is primarily responsible for determining the __________ of the electron and the size of the __________ occupied by the electron.

ITEM 43

For instance, s electrons shield p electrons from the attraction of the nucleus. Thus, for a given value of n, p electrons have slightly __________ energy than s electrons.

ITEM 56

For a given value of n there is (are) only __________ s orbital(s), but there are __________ p orbitals, differing only in their __________ in space.

ITEM 69

The arrangement, or *configuration*, of the electrons in fluorine can also be represented by the notation $1s^2 2s^2 2p_x{}^2 2p_y{}^2 2p_z{}^1$, or $1s^2 2s^2 2p^5$, in which the superscripts indicate the number of electrons in that quantum state. At the right, indicate diagrammatically, and with this notation, the electronic configuration of sodium (at. no. = 11).

ANSWER 3 most probable distance from the nucleus
energy

ANSWER 16 1

ANSWER 29 electron
nucleus

ANSWER 42 l

ANSWER 55 d
five
m

ANSWER 68

___ ___ ___ ___ ___
___ ___ ___
___
$\underline{\uparrow\downarrow}$ $2p_x$ $\underline{\uparrow\downarrow}$ $2p_y$ $\underline{\uparrow}$ $2p_z$
$\underline{\uparrow\downarrow}$ $2s$
$\underline{\uparrow\downarrow}$ $1s$

ITEM 5

The first of the four quantum numbers that specify the ________ and the ________ of an electron is called the *principal quantum number* and is denoted by the symbol n.

ITEM 18

If energy is absorbed by a hydrogen atom in the ground state, the value of n will ________. Then the atom is in an *excited state.*

ITEM 31

The shape of the region in which the probability of the electron being located is high (i.e., the shape of the ________) is determined by the value of a second quantum number, the *azimuthal quantum number* l.

ITEM 44

Similarly, p electrons have somewhat lower energy than d electrons, and so forth. The energy is principally determined, however, by the value of ________.

ITEM 57

Since the energy of an electron in an atom is independent of the orientation of the orbital (so long as the atom is not subjected to an electric or magnetic field), electrons with different values of m have the same ________.

ITEM 70

The electronic configuration of oxygen (at. no. = 8) in the ground state is $1s^2 2s^2 2p_x^{\,2} 2p_y^{\,1} 2p_z^{\,1}$. (The repulsion between two electrons causes them to occupy two different orbitals of the same energy whenever possible; such unpaired electrons have the same spin.) Represent the ground state configuration of oxygen in the diagram at the right.

—— —— —— —— ——
—— —— ——
——
—— —— ——
——
——

ANSWER 4 quantum

ANSWER 17 1

ANSWER 30 1, 2, 3, . . . , ∞
energy
orbital

ANSWER 43 higher

ANSWER 56 one
three
direction

ANSWER 69

—— —— —— —— ——
—— —— ——
↑ 3s
⇅ $2p_x$ ⇅ $2p_y$ ⇅ $2p_z$
⇅ $2s$
⇅ $1s$

$1s^22s^22p_x^22p_y^22p_z^23s^1$ (or $1s^22s^22p^63s^1$)

ITEM 6

The principal quantum number, symbol _________, can have integral values from 1 to ∞. This is indicated by writing $n = 1, 2,$ _________.

ITEM 19

In the diagram at the right, draw a vertical arrow to represent the change in energy level when a hydrogen atom in the ground state absorbs 278.7 kcal mole^{-1} of energy.

ITEM 32

The solutions of the quantum mechanical equation that describes the behavior of an electron in a hydrogen atom show that l can have integral values from zero to one less than the value of n; that is, $l =$ _________.

ITEM 45

For atoms with several electrons, each state with n greater than one is split into different levels depending on the number of l values possible. Complete the labeling of the diagram at the right.

ITEM 58

The energy level diagram at the right has lines corresponding to some of the states available to an electron in an atom. Complete the labeling.

ITEM 71

Using the notation s, p_x, and so on, write the electronic configurations of carbon (at. no. = 6) and nitrogen (at. no. = 7).

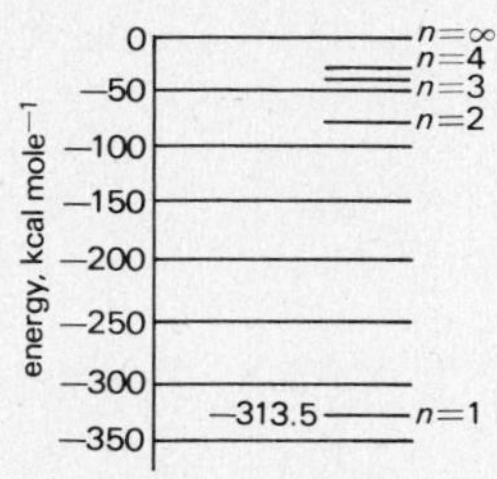

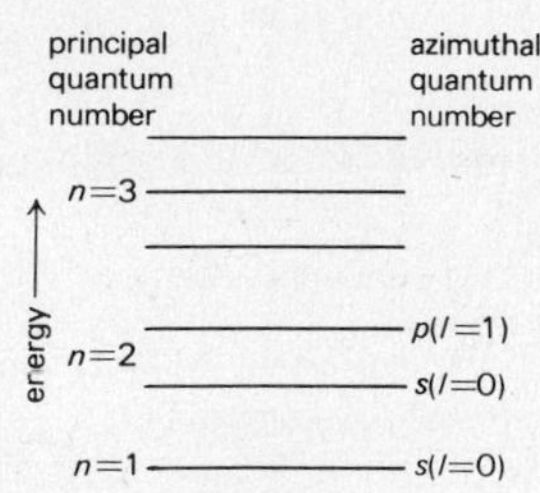

$3d_{xy}$ $3d_{xz}$ $3d_{yz}$ $3d_{x^2-y^2}$ $3d_{z^2}$

$n = 3$ ___ ___ ___

___

$2p_x$ ___ ___

$n = 2$ ___

$n = 1$ 1s

ANSWER 5 most probable distance from the nucleus
energy

ANSWER 18 increase

ANSWER 31 orbital

ANSWER 44 *n*

ANSWER 57 energy

ANSWER 70

___ ___ ___ ___ ___

___ ___ ___

___

⇅ $2p_x$ ↑ $2p_y$ ↑ $2p_z$

⇅ $2s$

⇅ $1s$

ITEM 7

If the state of an electron is that specified by a small value of n, the electron is relatively near the nucleus and has a relatively low __________.

ITEM 20

In the diagram at the right, indicate the change in energy level when a hydrogen atom having an electron in the $n = 3$ state emits 43.6 kcal mole^{-1}.

ITEM 33

Thus, when an electron has a principal quantum number of 1, l must be __________; if $n = 2$, l can be __________ or __________; if $n = 3$, l can be __________, __________, or __________.

ITEM 46

A convenient abbreviation of the orbital notation consists of writing 1s instead of $n = 1$, $l = 0$; 2s for $n = 2$, $l = 0$; 2p for $n = 2$, $l = 1$, and so forth. Use this abbreviated notation to label the energy levels at the right, as is done for the 1s level.

ITEM 59

All s orbitals have a __________ shape; the shape of all p orbitals can be represented by __________. The size of an orbital is dependent on the value of __________.

ITEM 72

The electronic configurations of heavier elements are obtained in the same way. Electrons are "placed" in orbitals in order of increasing energy. However, the simple energy level diagram must be modified, as shown at the right, when we include $n = 4$ levels. Indicate in the diagram at the right the configuration of potassium (at. no. = 19).

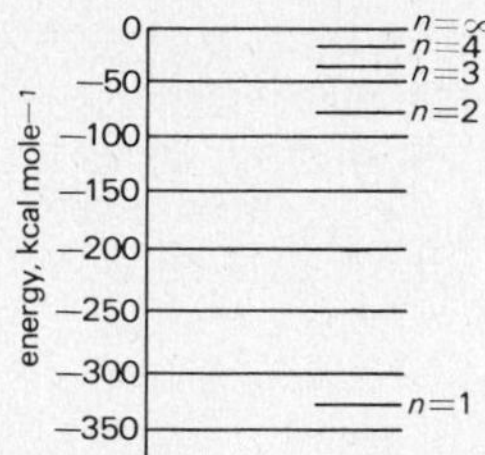

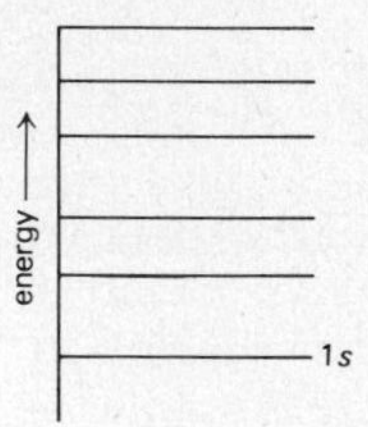

$4p$ ___ ___ ___
$3d$ ___ ___ ___ ___ ___
$4s$ ___
$3p$ ___ ___ ___
$3s$ ___
$2p$ ___ ___ ___
$2s$ ___
$1s$ ___

ANSWER 6 n
$3, \ldots, \infty$

ANSWER 19

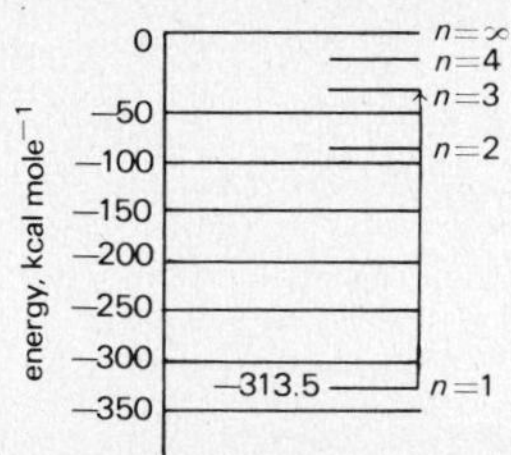

ANSWER 32 $0, 1, 2, \ldots, n-1$

ANSWER 45

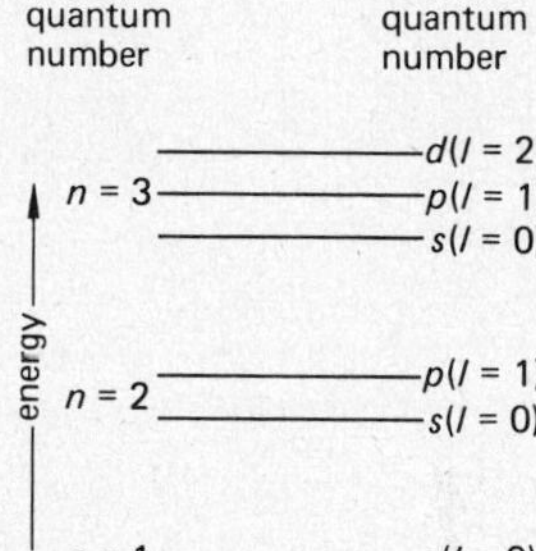

ANSWER 58

$n = 3$ $\begin{cases} \underline{3d_{xy}}\ \underline{3d_{xz}}\ \underline{3d_{yz}}\ \underline{3d_{x^2-y^2}}\ \underline{3d_{z^2}} \\ \underline{3p_x}\ \underline{3p_y}\ \underline{3p_z} \\ \underline{3s} \end{cases}$

$n = 2$ $\begin{cases} \underline{2p_x}\ \underline{2p_y}\ \underline{2p_z} \\ \underline{2s} \end{cases}$

$n = 1$ $\underline{1s}$

ANSWER 71 C $1s^2 2s^2 2p_x^1 2p_y^1$
N $1s^2 2s^2 2p_x^1 2p_y^1 2p_z^1$

ITEM 8

The state in which the electron has the lowest energy and is nearest the nucleus is that state specified by $n =$ __________.

ITEM 21

The only energies that an electron of a hydrogen atom can have are those calculated by using the formula $E = -313.5/n^2$ kcal mole^{-1}, in which n, known as the __________, is an integer. Thus, the smallest amount of energy that a ground state hydrogen atom can absorb is __________.

ITEM 34

Electrons that undergo transitions to states with $l =$ 0 emit energy and produce spectral lines that are sometimes described as *sharp.* The first letter, s, of this word is often used to indicate an electronic state where $l =$ __________.

ITEM 47

We have now dealt with two of the four quantum numbers necessary to describe the behavior of an electron in an atom. These are the __________ quantum number, having values __________, and the __________ quantum number, having values __________.

ITEM 60

In addition to the three quantum numbers introduced so far, the __________, the __________, and the __________ quantum numbers, a final one is necessary to describe the behavior of an electron in an atom.

ITEM 73

Show the electronic configuration of arsenic (at. no. = 33) on the diagram at the right and by means of the notation s, p_x, and so forth.

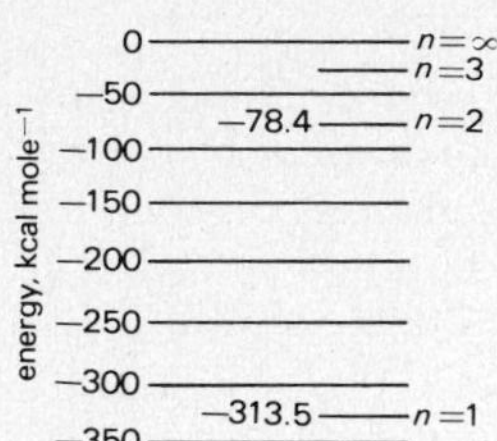

4*p* ___ ___ ___
3*d* ___ ___ ___ ___ ___
4*s* ___
3*p* ___ ___ ___
3*s* ___
2*p* ___ ___ ___
2*s* ___
1*s* ___

ANSWER 7 energy

ANSWER 20

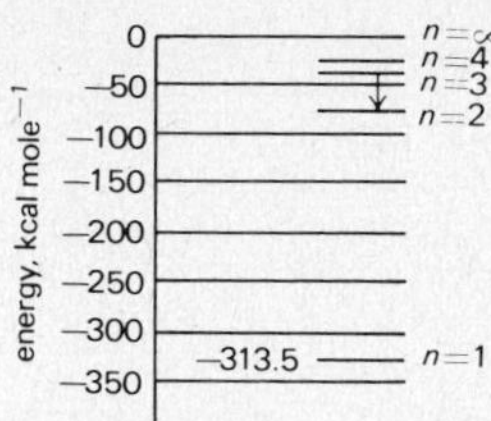

ANSWER 33 0
0, 1
0, 1, 2

ANSWER 46

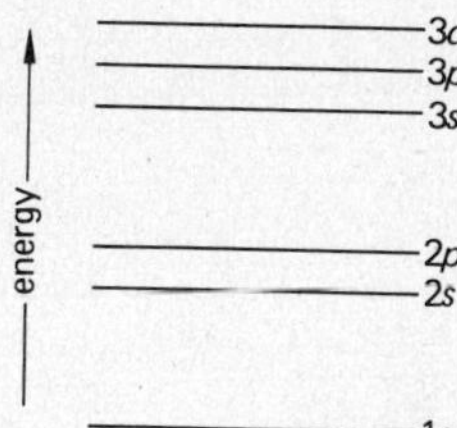

ANSWER 59 spherical
dumbbells
n

ANSWER 72

4*p* ___ ___ ___
3*d* ___ ___ ___ ___ ___
4*s* ↑
3*p* ⇅ ⇅ ⇅
3*s* ⇅
2*p* ⇅ ⇅ ⇅
2*s* ⇅
1*s* ⇅

ITEM 9

For larger values of the principal quantum number, the distance of the electron from the nucleus is __________, and the energy is __________.

ITEM 22

Using the formula $E = -313.5/n^2$ kcal mole^{-1} for the allowed energies of an electron of a hydrogen atom, calculate the energy emitted when an electron changes from a state with $n = 4$ to one with $n = 2$.

ITEM 35

Similarly, since l values of 1, 2, and 3 were related to series of spectral lines described as *principal, diffuse,* and *fundamental,* the letters __________, __________, or __________ generally are written to indicate l values of 1, 2, or 3, respectively.

ITEM 48

Although s orbitals are spherically symmetrical and cannot be oriented in any particular direction, it is clear that __________ orbitals (and, in fact, all orbitals with higher values of l) can be oriented along different directions.

ITEM 61

An electron is sometimes pictured as spinning around an axis through its center. With this picture, it can spin either counterclockwise, conventionally represented by an arrow pointing up, or it can spin __________, represented by ↓.

ITEM 74

All the orbitals of any one value of n are generally called, collectively, a *shell.* Helium and neon, elements that are not known to react chemically, are characterized by completed shells of electrons, as shown by their respective electronic configurations, __________ and __________. (*For the atomic numbers of He and Ne see the periodic table in the inside front cover.*)

ANSWER 8 1

ANSWER 21 principal quantum number
235.1 kcal $mole^{-1}$

ANSWER 34 0

s orbital

p orbital

ANSWER 47 principal
1, 2, 3, . . . , ∞
azimuthal
0, 1, 2, . . . , $n-1$

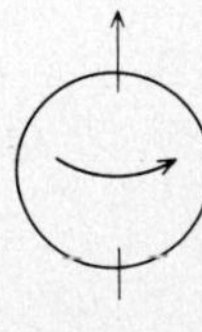

ANSWER 60 principal
azimuthal
magnetic

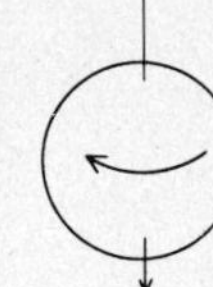

ANSWER 73

$4p$	↑	↑	↑		
$3d$	⇅	⇅	⇅	⇅	⇅
$4s$	⇅				
$3p$	⇅	⇅	⇅		
$3s$	⇅				
$2p$	⇅	⇅	⇅		
$2s$	⇅				
$1s$	⇅				

$1s^2 2s^2 2p^6 3s^2 3p^6 4s^2 3d^{10} 4p_x^1 4p_y^1 4p_z^1$

ITEM 10

At the right is a set of lines representing the energies of the lowest allowed states of an electron of an atom. Label these with appropriate values of the principal quantum number.

ITEM 23

At room temperature, atoms are generally in their ground states, which means that states with the __________ possible values of n will be occupied by electrons.

ITEM 36

If the principal quantum number of an electron is 4, the value of the azimuthal quantum number will be designated by one of the integers __________, or by one of the letters __________.

ITEM 49

The energy of an electron of an atom placed in a magnetic field depends, to some extent, on the direction of its orbital relative to the field. The different possible orientations are designated by the quantum number, m, known as the *magnetic quantum number.* It determines the __________ in which an orbital lies.

ITEM 62

In speaking of the *spin* of an electron we refer to its intrinsic angular momentum quantum number, which has a fixed value of $\frac{1}{2}$. From the preceding item, we know that there are __________ possible orientations of spin, one designated as up and __________ designated as __________.

ITEM 75

This special chemical stability apparently is characteristic of electronic configurations in which eight electrons (completing both s and p orbitals) are placed in the outermost occupied shell. Thus, the elements Ar and Kr have only a slight tendency to react chemically. Their electronic configurations are __________ and __________.

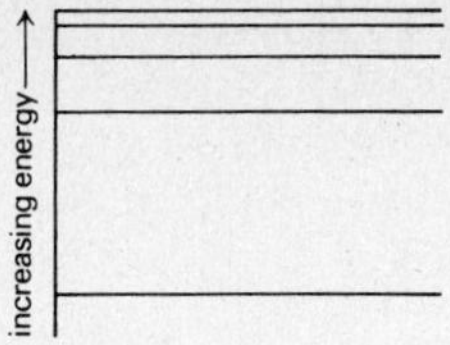

ANSWER 9 greater
higher

ANSWER 22

$E_{n=4} = -313.5/16 = -19.6$ kcal mole^{-1}
$E_{n=2} = -313.5/4 = -78.4$ kcal mole^{-1}
$E_{n=4} - E_{n=2} = 58.8$ kcal mole^{-1}

ANSWER 35 *p*
d
f

ANSWER 48 *p*

ANSWER 61 clockwise

ANSWER 74 $1s^2$
$1s^2 2s^2 2p^6$

ITEM 11

The energy of an electron for $n = \infty$ was assigned the value zero when the theory of atomic structure was worked out. Therefore, for smaller values of n, for which the energy is __________, the value of the energy will have a __________ sign.

ITEM 24

An atom in an excited state can return to the ground state by __________ energy.

ITEM 37

The shape of the orbital for any s electron ($l = 0$) is often represented by a diffuse sphere having the nucleus as its center. The size of the sphere is determined by the value of __________.

ITEM 50

The allowed values of m are 0 and the positive and negative integers that are less than or equal to the value of l. Complete the table of m values at the right.

ITEM 63

We will use the symbol s for the spin orientation quantum number. Then the quantum number for the "up" orientation has the value $+\frac{1}{2}$ and the quantum number for the "down" orientation has the value __________. Thus we write __________ $= \pm\frac{1}{2}$.

ITEM 76

Fluorine, chlorine, and bromine are chemically alike in that they easily add one electron to form a negative ion. Their electronic configurations suggest that this tendency is related to their needing only one electron to complete a p orbital. The configurations of the atoms are F __________; Cl __________; Br __________.

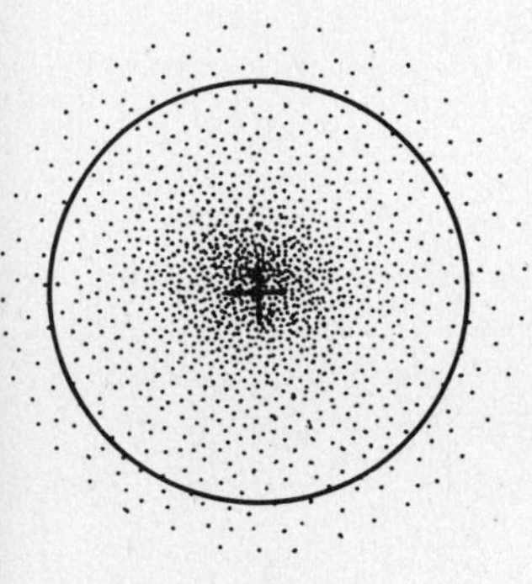

l		*m*			
0			0		
1		−1	0	+1	
2	___	___	0	___	___

ANSWER 10

ANSWER 23 smallest

ANSWER 36 0, 1, 2, 3
s, p, d, f

ANSWER 49 direction

ANSWER 62 two
one
down

ANSWER 75 $1s^22s^22p^63s^23p^6$
$1s^22s^22p^63s^23p^64s^23d^{10}4p^6$

ITEM 12

For some atoms only one electron need be considered. The energy, E, of this electron is given by $E = -k/n^2$, in which the constant, k, depends on the atom being considered. For $n = 1$, $E =$ __________; for $n = 2$, $E =$ __________, and for $n = \infty$, $E =$ __________.

ITEM 25

The principal quantum number n determines not only the __________ of an electron but also its __________ from the nucleus.

ITEM 38

The region in which there is a relatively high probability of finding a p electron ($l =$ __________) is dumbbell shaped, as shown at the right. (This representation is the easiest to draw but a more correct representation is shown in Figure 8–21 of *Chemical Principles*.)

ITEM 51

For a p electron ($l = 1$), there are three possible values of m: __________, __________, or __________.

ITEM 64

The allowed values of the four quantum numbers that describe electron behavior can be arranged as in the table at the right. Extend this table to show all the possible quantum number combinations for $n = 2$ and $n = 3$.

ITEM 77

Similarly, oxygen and sulfur easily add two electrons to form negative ions. Compare the electronic configurations of neutral oxygen and sulfur with those of the doubly charged negative ions.

O __________ S __________

O^{2-} __________ S^{2-} __________

$E = -k/n^2$

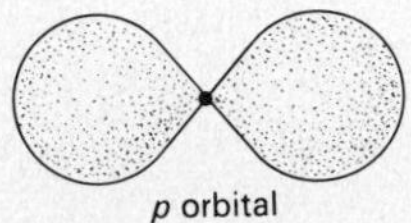

$n = 1 \quad l = 0 \quad m = \ \ 0 \quad s = \pm\frac{1}{2}$
$n = 2 \quad l = 0 \quad m = \ \ 0 \quad s =$
$\qquad\quad l = 1 \quad m = +1 \quad s =$
$\qquad\quad l = 1 \quad m = \qquad\ \ s =$
$\qquad\quad l = 1 \quad m =$
$n = 3 \quad l = 0$
$\qquad\quad l =$

ANSWER 11 lower
negative

ANSWER 24 emitting (or losing)

ANSWER 37 n

ANSWER 50 $-2, -1, 0, +1, +2$

ANSWER 63 $-\frac{1}{2}$
s

ANSWER 76 $1s^22s^22p^5$
$1s^22s^22p^63s^23p^5$
$1s^22s^22p^63s^23p^64s^23d^{10}4p^5$

ITEM 13

In the diagram at the right, draw horizontal lines to represent the energies of states corresponding to $n = 2, 3, 4$, and ∞.

ITEM 26

The region in space occupied by an electron is called its *orbital*. More specifically, an orbital is the region in space where there is appreciable probability of finding the __________.

ITEM 39

A p electron most likely would be located in the regions suggested by the lobes of the dumbbell. The probability of its being located in regions perpendicular to the axis of the dumbbell would be relatively __________.

ITEM 52

Similarly, a d electron can have the values of m: __________, that is, a total of __________ different values.

ITEM 65

The total number of possible combinations of quantum numbers with $n = 1$ is 2; the number of combinations with $n = 2$ is __________ and with $n = 3$ is __________.

ITEM 78

Conversely, the elements Li, Na, and K easily lose one electron to form a positive ion, as might be inferred from their electronic configurations: Li __________; Na __________; K __________.

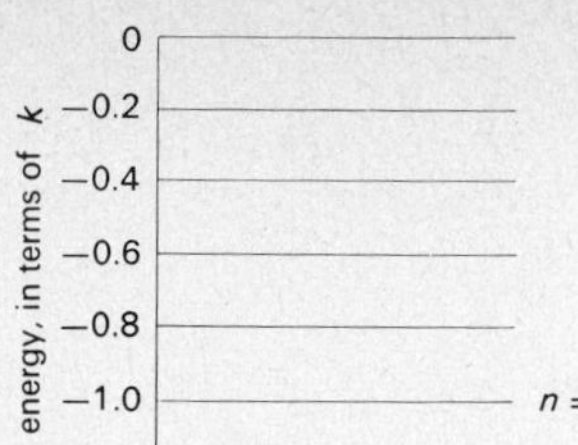

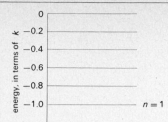

ANSWER 12 $-k$
$-\frac{1}{4}k$
0

ANSWER 25 energy
most probable distance

ANSWER 38 1

ANSWER 51 −1, 0, +1

ANSWER 64

$n=1$	$l=0$	$m=0$	$s=\pm\frac{1}{2}$
$n=2$	$l=0$	$m=0$	$s=\pm\frac{1}{2}$
	$l=1$	$m=+1$	$s=\pm\frac{1}{2}$
	$l=1$	$m=0$	$s=\pm\frac{1}{2}$
	$l=1$	$m=-1$	$s=\pm\frac{1}{2}$
$n=3$	$l=0$	$m=0$	$s=\pm\frac{1}{2}$
	$l=1$	$m=+1$	$s=\pm\frac{1}{2}$
	$l=1$	$m=0$	$s=\pm\frac{1}{2}$
	$l=1$	$m=-1$	$s=\pm\frac{1}{2}$
	$l=2$	$m=+2$	$s=\pm\frac{1}{2}$
	$l=2$	$m=+1$	$s=\pm\frac{1}{2}$
	$l=2$	$m=0$	$s=\pm\frac{1}{2}$
	$l=2$	$m=-1$	$s=\pm\frac{1}{2}$
	$l=2$	$m=-2$	$s=\pm\frac{1}{2}$

ANSWER 77

O	$1s^22s^22p^4$
O^{2-}	$1s^22s^22p^6$
S	$1s^22s^22p^63s^23p^4$
S^{2-}	$1s^22s^22p^63s^23p^6$

ANSWER 13

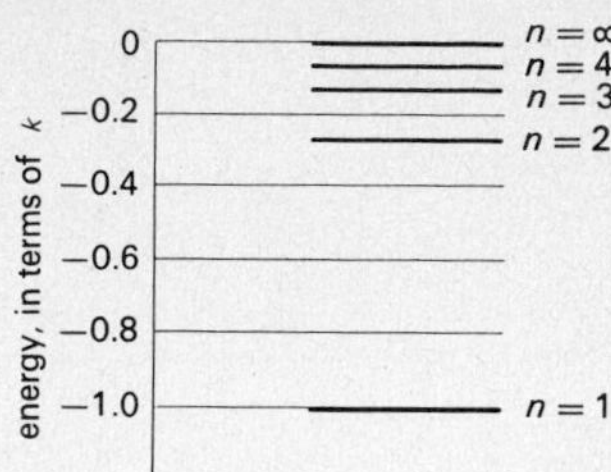

(*Item 14 is on page 408.*)

ANSWER 26 electron

(*Item 27 is on page 408.*)

ANSWER 39 low

(*Item 40 is on page 408.*)

ANSWER 52 −2, −1, 0, +1, +2
five

(*Item 53 is on page 408.*)

ANSWER 65 8
18

(*Item 66 is on page 408.*)

ANSWER 78 Li $1s^2 2s^1$
Na $1s^2 2s^2 2p^6 3s^1$
K $1s^2 2s^2 2p^6 3s^2 3p^6 4s^1$

REVIEW 9 THE TRANSITION METALS

In Review 6 you built a periodic table for the representative elements. As you did this, you learned how to write the formulas and names of the ionic, or mostly ionic, compounds formed from these elements. In Review 9 you will use electronic configurations as a basis for learning about the most important of the transition metals, the elements that involve the filling of the *d* orbitals. To do this review you must be familiar with the material of Reviews 6 and 8 so that you know the ground rules for naming compounds and for writing electronic configurations.

At the end of the review you will be able to predict formulas and write names for the compounds of the first row of transition elements in the periodic table and for silver, cadmium, and mercury. Although Stock nomenclature is by far the most convenient, you will learn the common names for the ions, too. Since there may be as many as three choices for the name of a given compound, you should try to give at least two names, where appropriate. The important thing is that you can write at least one correct name.

This material is a review of Table 7–2 and the relevant parts of Sections 9–1 and 9–4 in *Chemical Principles*. If you can list the important oxidation states of the transition metals and write formulas and names for the metal oxides, you are already sufficiently familiar with the transition metals and can skip this review.

ITEM 1

The order of filling orbitals with electrons is

$1s2s2p3s3p4s3d4p5s4d5p6s4f5d6p \ldots$

The *s* orbitals contain a maximum of ________ electrons, the *p* orbitals a maximum of ________ electrons, and the *d* orbitals a maximum of ________ electrons.

ITEM 12

The energy required to remove completely four electrons from an atom is exorbitant, so the free Ti^{4+} ion is not formed. However, in the compound TiO_2, named ________, titanium has an oxidation number of ________.

ITEM 23

Name the following compounds: CrO_3, Cr_2O_3, V_2O_5, $Sc(NO_3)_3$, TiO.

ITEM 34

The most important oxidation state of iron is +3 and exists as the *ferric* ion, Fe^{3+}. Write the names and formulas of the sulfate salts of iron(II) and iron(III).

ITEM 45

Element number 29 is copper. The electronic configuration is $1s^22s^22p^63s^23p^64s^13d^{10}$. Again there is an irregularity in the filling of orbitals. We could expect copper to have an oxidation number of ________ so the $3d$ orbital would remain completely filled.

ITEM 56

The symbols for the zinc, cadmium, and both mercury ions are ________, ________, ________, and ________.

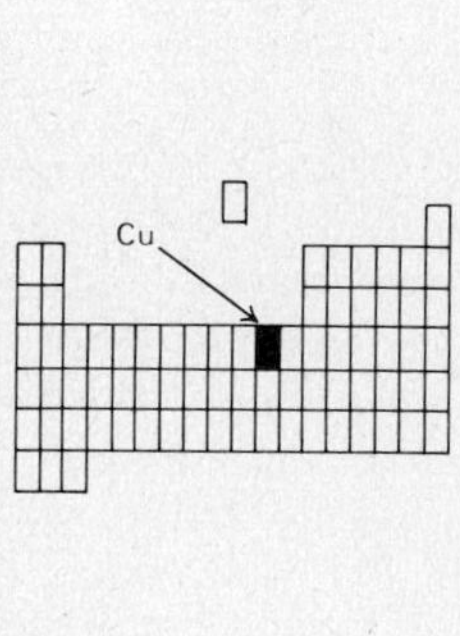
Cu

ITEM 2

Calcium, element number 20, has the electronic configuration __________. It is not a transition element, but a representative element, because there are no electrons partially filling a __________ orbital.

ITEM 13

In the compound TiO, named __________, titanium has an oxidation number of __________. The electrons considered to be transferred or partially transferred are the ones from the __________ orbital(s).

ITEM 24

The twenty-fifth element, manganese, has the electronic configuration $1s^22s^22p^63s^23p^64s^23d^5$. (Note that the $4s$ orbital again has two electrons.) The maximum oxidation state for manganese is __________.

ITEM 35

The maximum oxidation state of a transition element is equal to the number of unpaired electrons in the outermost occupied *d* orbital plus the number of electrons in the outermost occupied *s* orbital. Thus, the maximum oxidation number of iron is __________.

ITEM 46

Compounds containing the *cuprous* ion Cu^+ are known (and include the principal ores of copper), but the principal oxidation state of copper is +2, which occurs as the *cupric* ion Cu^{2+}. Write the common names and formulas for the chlorides and oxides of copper(I) and copper(II).

ITEM 57

Write the formulas of the following compounds: cadmium oxide, __________; zinc nitrate, __________; mercurous fluoride, __________; mercuric sulfide, __________; chromous chloride, __________.

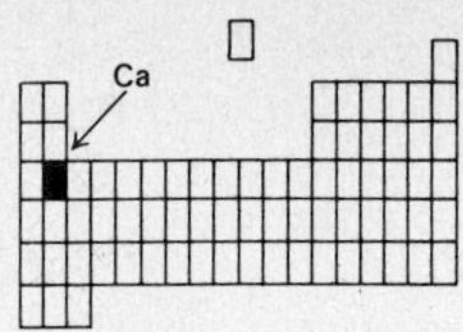

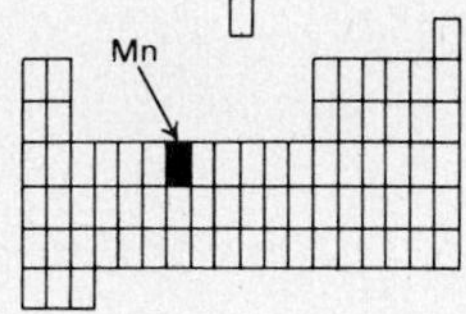

ANSWER 1 two
six
ten

ANSWER 12 *titanium dioxide,*
titanium(IV) oxide,
or *titanic oxide*
+4

ANSWER 23

chromium trioxide or chromium(VI) oxide
chromic oxide or chromium(III) oxide
vanadium pentoxide or vanadium(V) oxide
scandium trinitrate or scandium(III) nitrate
titanium monoxide, titanium(II) oxide,
or titanous oxide

ANSWER 34

ferrous sulfate [iron(II) sulfate] $FeSO_4$
ferric sulfate [iron(III) sulfate] $Fe_2(SO_4)_3$

ANSWER 45 +1

ANSWER 56 Zn^{2+}
Cd^{2+}
Hg^{2+}
Hg_2^{2+}

ITEM 3

Atoms of calcium usually react by donating electrons of the __________ orbital to elements of high electron affinity. The resulting ion has the symbol __________.

ITEM 14

Name the compounds $TiCl_4$, TiF_2, and $ScBr_3$.

ITEM 25

Manganese has all oxidation numbers from 0 through +7, but the principal oxidation number is +7, found in the *permanganate* anion, MnO_4^-. The corresponding acid has the formula __________ and the name __________.

ITEM 36

Write the formula of each of the following: ferric chloride, __________; ferrous sulfite, __________; chromic oxide, __________; lithium chromate, __________; ferric sulfide, __________.

ITEM 47

Silver, atomic number 47, is beneath copper in the periodic table and has an analogous electronic configuration: ten electrons in the $4d$ and one electron in the $5s$ orbital. The principal ion is Ag^+ and is formed by __________.

ITEM 58

Name the following compounds: $HgSO_4$, $AgNO_3$, FeO, $Cu(OH)_2$, TiO_2, H_2CrO_4.

ANSWER 2 $1s^2 2s^2 2p^6 3s^2 3p^6 4s^2$
d

ANSWER 13 *titanium monoxide,*
titanium(II) oxide,
or *titanous oxide*
+2
$4s$

ANSWER 24 +7

ANSWER 35 +6 (It occurs only in the rare ferrate ion, FeO_4^{2-}.)

ANSWER 46 cuprous chloride, $CuCl$
cupric chloride, $CuCl_2$
cuprous oxide, Cu_2O
cupric oxide, CuO

ANSWER 57 CdO
$Zn(NO_3)_2$
Hg_2F_2
HgS
$CrCl_2$

ITEM 4

The electronic configuration of Ca^{2+}, with 18 electrons, is ________. This is also the configuration of the element ________, which belongs to the family of ________ elements.

ITEM 15

Element number 23, vanadium, has the electronic configuration ________. The maximum oxidation number is expected to be ________.

ITEM 26

One important compound of manganese is MnO_2, named ________, in which manganese has the oxidation number ________.

ITEM 37

Cobalt has the atomic number 27 and the electronic configuration ________.

ITEM 48

Write formulas for the four important silver halides.

ITEM 59

Write the principal oxidation states of the following elements: iron, ________; copper, ________; mercury, ________; cobalt, ________; nickel, ________; zinc, ________; silver, ________.

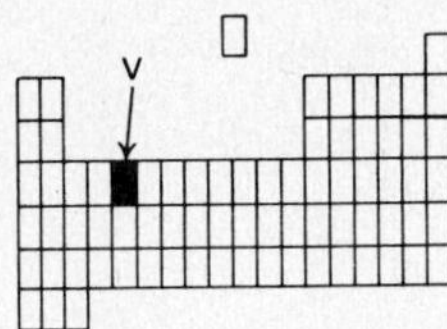

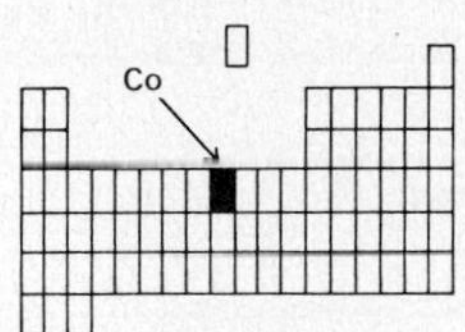

ANSWER 3 $4s$
Ca^{2+}

ANSWER 14

titanium tetrachloride, titanium(IV) chloride, or titanic chloride
titanium difluoride, titanium(II) fluoride, or titanous fluoride
scandium tribromide or scandium(III) bromide

ANSWER 25 $HMnO_4$
permanganic

ANSWER 36 $FeCl_3$
$FeSO_3$
Cr_2O_3
Li_2CrO_4
Fe_2S_3

ANSWER 47 donation of the $5s$ electron to an element of high electron affinity

ANSWER 58

mercuric sulfate or mercury(II) sulfate
silver nitrate
ferrous oxide or iron(II) oxide
cupric hydroxide or copper(II) hydroxide
titanium dioxide, titanium(IV) oxide, or titanic oxide
chromic acid

ITEM 5

Atoms or ions are especially stable when the valence shell contains ________ electrons, thus filling the outermost occupied ________ and ________ orbitals.

ITEM 16

The principal oxidation number of vanadium is +5. It occurs in the *vanadate* anion, VO_4^{3-}, and in vanadium pentoxide, which has the formula ________.

ITEM 27

The *manganate* ion is MnO_4^{2-}, the only ion in which manganese has an oxidation number of ________.

ITEM 38

In the cobalt atom there are ________ unpaired electrons in the *d* orbitals. The maximum oxidation number would be ________.

ITEM 49

The principal oxidation state of copper is ________, and that of silver is ________. Write formulas for the sulfides corresponding to these oxidation states and name them.

ITEM 60

Write formulas for the following compounds: scandium oxide, ________; potassium permanganate, ________; vanadium pentoxide, ________; nickel sulfate, ________; manganese dioxide, ________.

ANSWER 4 $1s^22s^22p^63s^23p^6$
argon
rare gas
(noble gas or inert gas)

ANSWER 15 $1s^22s^22p^63s^23p^64s^23d^3$
+5

ANSWER 26 manganese dioxide
or manganese(IV) oxide
+4

ANSWER 37 $1s^22s^22p^63s^23p^64s^23d^7$

ANSWER 48 AgF
AgCl
AgBr
AgI

ANSWER 59 +2 and +3
+2
+1 and +2
+3
+2
+2
+1

ITEM 6

Element number 21, scandium, has the electronic configuration __________.

ITEM 17

The electronic configuration of chromium, element 24, is $1s^2 2s^2 2p^6 3s^2 3p^6 4s^1 3d^5$. How is it different than what would have been expected from the configurations of Sc, Ti, and V?

ITEM 28

Name the following compounds: TiO_2, Na_2CrO_4, $Ca(MnO_4)$, $Mg(MnO_4)_2$, $K_2Cr_2O_7$.

ITEM 39

However, the principal oxidation state of cobalt occurs in the *cobaltic* ion Co^{3+}. The *cobaltous* ion Co^{2+} is found in cobaltous oxide, which has the formula __________, and in a few other compounds.

ITEM 50

Zinc has 30 electrons in the electronic configuration __________.

ITEM 61

Write formulas and names for the sulfides of silver, iron, cadmium, nickel, and zinc.

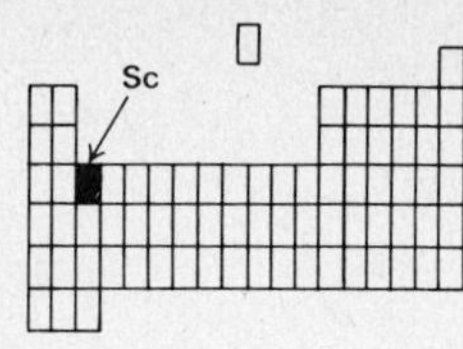

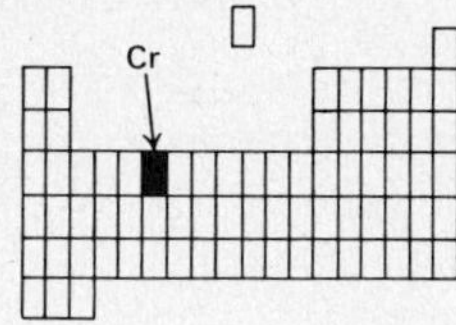

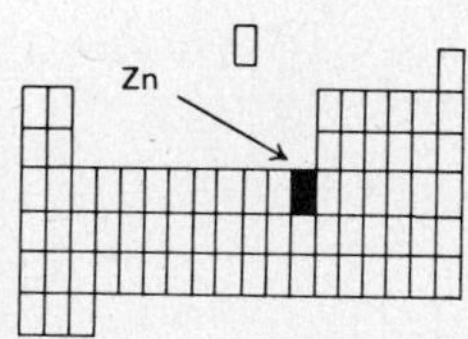

ANSWER 5 eight
s
p

ANSWER 16 V_2O_5

ANSWER 27 +6

ANSWER 38 3
+5

ANSWER 49 +2
+1
CuS, cupric sulfide
or copper(II) sulfide
Ag_2S, silver sulfide
or silver(I) sulfide

ANSWER 60 Sc_2O_3
$KMnO_4$
V_2O_5
$NiSO_4$
MnO_2

ITEM 7

Scandium can be expected to donate ______ electrons from the orbitals ______ and ______ to yield an ion with a charge of ______.

ITEM 18

The maximum oxidation number of chromium is ______. This is the oxidation number assigned to chromium in two important anions: CrO_4^{2-}, named *chromate*, and $Cr_2O_7^{2-}$, named *dichromate*.

ITEM 29

Write the formulas of the following compounds: manganese dioxide, chromic oxide, vanadium pentoxide, sodium dichromate, sodium permanganate.

ITEM 40

The common name and formula of cobalt(III) sulfate is ______.

ITEM 51

The removal of ______ electrons from a zinc atom leaves an ion in which the $n = 3$ level is filled and higher levels are empty. It is not surprising that the zinc ion has a charge of ______.

ITEM 62

In the first row of transition elements, electrons fill the ______ orbitals from scandium to zinc. Can you name all the first-row transition elements?

ANSWER 6 $1s^2 2s^2 2p^6 3s^2 3p^6 4s^2 3d^1$

ANSWER 17 The $3d$ orbital contains not one but two additional electrons because one has been transferred from the $4s$ orbital.

ANSWER 28 titanium dioxide, titanium(IV) oxide, or titanic oxide
sodium chromate
calcium manganate
magnesium permanganate
potassium dichromate

ANSWER 39 CoO

ANSWER 50 $1s^2 2s^2 2p^6 3s^2 3p^6 4s^2 3d^{10}$

ANSWER 61

Ag_2S, silver sulfide or silver(I) sulfide
FeS, ferrous sulfide or iron(II) sulfide
Fe_2S_3, ferric sulfide or iron(III) sulfide
CdS, cadmium sulfide
NiS, nickel sulfide
ZnS, zinc sulfide

ITEM 8

The oxidation number +3, as found in the ion Sc^{3+}, is by far the most common state of scandium. The ion has 18 electrons in the configuration ________.

ITEM 19

Chromium has an oxidation number of +6 not only in the two anions, CrO_4^{2-}, named ________, and $Cr_2O_7^{2-}$, named ________, but also in the compound CrO_3, named *chromium trioxide* or, in Stock nomenclature, ________.

ITEM 30

There are ________ *d* orbitals for any value of *n* greater than ________. Each of the orbitals can contain a maximum of two electrons, one with spin quantum number ________ and the other with spin quantum number ________.

ITEM 41

The properties of iron, cobalt, and nickel are quite similar. The usual oxidation state of nickel (atomic number 28, electronic configuration ________) is +2, as in the *nickelous* ion, Ni^{2+}.

ITEM 52

Similar electronic configurations for cadmium $(\ldots 5s^2 4d^{10})$ and mercury $(\ldots 6s^2 5d^{10})$ suggest that the principal oxidation number of cadmium should be ________, and that of mercury should be ________.

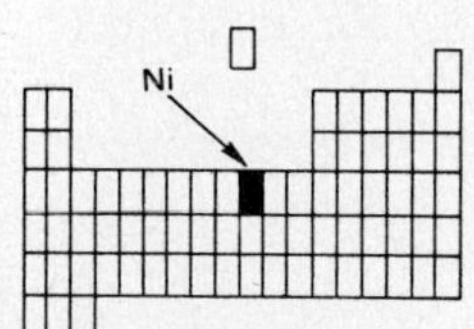

Zn
Cd
Hg

ANSWER 7 three
$4s$
$3d$
$3+$

ANSWER 18 +6

ANSWER 29 MnO_2
Cr_2O_3
V_2O_5
$Na_2Cr_2O_7$
$NaMnO_4$

ANSWER 40 cobaltic sulfate, $Co_2(SO_4)_3$

ANSWER 51 2
$2+$

ANSWER 62 $3d$

scandium	iron
titanium	cobalt
vanadium	nickel
chromium	copper
manganese	zinc

ITEM 9

Elements do not always lose enough electrons to yield an ion with a rare gas configuration, in which the outermost occupied __________ and __________ orbitals are __________.

ITEM 20

Write the formulas for the compounds potassium dichromate, potassium chromate, chromium trioxide, vanadium pentafluoride, and chromic acid.

ITEM 31

As electrons fill the orbitals, from scandium to manganese, each new electron goes in a different *d* orbital. In manganese, all the $3d$ orbitals contain __________ electron(s).

ITEM 42

The principal oxidation numbers of iron are __________ and __________. The principal oxidation number of cobalt is __________, and that of nickel is __________.

ITEM 53

In fact, the prediction is correct: Cadmium has oxidation states only of 0 and __________; mercury shares or donates __________ electrons per atom in all compounds.

ANSWER 8 $1s^22s^22p^63s^23p^6$

ANSWER 19 chromate
dichromate
chromium(VI) oxide

ANSWER 30 5
2
$+\frac{1}{2}$
$-\frac{1}{2}$

ANSWER 41 $1s^22s^22p^63s^23p^64s^23d^8$

ANSWER 52 +2
+2

ITEM 10

In general, the first electrons to be lost from an atom are those with the highest value of n. The rare Sc^{2+} ion has the electronic configuration ________.

ITEM 21

There are compounds in which chromium has oxidation states from 0 to +6. Two important ions are Cr^{3+}, named *chromic*, and Cr^{2+}, named *chromous.* Write the formulas for chromic oxide and chromous chloride.

ITEM 32

The electronic configuration of iron, element number 26, is ________. You know that one of the d orbitals will contain ________ electron(s) and four of them will contain ________ electron(s).

ITEM 43

Name the following compounds: $Ni(NO_3)_2$, Fe_2O_3, $Co(OH)_3$, $FeBr_2$, Cr_2O_3.

ITEM 54

However, mercury occurs in two oxidation states. *Mercuric* is the name of the +2 state. The +1 state is named ________.

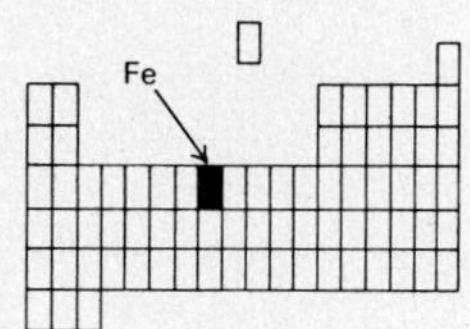

ANSWER 9 *s*
p
filled

ANSWER 20 $K_2Cr_2O_7$
K_2CrO_4
CrO_3
VF_5
H_2CrO_4

ANSWER 31 one

ANSWER 42 +2
+3
+3
+2

ANSWER 53 +2
two

ITEM 11

Titanium, element number 22, has the electronic configuration $1s^22s^22p^63s^23p^64s^23d^2$. The maximum oxidation number of this ion is expected to be ________ by transferral of electrons from the ________ and ________ orbitals.

ITEM 22

The positive oxidation states possible for chromium are ________. The +6 state is found in the anion ________, which is named ________, and in the ion ________, which is named ________.

ITEM 33

The first electrons to be donated from an iron atom to an element of high electron affinity are the ones in the ________ orbital. The ion that results has the symbol ________ and is called the *ferrous* ion.

ITEM 44

Write the formulas corresponding to ferrous carbonate, ________; ferric hydroxide, ________, nickelous sulfide, ________; cobaltic chloride, ________.

ITEM 55

The mercurous ion is actually Hg_2^{2+}, in which there is a covalent bond between the two atoms. The formula for mercurous chloride is therefore ________.

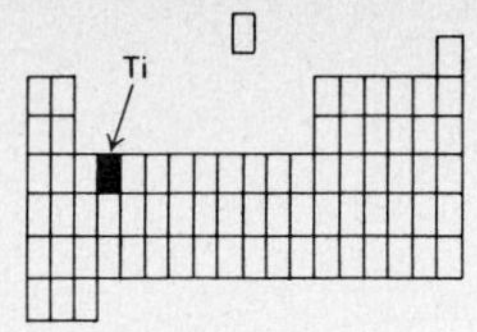

ANSWER 10 $1s^2 2s^2 2p^6 3s^2 3p^6 3d^1$

ANSWER 21 Cr_2O_3
$CrCl_2$

ANSWER 32 $1s^2 2s^2 2p^6 3s^2 3p^6 4s^2 3d^6$
two
one

ANSWER 43

nickel nitrate, nickel dinitrate,
nickel(II) nitrate or nickelous nitrate
ferric oxide or iron(III) oxide
cobaltic hydroxide, cobalt(III) hydroxide,
or cobalt trihydroxide
ferrous bromide, iron(II) bromide,
or iron dibromide
chromic oxide or chromium(III) oxide

ANSWER 54 mercurous

ANSWER 11 +4
4*s*
3*d*

(*Item 12 is on page 438.*)

ANSWER 22 +1 through +6
CrO_4^{2-}
chromate
$Cr_2O_7^{2-}$
dichromate

(*Item 23 is on page 438.*)

ANSWER 33 4*s*
Fe^{2+}

(*Item 34 is on page 438.*)

ANSWER 44 $FeCO_3$
$Fe(OH)_3$
NiS
$CoCl_3$

(*Item 45 is on page 438.*)

ANSWER 55 Hg_2Cl_2 (The empirical formula is HgCl, but the formula Hg_2Cl_2 is a better representation of the structure.)

(*Item 56 is on page 438.*)

REVIEW 10 THE STRUCTURE OF MOLECULES

"What holds the atoms together in molecules?" has been, and is, one of the most fascinating questions in chemistry. The answer to that question is far beyond the scope of this book, but Lewis diagrams and orbital diagrams are extremely useful in answering the question on a pictorial level. Section 10–1 will describe the bonding in ions and molecules by Lewis diagrams. Section 10–2 will use orbital diagrams and electronic configurations to describe the simplest diatomic molecules. Section 10–3 will teach you how to draw orbital diagrams for polyatomic molecules composed of the representative elements.

10–1 LEWIS DIAGRAMS

The chemical properties of an element are determined primarily by the number of electrons in the valence shell of the atom. A convenient method for representing these valence electrons and their participation in chemical bonding is the simple *Lewis diagram,* named after the American chemist G. N. Lewis. The explanation of the way in which these diagrams are drawn and the way in which they indicate the role of the valence electrons in the formation of ions or molecules will be a review of Section 10–1 of *Chemical Principles.* An understanding of these simple diagrams is of value because they provide a convenient bookkeeping system for electrons. The diagrams show the role, bonding or nonbonding, of electrons in many molecules, and they suggest the number of atoms that can be bound by covalent or ionic bonds in these molecules.

After studying the Lewis-diagram method for representing molecules and ions, you will be able to show that, for example, the simplest compound of nitrogen and hydrogen should be NH_3, rather than some other ratio of N to H, and that there is an unshared pair of valence electrons in the molecule. If you can do Items 94 through 97, you probably already understand how to draw Lewis diagrams.

ITEM 1

Two electrons can occupy a 1*s* orbital. A filled 1*s* orbital, indicated by $1s^2$, is the ground state configuration of the element ________.

ITEM 17

The fluorine atom in its ground state has one of its electrons unpaired, as is evident from its electronic configuration ________ or its Lewis diagram ________.

ITEM 33

The term *covalent* is used for bonds whose pairs of electrons are ________ between atoms.

ITEM 49

A molecule of HCN has a triple bond. Its Lewis diagram is ________, which can be simplified to ________.

ITEM 65

The Lewis diagram for water is ________. The formal charge on each atom is ________, and the sum of all these formal charges is ________.

ITEM 81

Boron trifluoride reacts with ions or molecules that have unshared pairs of electrons. For instance, BF_3 reacts with fluoride ion to form $(BF_4)^-$. Draw the Lewis diagram for this ion and insert formal charge signs where needed.

ITEM 97

Draw Lewis diagrams and show formal charges for the following compounds. Assume that only *s* and *p* orbitals are involved in the bonding.

dimethyl sulfone, $H_3C—\overset{\displaystyle O}{\underset{\displaystyle O}{S}}—CH_3$

dimethyl sulfoxide, $H_3C\overset{\displaystyle O}{S}CH_3$
aluminum tetrachloride ion, $[Al(Cl_4)]^-$

ITEM 2

The next noble gas elements below helium are neon and argon. Their electronic configurations are ________ and ________, respectively.

ITEM 18

Since the remaining halogens appear under fluorine in the periodic table (inside front cover), each also has ________ valence electron(s).

ITEM 34

The Lewis diagrams for hydrogen and oxygen atoms, ________ and ________, can show that H_2O forms as a result of covalent bonding between hydrogen and oxygen.

ITEM 50

In each of the molecules considered so far, every covalent bond between two atoms has been a shared pair of electrons. Each atom is considered to have contributed ________ electron(s) to the pair.

ITEM 66

In the presence of a strong base, a hydrogen ion, H^+, can be removed from H_2O. The Lewis diagram of the resulting ion is ________.

ITEM 82

A similar reaction of BF_3 occurs with NH_3 to form the molecule H_3NBF_3. Draw the Lewis diagram and insert formal charge signs where necessary.

ANSWER 1 He

ANSWER 17

$1s^2 2s^2 2p_x^2 2p_y^2 2p_z^1$ (or $1s^2 2s^2 2p^5$)

$:\ddot{\underset{..}{F}}\cdot$ (the unpaired electron can be written in any position)

ANSWER 33 shared

ANSWER 49 H : C : : : N : or H : C ⋮ ⋮ N :

H—C≡N :

ANSWER 65 $H : \ddot{\underset{..}{O}} : H$ or $H—\ddot{\underset{..}{O}}—H$

zero

zero

ANSWER 81

```
        ..
       :F:
        |
  ..    |      ..
 :F—B⊖—F:
  ..    |      ..
        |
       :F:
        ..
```

ANSWER 97

```
        H      ..     H
        |     :O:⊖    |
        |      |      |
  H—C———S(2+)—C—H
        |      |      |
        H     :O:⊖    H
               ..

        H      ..     H
        |     :O:⊖    |
        |      |      |
  H—C———S⊕———C—H
        |      ..     |
        H             H

               ..
              :Cl:
               |
  ..           |       ..
 :Cl—Al⊖—Cl:
  ..           |       ..
               |
              :Cl:
               ..
```

ITEM 3

If the term *valence electrons* implies those with the highest value of n in an atom, we see that to fill the s and p orbitals of the valence shell, as occurs for Ne and Ar, takes __________ electrons.

ITEM 19

By analogy with the diagram for fluorine, the Lewis diagrams of chlorine, bromine, iodine, and astatine can be written, without determining their electronic configurations, as __________, __________, __________, and __________.

ITEM 35

The Lewis diagram for H_2O is __________.

ITEM 51

For instance, the combination of four hydrogen atoms and one carbon atom gives CH_4 (Lewis diagram __________). Each bond is formed by one electron from __________ and one electron from __________.

ITEM 67

In the OH^- ion the number of valence electrons assigned to oxygen is __________. The neutral oxygen atom has __________ valence electrons. The formal charge on oxygen in OH^- is __________.

ITEM 83

Although you may have never encountered a compound of phosphorus and bromine, you could combine the Lewis diagrams at the right to predict that a possible compound of these elements might be __________.

$:\dot{\underset{\cdot}{P}}\cdot \quad :\ddot{\underset{\cdot\cdot}{Br}}\cdot$

ANSWER 2 $1s^22s^22p^6$
$1s^22s^22p^63s^23p^6$

ANSWER 18 seven

ANSWER 34 $H\cdot$
$\cdot\ddot{\underset{\cdot\cdot}{O}}\cdot$

ANSWER 50 one

ANSWER 66 $\left[:\ddot{\underset{\cdot\cdot}{O}}:H\right]^-$

ANSWER 82

$$\begin{array}{ccc} :\ddot{F}: & & H \\ | & & | \\ :\ddot{\underset{\cdot\cdot}{F}}-B^{\ominus} & - & {}^{\oplus}N-H \\ | & & | \\ :\underset{\cdot\cdot}{F}: & & H \end{array}$$

ITEM 4

Lack of chemical reactivity is a characteristic of atoms with a complete valence *octet* of *s* and *p* electrons, as illustrated by the noble gases __________ and __________.

ITEM 20

The Lewis diagram for oxygen can be written by deriving its electronic configuration, which is __________, or by noting that it comes before fluorine in the periodic table and, therefore, must have __________ valence electrons.

ITEM 36

Fluorine can achieve an octet by the formation of a covalent bond, as in F_2, having the Lewis diagram __________, or by accepting an electron (as from a sodium atom) with the formation of an __________ bond. The Lewis diagram of NaF is __________.

ITEM 52

From the Lewis diagram of CH_4, assign to each atom one electron from each pair it shares with another atom. This formal assignment of shared electrons gives carbon __________ valence electrons and each hydrogen __________.

ITEM 68

The Lewis diagram of CH_3CH_2Cl is __________.

ITEM 84

PBr_3 is one compound of phosphorus and bromine. However, PBr_5 is also a known compound. In PBr_5, the bonding probably involves the 3*d* orbitals as well as __________ and __________ orbitals.

P $1s^2 2s^2 2p^6 3s^2 3p^3$

ANSWER 3 eight

ANSWER 19

```
 ..    ..    ..    ..
:Cl·  :Br·  :I·   :At·
 ..    ..    ..    ..
```

ANSWER 35

```
  ..        H
H:O:H (or  :O:H etc., since
  ..        ..
```

Lewis diagrams are not really intended to show molecular shapes)

ANSWER 51

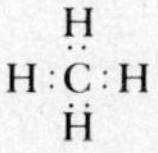

```
  H
  ..
H:C:H
  ..
  H
```

carbon
hydrogen

ANSWER 67 seven
six
1−

ANSWER 83

```
 ..   ..   ..
:Br—P—Br:
 ..   |    ..
     :Br:
      ..
```

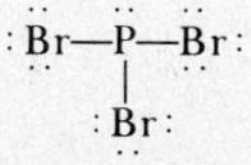

ITEM 5 In He, only *s* electrons can be accommodated and only ________ valence electrons are present.

ITEM 21 Lewis diagrams for atoms can be drawn either by first writing the ________ for the atom or by noting the position of the element in the ________.

ITEM 37 The sodium atom can achieve an octet of valence electrons by ________ one electron.

ITEM 53 The Lewis diagram for NH_3 is ________.

ITEM 69 The Lewis diagram of chloride ion, Cl^-, is ________.

ITEM 85 When *d* orbitals participate in bonding, Lewis diagrams are not particularly helpful in predicting the existence of compounds since the number of electrons, including shared electrons, in the valence shell of an atom can be more than ________.

ANSWER 4 Ne
Ar
(to a lesser extent, Kr and Xe also)

ANSWER 20 $1s^2 2s^2 2p_x^2 2p_y^1 2p_z^1$
six

ANSWER 36 $:\ddot{\underset{..}{F}}:\ddot{\underset{..}{F}}:$
ionic (or electrovalent)
$[Na]^+ \; [:\ddot{\underset{..}{F}}:]^-$

ANSWER 52 four (for carbon)
one (for hydrogen)

ANSWER 68

$$\begin{array}{c}H\;H\\ H:\ddot{C}:\ddot{C}:\ddot{\underset{..}{Cl}}:\\ H\;H\end{array} \quad \text{or} \quad \begin{array}{c}H\quad H\\ |\quad\; |\\ H\text{—}C\text{—}C\text{—}\ddot{\underset{..}{Cl}}:\\ |\quad\; |\\ H\quad H\end{array}$$

ANSWER 84 $3s$
$3p$

ITEM 6

The electronic configurations of lithium, sodium, and potassium (at right) show that each of these elements has ________ valence electron(s).

ITEM 22

Lewis diagrams show the number of outer *s* and *p* electrons. Therefore, these diagrams are not generally used for representing atoms that have electrons in partially filled ________ or ________ orbitals.

ITEM 38

In contrast, a sodium atom does not form an octet by sharing electrons. This is consistent with the fact that sodium does not form strong ________ bonds.

ITEM 54

From the Lewis diagram at the right for NH_3, assign to each atom half of the electrons it shares with other atoms plus all its unshared valence electrons. On this basis, the nitrogen atom has ________ valence electrons and each hydrogen has ________.

ITEM 70

If Cl^- is removed from the molecule CH_3CH_2Cl, the Lewis diagram of the resulting ion is ________.

ITEM 86

Lewis diagrams, then, are used mostly for molecules containing "light" elements in which the bonding can be described by a valence shell with no more than either ________ or ________ electrons.

Li $1s^2 2s^1$
Na $1s^2 2s^2 2p^6 3s^1$
K $1s^2 2s^2 2p^6 3s^2 3p^6 4s^1$

$\mathrm{H:\ddot{N}:H}$ (with H below N) or H—N̈—H (with H bonded below N)

ANSWER 5 two

ANSWER 21 electronic configuration
periodic table

ANSWER 37 donating
(or transferring)

ANSWER 53 $\mathrm{H:\ddot{N}:H}$ (with H below N) or H—N̈—H (with H bonded below N)

ANSWER 69 $[:\ddot{\underset{\cdot\cdot}{Cl}}:]^-$

ANSWER 85 eight

ITEM 7

A useful convention is to let the chemical symbol stand for the nucleus and all inner, nonvalence electrons. Then the outer electrons are represented by dots. Such a diagram, an example of a *Lewis diagram,* for lithium is Li·. Corresponding diagrams for sodium and potassium are __________ and __________.

ITEM 23

Atoms often react with other atoms to achieve a complete outer shell, that is, a complete octet. For example, a sodium atom, with the Lewis diagram shown at the right, tends to __________ one electron, thereby becoming the ion Na^{+}.

ITEM 39

From the Lewis diagram for nitrogen at the right it is apparent that a nitrogen atom needs to share __________ pairs of electrons to achieve an octet.

ITEM 55

This *formal* assignment of electrons to atoms of a molecule or ion, in which we assign __________ electron from each shared pair plus __________ unshared valence electrons, is the first step in determining the *formal charge* of an atom.

ITEM 71

Add (+) or (−) signs to the diagram at the right (the ion treated in Item 70) to indicate the formal charges.

ITEM 87

For instance, Lewis diagrams predict correctly that helium and neon will not donate or accept electrons because the valence shells, holding __________ and __________ electrons, respectively, are filled.

Li $1s^2 2s^1$
Na $1s^2 2s^2 2p^6 3s^1$
K $1s^2 2s^2 2p^6 3s^2 3p^6 4s^1$

$Na\cdot$

$:\dot{\underset{\cdot}{N}}\cdot$

$$\begin{array}{c} \;\;\;H\;\;\;H \\ \;\;\;|\;\;\;\;| \\ H—C—C \\ \;\;\;|\;\;\;\;| \\ \;\;\;H\;\;\;H \end{array}$$

ANSWER 6 one

ANSWER 22 *d*
f

ANSWER 38 covalent

ANSWER 54 five (for nitrogen)
one (for hydrogen)

ANSWER 70

$$\left[\begin{array}{c} \;\;H\;\;H \\ H:\ddot{\underset{\cdot\cdot}{C}}:\ddot{\underset{\cdot\cdot}{C}} \\ \;\;H\;\;H \end{array}\right]^+ \quad \text{or} \quad \left[\begin{array}{c} \;\;\;H\;\;\;H \\ \;\;\;|\;\;\;\;| \\ H—C—C \\ \;\;\;|\;\;\;\;| \\ \;\;\;H\;\;\;H \end{array}\right]^+$$

ANSWER 86 two
eight

ITEM 8

The electronic configurations of the alkaline earth elements (at right) indicate that each has __________ valence electron(s).

ITEM 24

Similarly, a halogen such as fluorine, with Lewis diagram __________, might acquire one electron to become the fluoride ion, for which the Lewis diagram is __________.

ITEM 40

The octet is completed if nitrogen forms covalent compounds with other elements. For example, the Lewis diagram of NH_3 is __________ and of NF_3 is __________.

ITEM 56

If an atom is assigned more or less valence electrons than it possesses as a neutral atom, it carries a positive or negative formal charge. If it is assigned fewer electrons than it has as a neutral atom, its formal charge is positive; if it has more electrons, its formal charge is __________.

ITEM 72

A chemical reaction might remove an H^+ ion from a hydrocarbon molecule. If this happened to an ethane molecule, C_2H_6, the Lewis diagram of the resulting ion, including the formal charge sign, would be __________.

ITEM 88

Conversely, such diagrams are worthless in predicting the existence or the composition of compounds of krypton or xenon, such as KrF_2 and XeO_3, for which Lewis diagrams would have to show more than __________ electrons in the outer shell of the central atom.

Be $1s^2 2s^2$
Mg $1s^2 2s^2 2p^6 3s^2$
Ca $1s^2 2s^2 2p^6 3s^2 3p^6 4s^2$

```
   H  H
   |  |
H—C—C—H
   |  |
   H  H
```

Kr $1s^2 2s^2 2p^6 3s^2 3p^6 4s^2 3d^{10} 4p^6$

Xe $1s^2 2s^2 2p^6 3s^2 3p^6 4s^2 3d^{10} 4p^6 5s^2 4d^{10} 5p^6$

ANSWER 7 Na·
K·

ANSWER 23 lose (or donate)

ANSWER 39 three

ANSWER 55 one
all

ANSWER 71

```
   H  H
   |  |
H—C—C⊕
   |  |
   H  H
```

(formal charges are often circled as ⊕ or ⊖)

ANSWER 87 two
eight

ITEM 9

Diagrams showing the valence electrons, that is, Lewis diagrams, of the alkaline earth elements beryllium, magnesium, and calcium are ·Be· or Be:, ________, and ________.

ITEM 25

A sodium atom and a fluorine atom both can achieve complete octets of valence electrons by the ________ of one electron from the sodium atom to the fluorine atom.

ITEM 41

The Lewis diagram for N_2 at the right shows another way that three pairs of electrons can be shared to give ________ electrons around each nitrogen atom.

ITEM 57

In the neutral N and H atoms there are ________ and ________ valence electrons, respectively.

ITEM 73

Neutral molecules also can have atoms with formal charges, for example, trimethylamine oxide, $(CH_3)_3NO$. Its Lewis diagram is shown at the right. The formal charge on the nitrogen is ________. The formal charge on the oxygen is ________.

ITEM 89

The ion I_3^- (in which there are a total of 22 outer electrons) is known. We conclude, since a Lewis diagram with octets cannot be drawn (verify this), that ________ orbitals may be involved in the bonding.

$:N\vdots\vdots N:$

```
 H    :Ö:    H
 |     |     |
H—C————N————C—H
 |     |     |
 H  H—C—H    H
       |
       H
```

ANSWER 8 two

ANSWER 24 $:\ddot{F}\cdot$

$:\ddot{\underset{\cdot\cdot}{F}}:^-$ or, as ions are usually written, $[:\ddot{\underset{\cdot\cdot}{F}}:]^-$

ANSWER 40 $H:\ddot{\underset{\cdot\cdot}{N}}:H$ over H (ammonia, NH_3) $\qquad :\ddot{\underset{\cdot\cdot}{F}}:\ddot{N}:\ddot{\underset{\cdot\cdot}{F}}:$ over $:\ddot{\underset{\cdot\cdot}{F}}:$ (NF_3)

ANSWER 56 negative

ANSWER 72

```
   H  H
   |  |
H—C—C:⊖
   |  |
   H  H
```

(any H might be removed)

ANSWER 88 eight

ITEM 10 The normal hydrogen atom, with the electronic configuration __________, has one valence electron and has the Lewis diagram __________.

ITEM 26 In a similar way, one calcium atom (Lewis diagram __________) must transfer __________ electrons to become the positive ion Ca^{2+}.

ITEM 42 In the ethylene molecule, C_2H_4, two pairs of electrons are shared between the carbon atoms. The Lewis diagram __________ can be drawn to show that each atom acquires a completed outer shell.

ITEM 58 In NH_3 the number of valence electrons assigned to N and H are __________ and __________, respectively. Thus, the formal charges of N and H in NH_3 are both __________.

ITEM 74 Draw Lewis diagrams and insert formal charge signs where needed for the following:

hydrogen cyanide, HCN

formamide, HCNH$_2$
O

formate ion, HCO$^-$
O

the trichloromethane anion, $^-CCl_3$

the amide ion, $^-NH_2$.

ITEM 90 A Lewis diagram for the sulfuric acid molecule, H_2SO_4, in which the atoms are arranged as

O
HOSOH
O

can be drawn so the sulfur atom need not use *d* orbitals. Draw this diagram and assign formal charges.

ANSWER 9 ·Mg· or Mg:
·Ca· or Ca:

ANSWER 25 transfer

ANSWER 41 eight

ANSWER 57 five
one

ANSWER 73 1+
1−

ANSWER 89 *d*

ITEM 11

The ground state electronic configuration of the carbon atom is __________. There are __________ valence electrons.

ITEM 27

Many ions can be recognized as resulting from the tendency of an atom to achieve a complete __________ of electrons by the transfer of electrons.

ITEM 43

In a molecule of formaldehyde, H_2CO, two pairs of electrons are shared between the carbon and the oxygen to give the Lewis diagram __________.

ITEM 59

A molecule of NH_3 can share both of its nonbonded valence electrons with a hydrogen ion. The ion $(NH_4)^+$ that is formed has the Lewis diagram __________.

ITEM 75

The Lewis diagrams written thus far have been for molecules or ions in which only *s* and __________ orbitals are needed to describe the bonding. And each of the atoms, through transfer or __________ of electrons, has attained a completed valence shell or octet.

ITEM 91

Other diagrams for sulfuric acid, such as the one at the right, are also reasonable because the __________ atom uses *d* orbitals and can accommodate more than __________ valence electrons.

$$H-\ddot{\underset{..}{O}}-\overset{\overset{:O:}{\|}}{\underset{\underset{:O:}{\|}}{S}}-\ddot{\underset{..}{O}}-H$$

ANSWER 10 $1s^1$

$H\cdot$

ANSWER 26 $\cdot Ca \cdot$ or $Ca:$

two

ANSWER 42

$$H:\overset{H}{\ddot{C}}::\overset{H}{\ddot{C}}:H$$

ANSWER 58 five

one

zero

ANSWER 74 $H-C\equiv N:$

$$H-\underset{\underset{:O:}{\|}}{C}-\ddot{N}(H)(H)$$

$$H-\underset{\underset{:O:}{\|}}{C}-\ddot{\underset{..}{O}}:^{\ominus}$$

$$:\ddot{\underset{..}{Cl}}-\overset{\ominus}{\underset{\underset{:\ddot{\underset{..}{Cl}}:}{|}}{C}}-\ddot{\underset{..}{Cl}}:$$

$$^{\ominus}:\underset{\underset{H}{|}}{\ddot{N}}-H$$

ANSWER 90

$$H-\ddot{\underset{..}{O}}-\overset{\overset{:\ddot{O}:^{\ominus}}{|}}{\underset{\underset{:\underset{..}{O}:^{\ominus}}{|}}{S^{2+}}}-\ddot{\underset{..}{O}}-H$$

ITEM 12

The Lewis diagram for carbon is $\cdot\dot{\underset{\cdot}{\mathrm{C}}}\cdot$. Here we notice that Lewis diagrams do not distinguish between *s* and __________ electrons.

ITEM 28

When electrons are transferred from one atom to another, the compounds that result contain charged species, that is, __________. The bond holding the groups together in such cases is referred to as an __________ bond.

ITEM 44

In the formaldehyde molecule each hydrogen atom has a share of two electrons. The total number of electrons that each carbon and each oxygen atom has, shared plus unshared, in its outer shell is __________.

ITEM 60

From the Lewis diagram for $(NH_4)^+$, assign one electron from each bond to each atom of the bond. Each hydrogen atom will be assigned __________ electron(s), and the nitrogen atom will be assigned __________ electron(s).

ITEM 76

The tendency of atoms to achieve a complete shell or octet can be useful for predicting the composition of compounds. For example, a nitrogen atom has the Lewis diagram __________, from which it is apparent that this atom must obtain a share of __________ more electrons to attain an octet.

ITEM 92

Thus, a kind of Lewis diagram can be drawn when __________ orbitals are involved, but it is not as useful because the octet rule need not be obeyed.

ANSWER 11 $1s^22s^22p^2$ (or $1s^22s^22p_x^12p_y^1$)
four

ANSWER 27 octet (or valence shell)

ANSWER 43

$$\begin{array}{c} H \\ \cdot\cdot\cdot\cdot \\ H:C::O: \end{array}$$

ANSWER 59

$$\left[\begin{array}{c} H \\ H:\ddot{\underset{\cdot\cdot}{N}}:H \\ H \end{array}\right]^+$$

ANSWER 75 *p*
sharing

ANSWER 91 sulfur
eight

ITEM 13

Fill in the remaining Lewis diagrams of the first-row elements in the table at the right.

ITEM 29

An alternative way in which completed outer shells or octets are attained is illustrated by two hydrogen atoms (Lewis diagram __________) that combine by sharing their electrons, thereby forming H_2, represented by H : H.

ITEM 45

A shared pair of electrons often is denoted by a line, as in H—H. With this modification the Lewis diagram for HF is __________.

ITEM 61

Since in $(NH_4)^+$ the nitrogen atom is assigned one less valence electron than it possesses as a neutral atom, the formal charge on nitrogen in $(NH_4)^+$ is __________.

ITEM 77

Therefore, the simplest neutral molecule having no unpaired electrons that can be formed from N atoms and H atoms has the Lewis diagram __________.

ITEM 93

According to the rule for writing Lewis diagrams, the two diagrams at the right are equally good. Lewis diagrams do not attempt to indicate the __________ of molecules.

Li	Be	B	C	N	O	F	Ne
Li·	·Be·	_	·Ċ·	_	_	_	:N̈e:

ANSWER 12 *p*

ANSWER 28 ions
ionic (or electrovalent)

ANSWER 44 eight

ANSWER 60 one (for hydrogen)
four (for nitrogen)

ANSWER 76 ·N̈·
three

H:Ö:H H
:Ö:H

ANSWER 92 *d*

ITEM 14

The maximum number of outer *s* and *p* electrons, as illustrated by neon, is ________. Its complete octet of electrons can be shown as dots arranged neatly in pairs on the "sides" of the symbol. Draw the Lewis diagram of neon.

ITEM 30

When a pair of electrons is shared between a chlorine atom $:\overset{..}{\underset{..}{\mathrm{Cl}}}\cdot$ and a hydrogen atom $\mathrm{H}\cdot$, the H atom has a share of two electrons and the Cl atom has a share of eight electrons. This arrangement is represented by the Lewis diagram of HCl, ________.

ITEM 46

Two shared pairs constitute a double bond, represented by two lines. With this notation the Lewis diagram for formaldehyde, shown at the right, becomes ________.

ITEM 62

Since the formal charge shows the difference between the number of electrons that an atom has in the molecule or ion and the number that the neutral atom possesses, the sum of the formal charges of all the atoms must be ________ for a molecule and equal to the net ________ on an ion.

ITEM 78

It is possible to combine Lewis diagrams and to predict that the simplest compound of oxygen and fluorine (in which each atom has a complete octet) is ________.

ITEM 94

Lewis diagrams do provide a bookkeeping system that shows, since only *s* and *p* orbitals are involved in the bonding, that the simplest molecule involving N and H will be ________ and that the molecule has an unshared ________ of electrons.

```
  H
  ..   ..
H:C: :O:
```

ANSWER 13

```
 .    .    .    ..
·B·  ·N·  :O·  :F·
      ..   ..   ..
```

ANSWER 29 H·

ANSWER 45

```
    ..
H—F:
    ..
```

ANSWER 61 1+

ANSWER 77

```
    ..
H—N—H
    |
    H
```

ANSWER 93 shapes

ITEM 15

The pairs of dots in Lewis diagrams can imply two electrons in a given orbital, in which case the two electrons must have opposite ________.

ITEM 31

Although chlorine atoms can achieve an octet by accepting one electron from an atom such as sodium, they can also achieve this octet by ________ a pair of electrons.

ITEM 47

The Lewis diagram for H_2O can be written as $H:\ddot{\underset{..}{O}}:H$ or simplified to ________.

ITEM 63

In NH_3 the formal charge on the N atom is ________; on the H atoms it is ________. The sum is ________, as expected for a neutral molecule.

ITEM 79

A reactive compound of boron is BF_3. Combine the Lewis diagrams of B and F to show the Lewis diagram of BF_3. The formal charge on B is ________ and on F is ________.

ITEM 95

Draw Lewis diagrams for the single-bonded compounds:
ethane, C_2H_6
hydrazine, N_2H_4
hydrogen selenide, H_2Se.

B $1s^22s^22p^1$

ANSWER 14 eight

$:\ddot{\underset{..}{Ne}}:$

ANSWER 30 $H:\ddot{\underset{..}{Cl}}:$

ANSWER 46

```
     H
     |      ..
H—C═O:
```

ANSWER 62 zero
charge

ANSWER 78 $:\ddot{\underset{..}{F}}—\ddot{\underset{..}{O}}—\ddot{\underset{..}{F}}:$

ANSWER 94

```
                 H
                 |
NH3   or   :N—H
                 |
                 H
```

pair

ITEM 16

Since electrons repel one another, the electrons of an atom avoid pairing in the same orbital whenever there are other available orbitals of similar energy. Thus, the normal oxygen atom has two unpaired electrons. The dots of the Lewis diagram for the oxygen atom can show this if they are arranged as __________.

ITEM 32

The Lewis diagram for Cl_2 is __________.

ITEM 48

The CO_2 molecule is described as having two double bonds. Its Lewis diagram is __________, which can be simplified to __________.

ITEM 64

In $(NH_4)^+$ the formal charge on each hydrogen atom is __________; the formal charge on the nitrogen atom is __________. The total of the formal charges is equal to __________, and this equals the charge of the ion.

ITEM 80

The incompletely filled valence shell of B in BF_3 is indicated clearly in the Lewis diagram. The boron atom needs to share __________ more electrons to fill its outer shell.

ITEM 96

Draw Lewis diagrams for the multiple-bonded compounds:

acetylene, C_2H_2

formic acid, HCOH
O

O
phosgene, ClCCl

allene, H_2CCCH_2.

ANSWER 15 spins

ANSWER 31 sharing

ANSWER 47

```
    ..              ..
H—O—H    or   :O—H
    ..               |
                    H
```

ANSWER 63 zero
zero
zero

ANSWER 79

```
   ..
 :F:
   |       ..
  B—F:
   |       ..
 :F:
   ..
```

zero

zero

ANSWER 95

```
     H   H
     |    |
H—C—C—H
     |    |
     H   H
```

```
     ..   ..
H—N—N—H
     |    |
     H   H
```

```
     ..
H—Se—H
     ..
```

ANSWER 16 $:\dot{\underset{\cdot\cdot}{O}}\cdot$ or $\cdot\ddot{\underset{\cdot\cdot}{O}}\cdot$ etc.

(*Item 17 is on page 466.*)

ANSWER 32 $:\ddot{\underset{\cdot\cdot}{Cl}}:\ddot{\underset{\cdot\cdot}{Cl}}:$

(*Item 33 is on page 466.*)

ANSWER 48 $:\ddot{O}::C::\ddot{O}:$

$:\ddot{O}{=}C{=}\ddot{O}:$

(*Item 49 is on page 466.*)

ANSWER 64 zero
1+
1+

(*Item 65 is on page 466.*)

ANSWER 80 two

(*Item 81 is on page 466.*)

ANSWER 96 $H{-}C{\equiv}C{-}H$

$$\begin{array}{c} H{-}C{-}\ddot{\underset{\cdot\cdot}{O}}{-}H \\ \| \\ :O: \end{array}$$

$$\begin{array}{c} :O: \\ \| \\ :\ddot{\underset{\cdot\cdot}{Cl}}{-}C{-}\ddot{\underset{\cdot\cdot}{Cl}}: \end{array}$$

$$\begin{array}{ccccccc} H{-}&C&=&C&=&C&{-}H \\ &| & & & &| & \\ &H& & & &H& \end{array}$$

(*Item 97 is on page 466.*)

10–2 BONDING IN HOMONUCLEAR DIATOMIC MOLECULES

One of the most exciting goals in chemistry is that of understanding and representing the bonding that holds atoms together in molecules. You have learned something about the energy and spatial arrangements of electrons in atoms, and you have learned a method (Lewis diagrams) for representing the involvement of pairs of electrons in bonding. Now you will learn to make more detailed diagrams, like those studied for atoms in Review 8, that indicate the spatial arrangement of the electrons in a molecule. In this section you will learn to make the diagrams for orbitals of very simple molecules and practice writing electronic configurations. This will be a review of Sections 10–3 and 10–4 of *Chemical Principles*.

It is important to remember that the treatment given here is only one of a number of approaches to the description of chemical bonds. This approach provides a useful and easily presented picture of bonding and predicts reasonably well the geometry and chemical properties of molecules whose bonding can be described on the basis of *s* and *p* orbitals.

When you are finished with this section, you should be able to write the electronic configuration of the N_2^+ ion, for example, and draw diagrams of the molecular orbitals used for bonding. If you can correctly answer Items 53 through 59, then you are ready to go on to Section 3.

ITEM 1

The electronic configuration of the hydrogen atom is ________.

ITEM 11

Draw a diagram of the 1*s* orbital of helium analogous to that of Item 2 for hydrogen.

ITEM 21

The electronic configuration of the lithium atom is ________.

ITEM 31

At the right are shown the 2*s* and 2*p* orbitals. Label the orbitals and place dots in the orbitals to represent the electronic configuration of boron.

ITEM 41

The two "half-bonds" arising by overlap of two half-filled *p* orbitals to form two half-filled π orbitals constitute the bonding of atoms in B_2. The electronic configuration is written as

B_2: *KK* ________

ITEM 51

The electronic configuration of O_2 is ________.

H

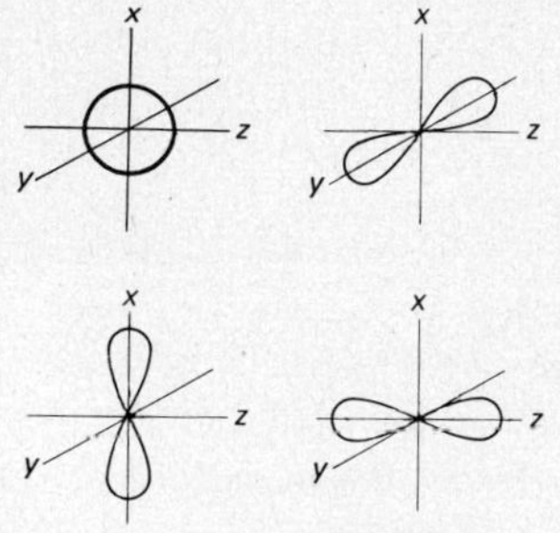
x
y
z
x
y
z
x
y
z
x
y
z

ITEM 2

The 1*s* orbital of hydrogen can be drawn as in the diagram at the right. The symbol represents the nucleus and the dot represents the electron. This is analogous to the Lewis diagram for a hydrogen atom, which is ________, but is intended to show that the electron density cloud has the shape of a ________.

ITEM 12

If two atoms of helium are brought close together, we could imagine two molecular orbitals being produced, one bonding and the other ________. Draw outlines of the two orbitals.

ITEM 22

Draw an atomic orbital diagram for lithium. Let the symbol Li stand for the nucleus and the completely filled 1*s* shell that is not involved in bonding.

ITEM 32

The orbital diagram for atomic boron is a composite of the diagrams in Answer 31. The symbol, which usually represents the nucleus and filled shells of electrons, is omitted for clarity. Draw a diagram showing the valence orbitals of boron.

ITEM 42

The electronic configuration of the carbon atom is ________. Draw an atomic orbital diagram of carbon.

ITEM 52

The electronic configuration of atomic fluorine is ________, and the electronic configuration of F_2 is written as ________.

H •

ANSWER 1 $1s^1$

ANSWER 11

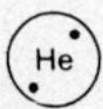

ANSWER 21 $1s^2 2s^1$

ANSWER 31

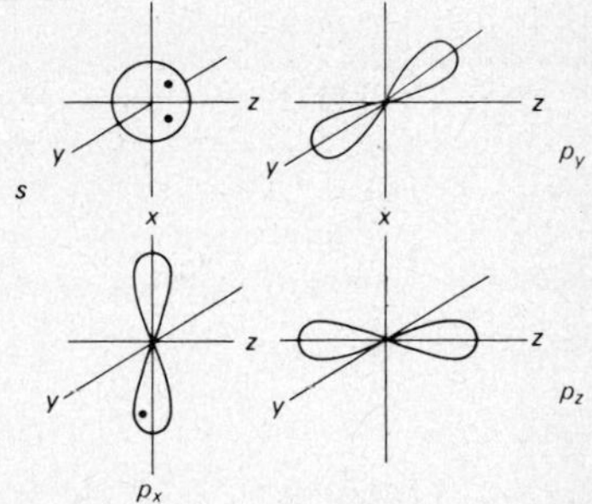

ANSWER 41 $(\sigma_s^b)^2(\sigma_s^*)^2(\pi_x^b)^1(\pi_y^b)^1$

ANSWER 51

$KK\,(\sigma_s^b)^2(\sigma_s^*)^2(\pi_x^b)^2(\pi_y^b)^2(\sigma_z^b)^2(\pi_x^*)^1(\pi_y^*)^1$

ITEM 3

The spherical 1s orbitals of two hydrogen atoms can overlap. The electrons can, to some extent, share the same region in space and form a bond. This is implied by the Lewis diagram for H_2, ________, or by the orbital diagram at the right.

ITEM 13

Since two helium atoms have a total of ________ electrons, ________ of these electrons would go into the bonding orbital and ________ of these electrons would go into the antibonding orbital.

ITEM 23

If two lithium atoms are brought to within bonding distance of one another, molecular orbitals can be formed by combination of the 2s atomic orbitals. Draw diagrams showing the two possible molecular orbitals and insert dots corresponding to the number of electrons expected to be in each.

ITEM 33

The molecule B_2 has two sigma orbitals formed by combinations of the 2s orbitals. The diagrams for these, including dots for the electrons, are

$\sigma_s{}^b$ ________ $\sigma_s{}^*$ ________

ITEM 43

Experiments have shown that the C_2 molecule has the electronic configuration $KK\ (\sigma_s{}^b)^2\ (\sigma_s{}^*)^2\ (\pi_x{}^b)^2\ (\pi_y{}^b)^2$. Draw orbital diagrams for C_2. Show the σ^b, σ^*, and π^b orbitals and use arrows to indicate the electron spins.

ITEM 53

Ne_2 is not likely to exist under ordinary conditions. The reason is ________.

ANSWER 2 H·
sphere

ANSWER 12 antibonding

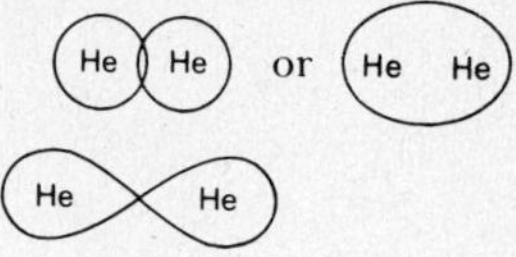

ANSWER 22

ANSWER 32

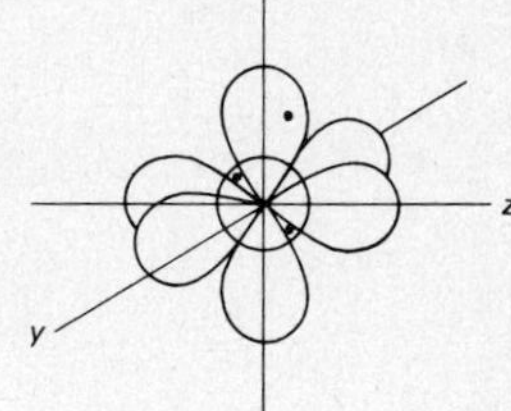

ANSWER 42 $1s^22s^22p_x^12p_y^1$

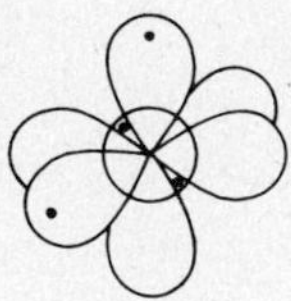

ANSWER 52 $1s^22s^22p_x^22p_y^22p_z^1$
$KK\ (\sigma_s^b)^2(\sigma_s^*)^2(\pi_x^b)^2$
$(\pi_y^b)^2(\sigma_z^b)^2(\pi_x^*)^2(\pi_y^*)^2$

ITEM 4

This type of combination of two atomic orbitals produces a *bonding molecular orbital* that has a watermelon shape. Draw a diagram indicating roughly the correct electron density of the bonding molecular orbital of H_2.

ITEM 14

The net effect of having equal numbers of electrons in bonding orbitals and in antibonding orbitals is that no bond is formed between the atoms. Therefore, a molecule of He_2 (would, would not) be expected to exist.

ITEM 24

Since the number of electrons in the __________ orbital is greater than the number of electrons in the __________ orbital, the molecule Li_2 can be expected to exist.

ITEM 34

Since the bonding electrons are canceled by the antibonding ones, these electrons play no part in holding together the B_2 molecule. The actual bonding must involve the __________ orbital electrons.

ITEM 44

The electronic configuration of atomic nitrogen is __________. Draw the orbital diagram.

ITEM 54

The electronic configuration of the ion Ne_2^+ is written as __________.

ANSWER 3 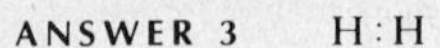

ANSWER 13 four
two
two

ANSWER 23

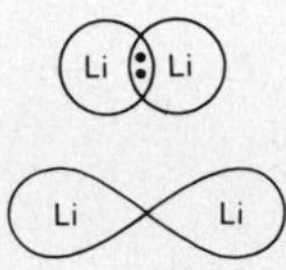

ANSWER 33

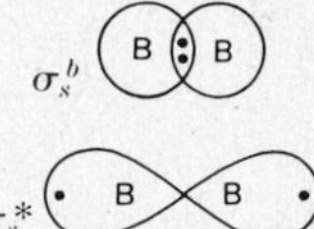

ANSWER 43

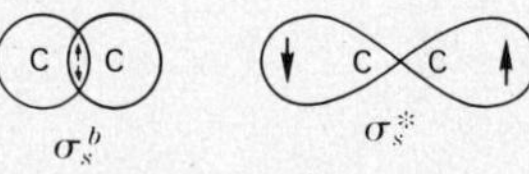

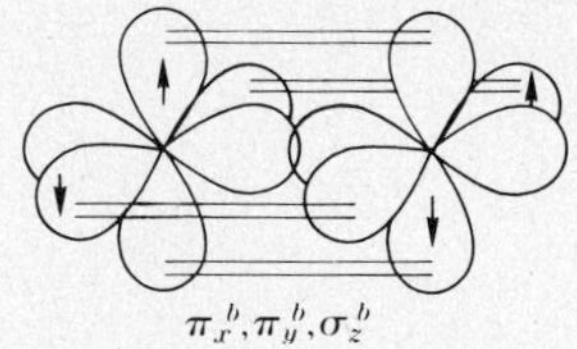

$\pi_x^b, \pi_y^b, \sigma_z^b$

ANSWER 53 that there would be equal numbers of electrons in bonding and in antibonding molecular orbitals.

ITEM 5

However, it is more convenient, especially for polyatomic molecules, to diagram bonds by showing the overlap of atomic orbitals. Draw this type of diagram for the H_2 molecule.

ITEM 15

Draw an orbital diagram for the ion He^+.

ITEM 25

The electronic configuration of Li_2 can be written $KK\ (\sigma_s^b)^2$, in which KK represents the electrons in the inner $n = 1$ shells. The electronic configuration of Li_2^+ is __________.

ITEM 35

The p orbitals can be combined to form bonding and __________ molecular orbitals. p orbitals can overlap to form bonding orbitals in two ways. If the $2p_z$ orbitals overlap, a σ_z^b molecular orbital is formed, as drawn at the right. Label these orbitals.

ITEM 45

The 10 valence electrons in a molecule of N_2 will fill these molecular orbitals: __________.

ITEM 55

Write the electronic configuration of F_2^-.

ANSWER 4

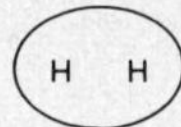

ANSWER 14 would not
(It has not been found experimentally.)

ANSWER 24 bonding (σ_s^b)
antibonding (σ_s^*)

ANSWER 34 p (or π)

ANSWER 44 $1s^2 2s^2 2p_x^1 2p_y^1 2p_z^1$

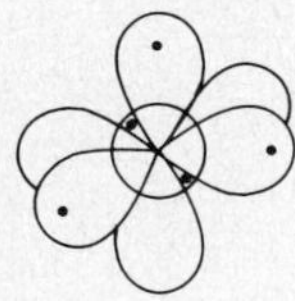

ANSWER 54 $KK\ (\sigma_s^b)^2(\sigma_s^*)^2(\pi_x^b)^2$
$(\pi_y^b)^2(\sigma_z^b)^2(\pi_x^*)^2(\pi_y^*)^2(\sigma_z^*)^1$

ITEM 6

The maximum number of electrons that can occupy a molecular orbital is two. Since they occupy the same bonding orbitals, the electrons forming the H—H bond must have opposite spins; that is, the spin quantum number s of one electron is $+\frac{1}{2}$; s for the other electron must be ________.

ITEM 16

If the atomic orbitals of He and He^+ were combined, the number of molecular orbitals formed would be ________.

ITEM 26

The electronic configuration of Li_2^- is ________.

ITEM 36

If the p_x orbitals overlap, an orbital given the symbol $\pi_x{}^b$ is formed. It has approximately the shape shown at the right. This type of molecular orbital, like all others, can hold a maximum of ________ electrons.

ITEM 46

The electronic configuration of N_2 is written ________.

ITEM 56

The electronic configuration of atomic sodium is ________. Would you expect that sodium could form Na_2 molecules?

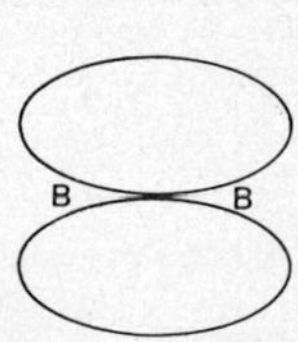

ANSWER 5

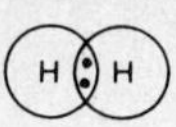

ANSWER 15

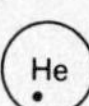

ANSWER 25 $KK\ (\sigma_s^b)^1$

ANSWER 35 antibonding

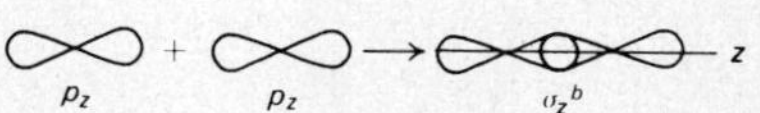

ANSWER 45 $\sigma_s^b \quad \sigma_s^* \quad \pi_x^b \quad \pi_y^b \quad \sigma_z^b$

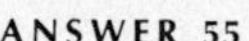

ANSWER 55

$KK\ (\sigma_s^b)^2(\sigma_s^*)^2(\pi_x^b)^2$
$(\pi_y^b)^2(\sigma_z^b)^2(\pi_x^*)^2(\pi_y^*)^2(\sigma_z^*)^1$

(Notice that Ne_2^+ and F_2^- are isoelectronic; that is, they contain the same number of electrons so the electronic configurations of the two are the same.)

ITEM 7

Two *s* atomic orbitals also can be combined so that an *antibonding orbital* is produced (as discussed in *Chemical Principles,* Section 10–3). The antibonding orbital can be shown as at the right. This is intended to indicate that the dumbbell-shaped cloud of electrons is concentrated away from the __________ of the molecule.

ITEM 17

The bonding molecular orbital formed from two *s* orbitals is given the symbol $\sigma_s{}^b$; the symbol $\sigma_s{}^*$ is given to the __________ orbital.

ITEM 27

The electronic configuration of the beryllium atom is __________. Draw the atomic orbital diagram.

ITEM 37

In this review, the formation of a π molecular orbital by overlap of p_y orbitals is indicated as in the diagram at the right. Draw an analogous diagram for the bonding combination of p_x orbitals.

ITEM 47

Draw the orbital diagrams for N_2 and include arrows to represent electrons.

ITEM 57

The electronic configuration of the ion $N_2{}^+$ is written __________.

ANSWER 6 $-\frac{1}{2}$

ANSWER 16 two

ANSWER 26 $KK\ (\sigma_s^b)^2(\sigma_s^*)^1$

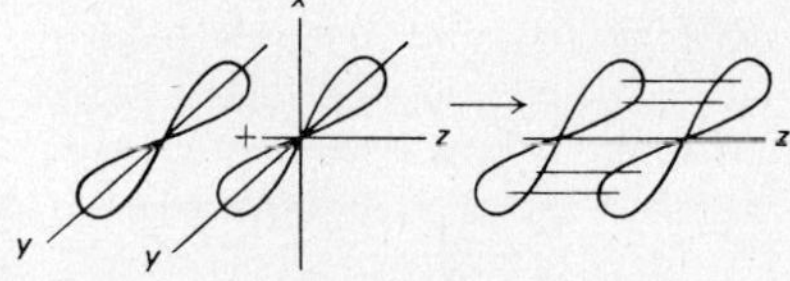

ANSWER 36 two

ANSWER 46 $KK\ (\sigma_s^b)^2(\sigma_s^*)^2(\pi_x^b)^2(\pi_y^b)^2(\sigma_z^b)^2$

ANSWER 56 $1s^22s^22p_x^22p_y^22p_z^23s^1$
Yes, by analogy with Li_2, Na_2 would have a σ_s^b orbital containing the two valence electrons.

ITEM 8

The energy of the bonding orbital formed by combination of two *s* orbitals is lower than that of the antibonding orbital. Therefore when electrons are "fed into" orbitals, they will go first into the __________ orbital, which will accommodate a maximum of __________ electrons.

ITEM 18

Draw diagrams of the two molecular orbitals of He_2^+, by analogy with those of H_2, and insert dots corresponding to the number of electrons expected to be in each.

σ_s^b __________ σ_s^* __________

ITEM 28

Draw the molecular orbitals that might be expected to form if two beryllium atoms were brought close together. Label them and include the number of electrons that would have to be in each orbital.

ITEM 38

Draw a diagram that shows σ^b and π^b overlap of the two sets of three *p* orbitals shown at the right.

ITEM 48

In N_2 all the bonding orbitals formed by combination of the 2*p* orbitals are filled. A molecule of O_2 has two more electrons than does N_2. Since the bonding orbitals are filled, two electrons in O_2 must occupy __________.

ITEM 58

Draw the molecular orbitals of the N_2^+ ion. Represent the electrons with arrows.

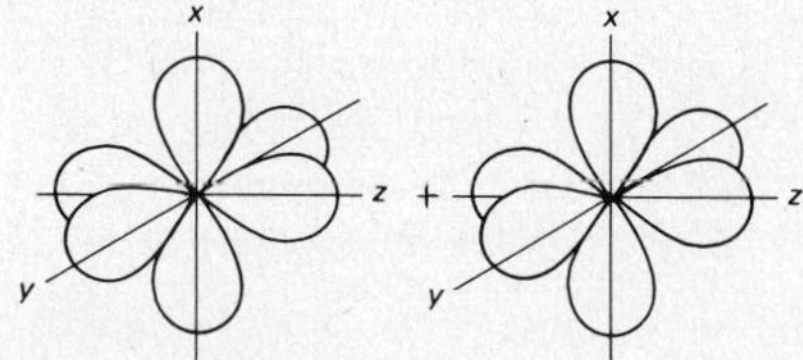

ANSWER 7 center

ANSWER 17 antibonding

ANSWER 27 $1s^2 2s^2$

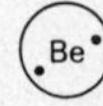

ANSWER 37

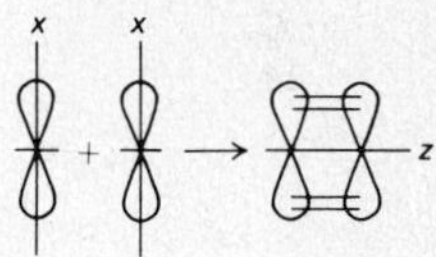

ANSWER 47

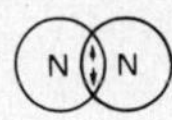

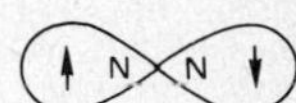

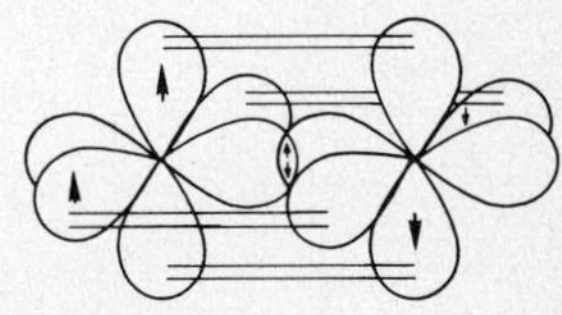

ANSWER 57 $KK\ (\sigma_s^b)^2(\sigma_s^*)^2(\pi_x^b)^2(\pi_y^b)^2(\sigma_z^b)^1$

ITEM 9

In a molecule of H_2, the number of electrons in the bonding orbital is ________ and the number of electrons in the antibonding orbital is ________.

ITEM 19

The number of electrons in the bonding molecular orbital exceeds that in the antibonding orbital. Would you expect that the ion He_2^+ could exist?

ITEM 29

As was true for He_2, the number of electrons in the bonding and antibonding orbitals would be ________. Would you expect a molecule of Be_2 ever to be formed?

ITEM 39

Of the three bonding orbitals, π_x^b, π_y^b, and σ_z^b, the two π orbitals have the same energy and σ_z^b is of higher energy. The orbitals in order of increasing energy are $\sigma_s^b < \sigma_s^* <$ ________ $=$ ________ $<$ ________.

ITEM 49

The antibonding molecular orbitals formed from $2p$ atomic orbitals have the following order of energy: $\pi_x^* = \pi_y^* < \sigma_z^*$. The maximum number of electrons that can occupy each orbital is ________. Sketches are shown at the right (the dots are nuclei).

ITEM 59

The electronic configuration of the O_2^{2-} ion is ________.

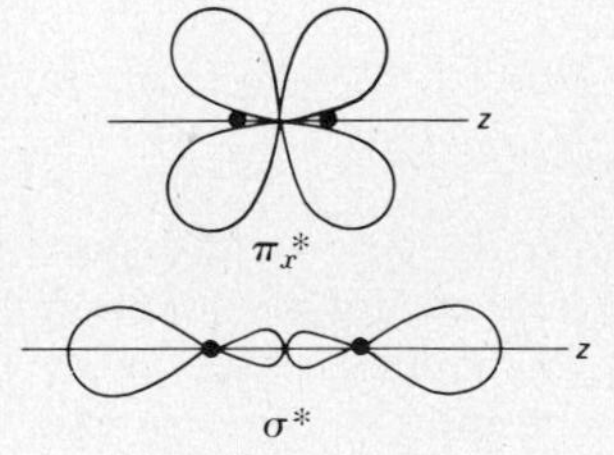

ANSWER 8 bonding
two

ANSWER 18

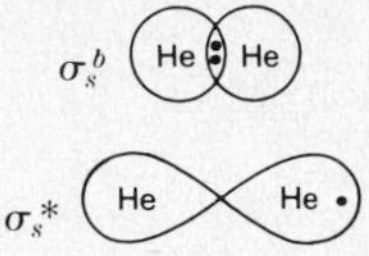

ANSWER 28

σ_s^b Be Be

σ_s^* Be Be

ANSWER 38

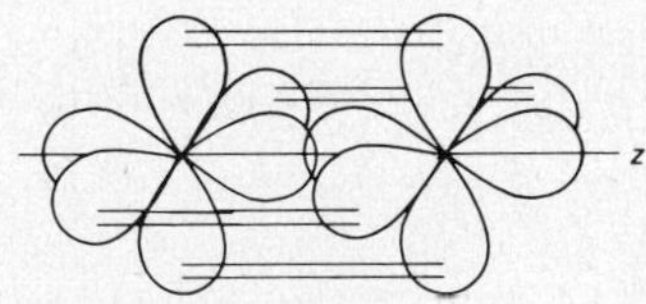

ANSWER 48 antibonding orbitals

ANSWER 58

(These drawings show electrons with different spins in the π orbitals and an electron with either spin in the σ_z orbital.)

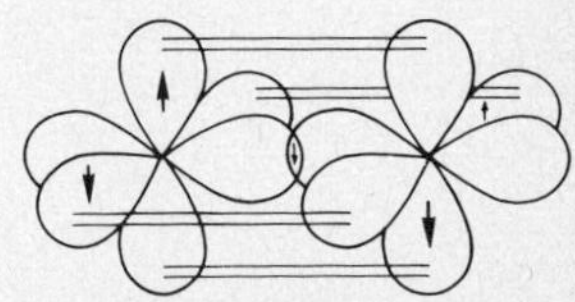

ITEM 10

The electronic configuration of the helium atom is ________.

ITEM 20

The bonding σ_s^b molecular orbital is lower in energy than the antibonding σ_s^* orbital. Therefore, the electronic configuration of the He_2^+ molecule is written $(\sigma_s^b)^2(\sigma_s^*)^1$. The electronic configuration of H_2 is written ________.

ITEM 30

The electronic configuration of boron is ________.

ITEM 40

As you discovered in Review 8 about atomic orbitals, the repulsion between two electrons causes them to occupy two different orbitals of the same energy whenever possible. Such unpaired electrons have the same spin. Add arrows to the diagram at the right to represent the π electronic configuration of B_2.

ITEM 50

By analogy with the B_2 molecule, we would expect that the two electrons of O_2 that occupy antibonding π orbitals will have the same spin, one electron in the ________ molecular orbital and the other in the ________ orbital.

ANSWER 9 two
zero

ANSWER 19 Yes. (It has been found experimentally that the bond is weak relative to that of H_2, but He_2^+ is known.)

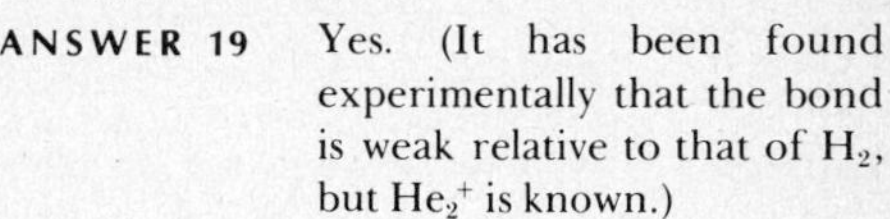

ANSWER 29 equal (two in each)
No.

ANSWER 39 $\pi_x^b = \pi_y^b < \sigma_z^b$

ANSWER 49 two

ANSWER 59

O_2^{2-} contains two more electrons than does O_2. These electrons are used to fill the $\pi_x{}^*$ and $\pi_y{}^*$ orbitals.

O_2^{2-}: $KK\,(\sigma_s^b)^2(\sigma_s{}^*)^2(\pi_x^b)^2(\pi_y^b)^2(\sigma_z^b)^2(\pi_x{}^*)^2(\pi_y{}^*)^2$

ANSWER 10 $1s^2$

(*Item 11 is on page 502.*)

ANSWER 20 $(\sigma_s^b)^2$

(*Item 21 is on page 502.*)

ANSWER 30 $1s^2 2s^2 2p_x^1$

(*Item 31 is on page 502.*)

ANSWER 40

(Any arrangement showing two electrons of the same spin, one in the π_x^b and the other in the π_y^b orbital, is correct.)

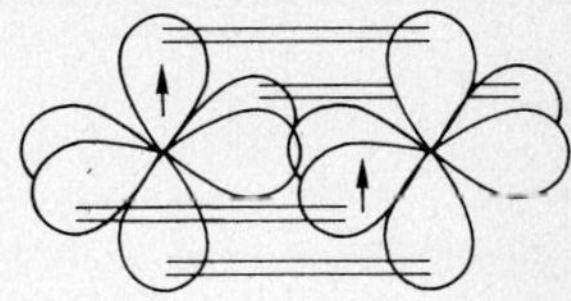

(*Item 41 is on page 502.*)

ANSWER 50 π_x^*
π_y^*

(*Item 51 is on page 502.*)

10-3 ORBITAL DIAGRAMS OF POLYATOMIC MOLECULES

The practice you have had in drawing orbital diagrams for diatomic molecules can be put to good use as we turn to more complex structures. In this section, we will look again at molecules in which the *s* and *p* electrons are used in bonding and will consider especially the relationship between molecular geometry and the molecular orbital theory of bonding. It is this structural aspect of Sections 10–5, 10–6, 10–7, and 10–8 of *Chemical Principles* that will be reviewed here.

When you have finished this review you will be able to draw diagrams that give information on the arrangement of the atoms and the most probable location of the electrons in many molecules. For example, you will be able to predict the planarity of a BH_3 molecule and the pyramid shape of NH_3. You will understand why an H_2O molecule is bent and a CO_2 molecule is linear.

ITEM 1

The electronic configuration of the hydrogen atom is __________, and the atomic orbital diagram for hydrogen is __________.

ITEM 15

The orbital diagram of Item 14 suggests that the angle between the two H—S bonds should be __________ degrees.

ITEM 29

Beryllium is the first of the alkaline earth elements. The Lewis diagrams of these elements, __________, __________, __________, __________, __________, and __________, suggest that they tend to form ions by the __________ of two electrons.

ITEM 43

The composite diagram at the right shows that the three sp^2 hybrid orbitals lie in a plane. The angle between any two of them would be expected to be __________.

ITEM 57

Arrange the components at the right to give an orbital diagram for methyl alcohol, CH_3OH.

ITEM 71

When only two atoms are bonded to a carbon atom, the bonding is described by first hybridizing the orbitals as $s, p_x, p_y, p_z \rightarrow sp, sp, p_y, p_z$. Label the orbitals in the diagram.

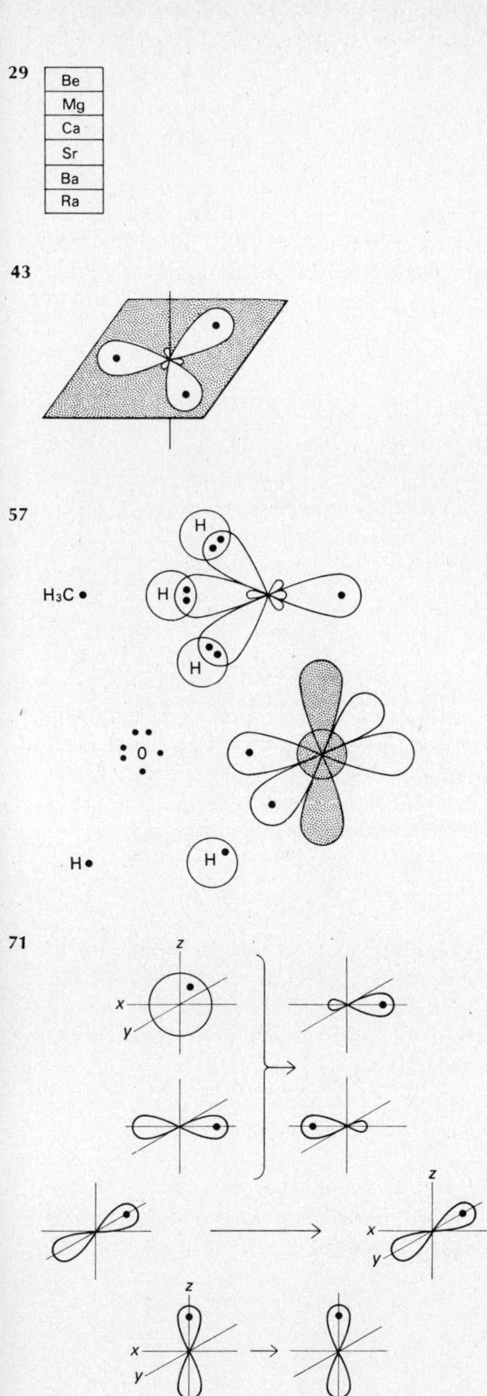
29
Be
Mg
Ca
Sr
Ba
Ra
43
57
$H_3C\bullet$
H
H
H
O
H•
H
71
z
x
y
z
x
y
z
x
y

ITEM 2

The electronic configuration of the lithium atom is __________, and the atomic orbital diagram is __________.

ITEM 16

The angle between the two H—S bonds has been determined experimentally to be 92°. Thus, we see that we can roughly predict molecular shapes on the basis of __________ diagrams, something we cannot do with __________ diagrams.

ITEM 30

In addition to forming ionic compounds, some alkaline earth metals form compounds, such as beryllium hydride, which are not ionic but have two __________ bonds.

ITEM 44

The molecule BH_3 can exist at high temperatures. The sharing of the three outer boron electrons with the electrons of the three hydrogen atoms can be represented by the orbital diagram __________.

ITEM 58

Although the bonding in H_2O can be described in terms of s, p_x, p_y, and p_z orbitals, an alternative description using four sp^3 hybrid orbitals for the six valence electrons of the oxygen atom is possible. Draw such an orbital diagram.

ITEM 72

Bonding in an acetylene molecule, H—C≡C—H, can be described by using the orbital diagrams of Item 71. Draw an orbital diagram for H—C≡C—H.

BeH_2

ANSWER 1 $1s^1$

ANSWER 15 90

ANSWER 29

Be: Mg: Ca: Sr: Ba: Ra:
loss (or transfer)

ANSWER 43 $360/3 = 120°$

ANSWER 57

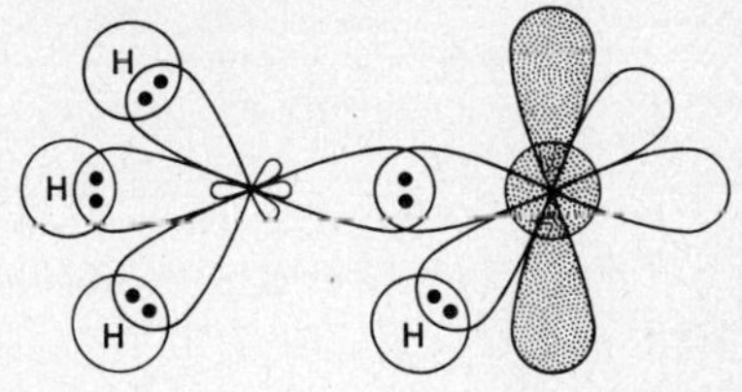

ANSWER 71

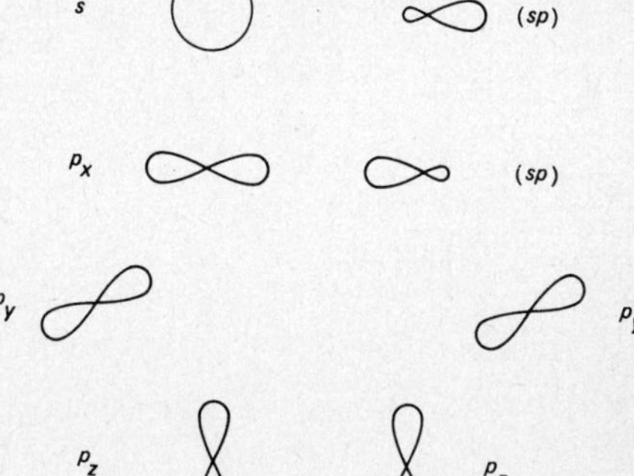

ITEM 3

A molecule of lithium hydride, LiH, considered as a covalent molecule, can be represented by the overlap of the two atomic orbitals as __________.

ITEM 17

The orbital diagram of the H_2O molecule is ________.

ITEM 31

If some energy were added to a beryllium atom, the process at the right could occur. The electronic configuration of the excited state would be __________, and __________ electrons would be available for bonding.

ITEM 45

The orbital diagram for the molecule BF_3 is ________.

ITEM 59

The angle predicted for the H_2O molecule by the sp^3 orbital description is __________. The angle that was predicted on the basis of s, p_x, p_y, and p_z orbitals was __________.

ITEM 73

In the diagram for Answer 72, the triple bond between the carbon atoms is described as involving one __________ bond and two __________ bonds.

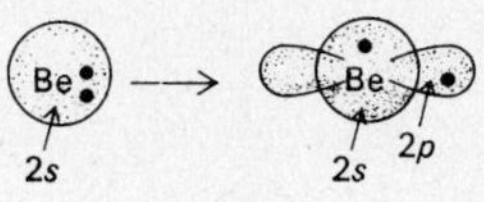

ANSWER 2 $1s^2 2s^1$

Li·

ANSWER 16 orbital
Lewis

ANSWER 30 covalent

ANSWER 44

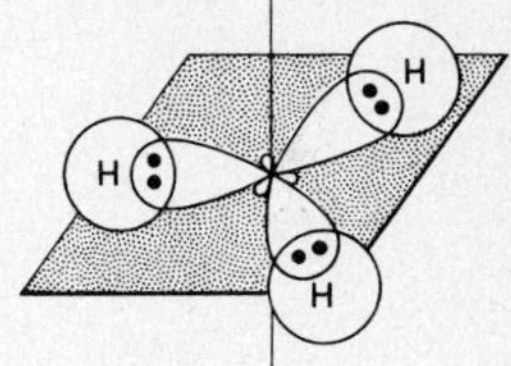

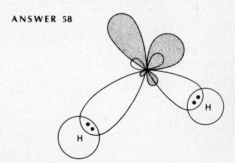

ANSWER 58

ANSWER 72

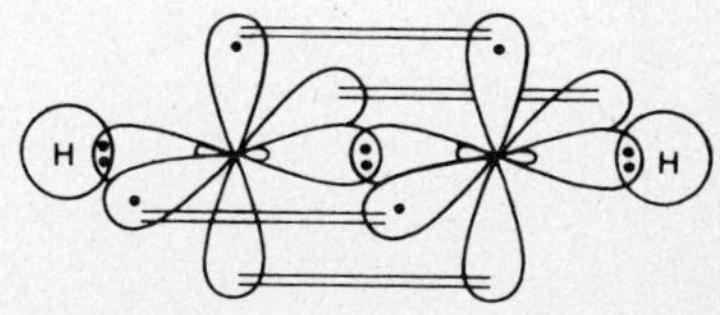

ITEM 4

This combination of two *s* orbitals gives rise to a *sigma bond.* The bonding orbital in LiH is represented by the symbol __________.

ITEM 18

On the basis of the orbital diagram of the H_2O molecule, the H—O—H angle is predicted to be __________. (In fact, it is 105°. The difference can be attributed to repulsion between the hydrogens and to hybridization – a term that will be discussed shortly.)

ITEM 32

For the description of the two bonds in BeH_2, which are found experimentally to be equivalent, it is assumed that the *s* and *p* orbitals mix or *hybridize,* as indicated at the right. Each new *sp* hybrid orbital contains __________ electron(s).

ITEM 46

On the basis of the diagram of Item 45, BF_3 is predicted to be a __________ molecule with angles between the B—F bonds of __________ degrees.

ITEM 60

The experimentally observed angle is 105°. Electron repulsion in so small a molecule is probably responsible for spreading out the electrons by hybridization of the *s* and *p* orbitals of oxygen to form __________ orbitals.

ITEM 74

Since two *sp* hybrid orbitals project linearly in opposite directions, the molecule acetylene must be __________ in shape.

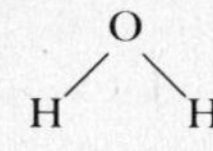

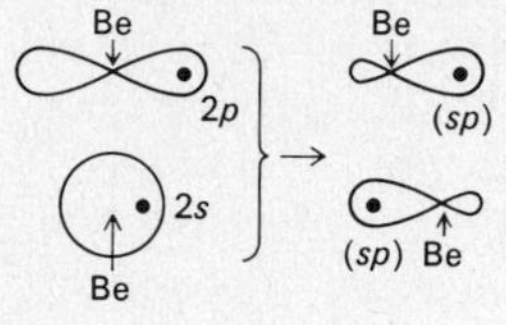

ANSWER 3

ANSWER 17

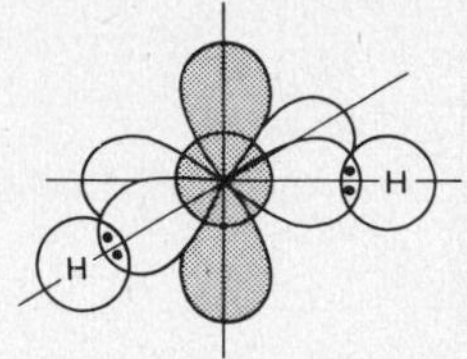

ANSWER 31 $1s^2 2s^1 2p^1$
two

ANSWER 45

ANSWER 59 109°28′
90°

ANSWER 73 sigma
pi

ITEM 5

In LiH there are ________ electrons of ________ spin in the bonding orbital and ________ electrons in the antibonding σ_s^* orbital.

ITEM 19

The electronic configuration of the nitrogen atom is ________. The Lewis diagram is ________.

ITEM 33

It is known that the compound BeH_2 has two equivalent covalent bonds. Yet the electronic configuration $1s^2 2s^2$ does not suggest covalent bonding. If one electron is *promoted* to a $2p$ orbital to give the configuration ________ and then the s and p orbitals are hybridized, the diagram at the right is obtained.

ITEM 47

The ground state electronic configuration of carbon is ________.

ITEM 61

The ground state electronic configuration of silicon is ________. The orbital diagram for SiH_4 is ______.

ITEM 75

The Lewis diagram for the hydrogen cyanide molecule, HCN, is ________.

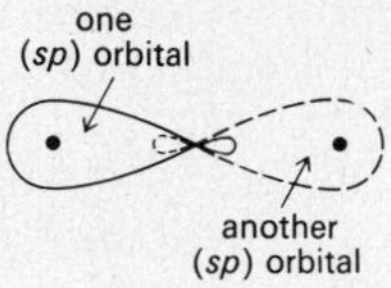

one
(*sp*) orbital
another
(*sp*) orbital

ANSWER 4 σ_s^b

ANSWER 18 90°

ANSWER 32 one

ANSWER 46 planar (or flat)
120
(This structure has been found experimentally.)

ANSWER 60 sp^3

ANSWER 74 linear

ITEM 6

The electronic configuration of the fluorine atom is __________. Draw an orbital diagram of the fluorine atom and shade with pencil those orbitals that are filled with two electrons and not available for bonding. Place dots in orbitals with single electrons.

ITEM 20

The orbital diagram for the nitrogen atom is ________.

ITEM 34

The two *sp* hybrid orbitals are arranged linearly as drawn at the right. Like other atomic orbitals, these hybrid orbitals can be combined with orbitals of other atoms to produce two kinds of molecular orbitals: __________ and __________.

ITEM 48

To explain the four equivalent bonds that occur in such molecules as CH_4, we say that one electron must be __________ from the $2s$ orbital to the $2p_z$ to give the configuration __________.

ITEM 62

When a carbon atom is bonded to three other atoms, three sp^2 orbitals, rather than four sp^3 orbitals, are used to form three sigma bonds. The sp^2 orbitals in carbon, as in boron, lie in a __________ and project at angles of __________ from one another.

ITEM 76

Draw an orbital diagram for HCN.

ANSWER 5 two
opposite
no

ANSWER 19 $1s^2 2s^2 2p_x^1 2p_y^1 2p_z^1$

$:\dot{\underset{\cdot}{N}}\cdot$

ANSWER 33 $1s^2 2s^1 2p^1$

ANSWER 47 $1s^2 2s^2 2p_x^1 2p_y^1$

ANSWER 61 $1s^2 2s^2 2p^6 3s^2 3p_x^1 3p_y^1$

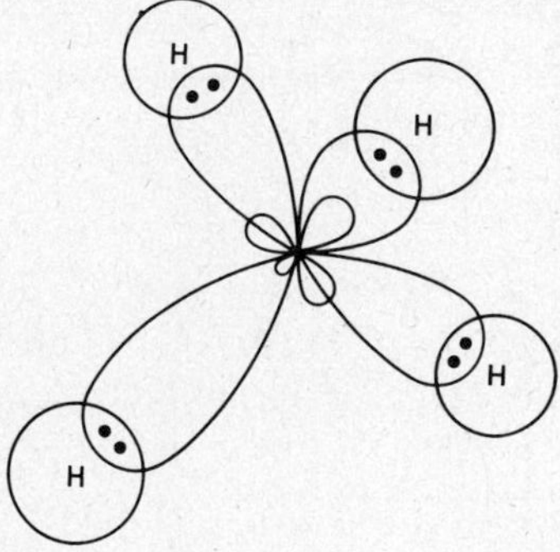

ANSWER 75 H—C≡N:

ITEM 7

A hydrogen atom and a fluorine atom each have one orbital containing an electron that is available for bonding. Combination of the *s* orbital of hydrogen and the *p* orbital of fluorine produces a bonding and a(n) __________ molecular orbital.

ITEM 21

The Lewis diagram for NH_3 is __________, and the orbital diagram is __________.

ITEM 35

In this section we will treat only molecules in their ground state. This means that for the molecules considered only bonding molecular orbitals will be needed. The bonding orbital diagram for BeH_2 is __________, and the shape of the molecule is predicted to be __________.

ITEM 49

The four orbitals $2s$, $2p_x$, $2p_y$, and $2p_z$ can be mixed, or __________, to give four equivalent orbitals.

ITEM 63

If one *s* orbital and two *p* orbitals of the carbon configuration $1s^2 2s^1 2p_x^1 2p_y^1 2p_z^1$ are hybridized to form three sp^2 orbitals, one electron remains in a __________ orbital.

ITEM 77

The bonding in a molecule of carbon dioxide, CO_2, also can be described in terms of sp, sp, p_y, and p_z orbitals on the carbon atom. Draw a Lewis and an orbital diagram for CO_2.

ANSWER 6 $1s^2 2s^2 2p_x^2 2p_y^2 2p_z^1$

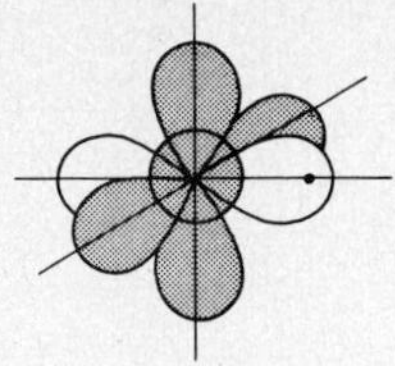

ANSWER 20

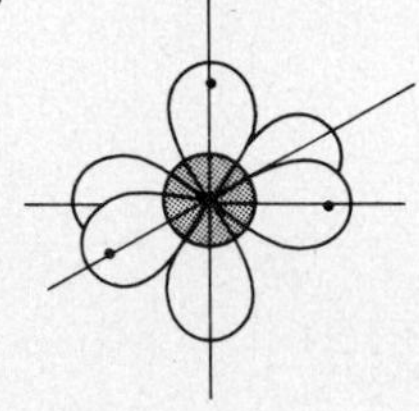

ANSWER 34 bonding
antibonding

ANSWER 48 promoted
$1s^2 2s^1 2p_x^1 2p_y^1 2p_z^1$

ANSWER 62 plane
120°

ANSWER 76

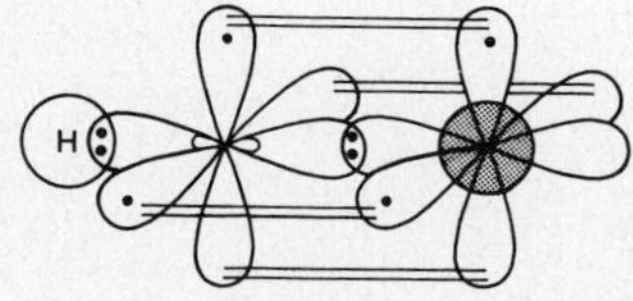

ITEM 8

Draw an orbital diagram for hydrogen fluoride, HF, showing the overlap of the *s* and *p* orbitals to form a bond.

ITEM 22

The NH_3 molecule is predicted to have a pyramidal rather than a planar structure. This has been verified experimentally. The angle between any two N—H bonds, however, is 107°, rather than the predicted value of __________.

ITEM 36

It has been found experimentally that BeH_2 is linear, as is the molecule $BeCl_2$. The orbital diagram for $BeCl_2$ is __________.

ITEM 50

The new sp^3 hybrid orbitals have the same general shape as sp or sp^2 orbitals; that is, they can be represented by the drawing __________.

ITEM 64

The *p* orbital is perpendicular to the plane of the three sp^2 orbitals. As in the C_2 and N_2 molecules, the two lobes of the *p* orbital can overlap with the two lobes of a *p* orbital of another atom to form a __________ bond.

ITEM 78

Each double bond of the CO_2 molecule is described as consisting of one __________ and one __________ bond.

ANSWER 7 antibonding

ANSWER 21

H
:N̈:H
H

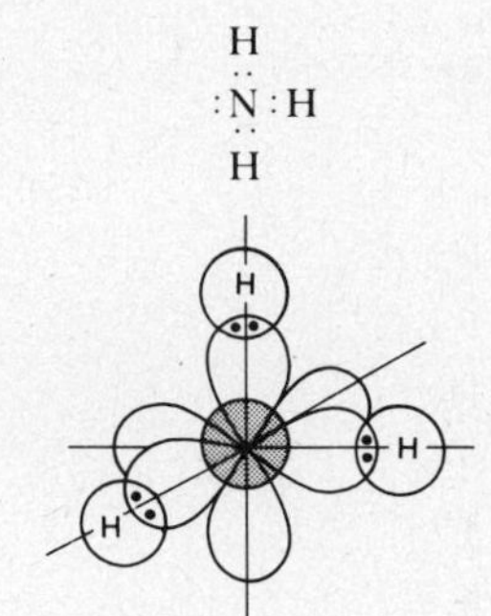

ANSWER 35

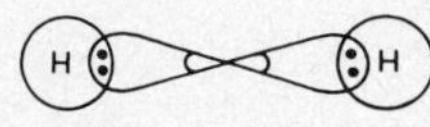

linear

ANSWER 49 hybridized

ANSWER 63 *p*

ANSWER 77 :Ö═C═Ö:

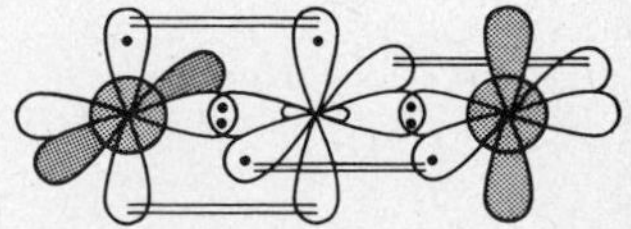

ITEM 9

The number of electrons in the σ^b molecular orbital that forms the bond in HF is ________. The number of electrons in the antibonding σ^* orbital is ________.

ITEM 23

In Section 10–2 you learned that *p* orbitals can overlap to form bonding orbitals in two ways. Label the orbitals at the right.

ITEM 37

Unlike $BeCl_2$, $CaCl_2$ is essentially ionically bonded. The bonding in $CaCl_2$ results from a ________ of electrons rather than a ________ of electrons.

ITEM 51

The four sp^3 hybrid orbitals project at angles of 109°28′, which can be drawn as at the right. Draw sp^3 orbitals along these directions.

ITEM 65

The Lewis diagram for ethylene, C_2H_4, is ________.

ITEM 79

On the basis of the orbital diagram, the shape of the CO_2 molecule is predicted to be ________.

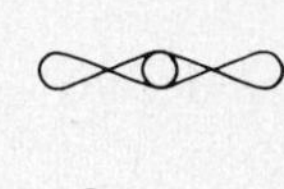

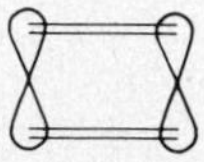

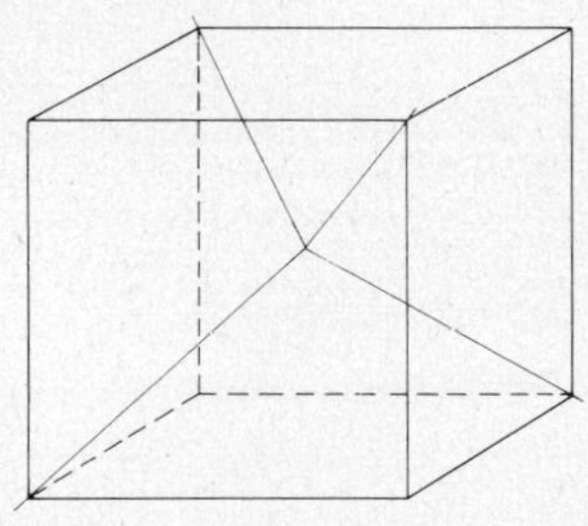

ANSWER 8

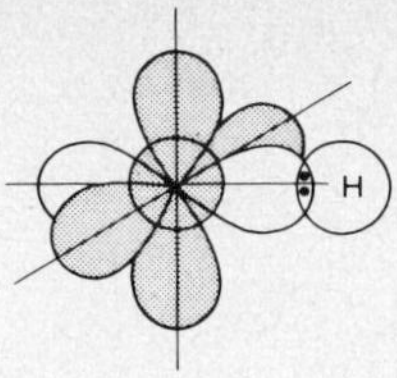

H

ANSWER 22 90°

ANSWER 36

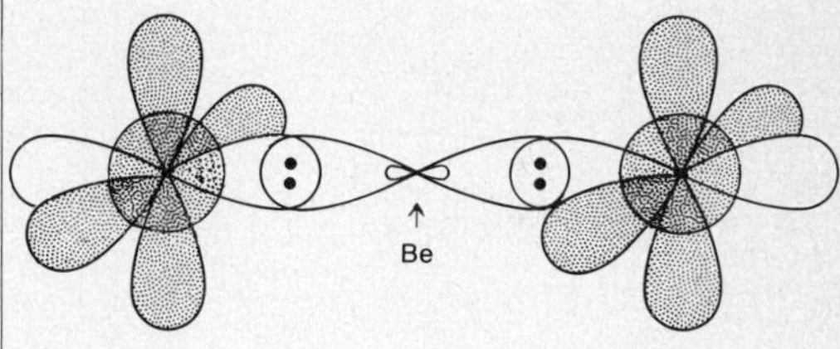

ANSWER 50

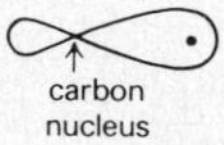

carbon nucleus

ANSWER 64 pi

ANSWER 78 sigma
pi

ITEM 10

The Lewis diagram of the chlorine atom is ________, and the orbital diagram (use shading for the filled orbitals) is ________.

ITEM 24

Draw diagrams for all the orbitals formed by combinations of the *s* and *p* orbitals of two nitrogen atoms in N_2 and use dots to represent the electrons.

ITEM 38

The covalent compound magnesium hydride, MgH_2, has the Lewis diagram ________ and the orbital diagram ________.

ITEM 52

In the carbon atom there are ________ valence electrons. If the bonding can be accounted for by sp^3 hybrid orbitals, there will be ________ bonds at angles of 109°28′.

ITEM 66

The orbital diagram for ethylene (at the right) shows that the carbon atoms are joined by one ________ bond and one ________ bond.

ITEM 80

Recall, by contrast, that the molecule H_2O is bent, as predicted by the orbital diagram for H_2O, which is ________.

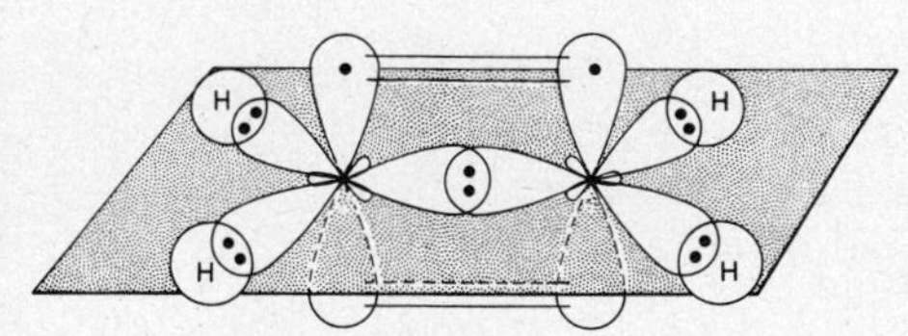

ANSWER 9 two
zero

ANSWER 23

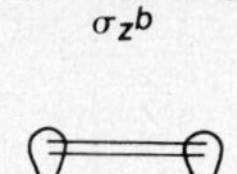

$\pi_x{}^b$

ANSWER 37 transfer
sharing

ANSWER 51

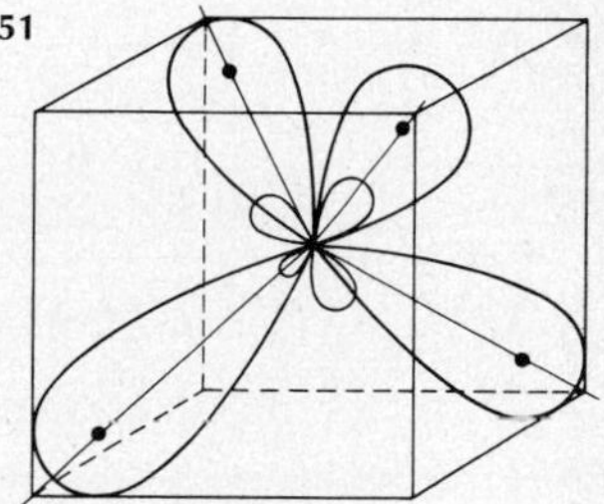

ANSWER 65

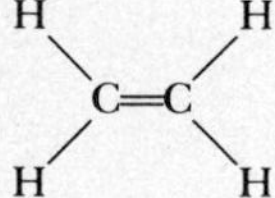

ANSWER 79 linear

ITEM 11

The orbital diagram of HCl is __________.

ITEM 25

Since there is an equal number of electrons in the σ_s^b and the σ_s^* orbitals, the *s* electrons (do, do not) contribute to the bonding of N_2.

ITEM 39

The predicted angle between the two Mg—H bonds is __________ degrees.

ITEM 53

Draw an orbital diagram for a molecule of methane, CH_4.

ITEM 67

The three sigma bonds of each carbon atom in ethylene result from combinations of hybrid orbitals that are described as __________ hybrids. The pi (π) bond is formed from __________ orbitals.

ITEM 81

Shortcomings of the simple atomic orbital method were indicated by its prediction for NH_3. An H—N—H angle of 90° was predicted if *p* orbitals were used for bonding. Since the actual H—N—H angle (107°) is closer to the one found in a tetrahedral molecule such as CH_4, it is possible that the bonding actually involves __________ orbitals.

ANSWER 10 $:\ddot{\underset{..}{\mathrm{Cl}}}\cdot$

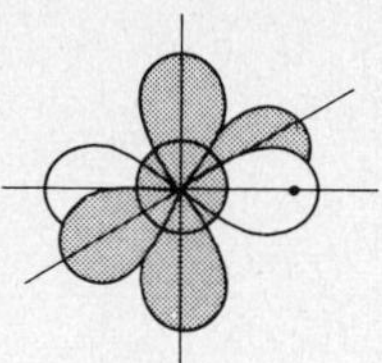

ANSWER 24

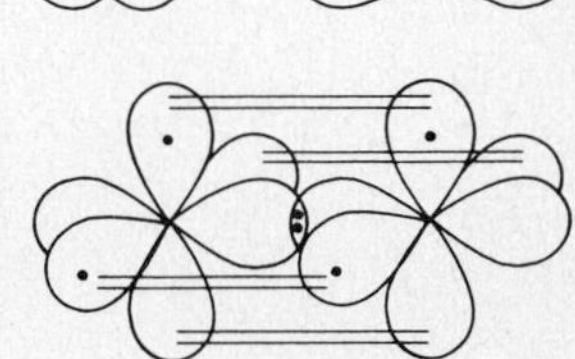

ANSWER 38 H : Mg : H

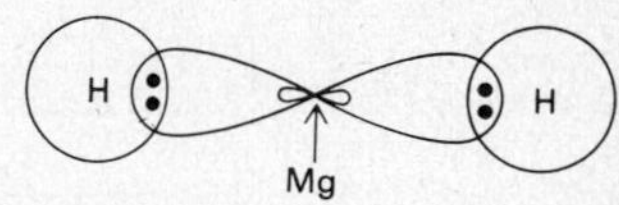

ANSWER 52 four
four

ANSWER 66 sigma
pi

ANSWER 80

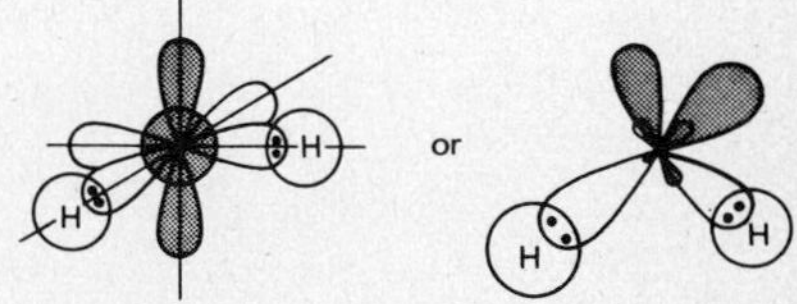

ITEM 12 The electronic configuration of sulfur and the Lewis diagram for the sulfur atom are __________.

ITEM 26 The description of the bonding in the N_2 molecule is that the nitrogen atoms are bonded by one __________ bond and two __________ bonds.

ITEM 40 The ground state electronic configuration of boron is __________.

ITEM 54 The solidly shaded figure at the right is a tetrahedron. Instead of saying that the angles in CH_4 are __________, we can say that the CH_4 molecule is tetrahedral.

ITEM 68 The Lewis diagram for formaldehyde, H_2CO, is __________.

ITEM 82 Draw an orbital diagram for NH_3 and show bonding by sp^3 orbitals.

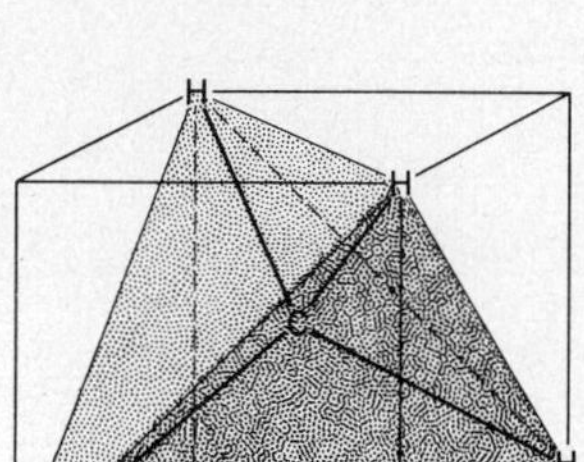

ANSWER 11

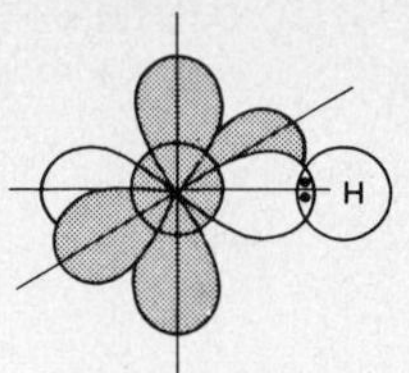

ANSWER 25 do not

ANSWER 39 180

ANSWER 53

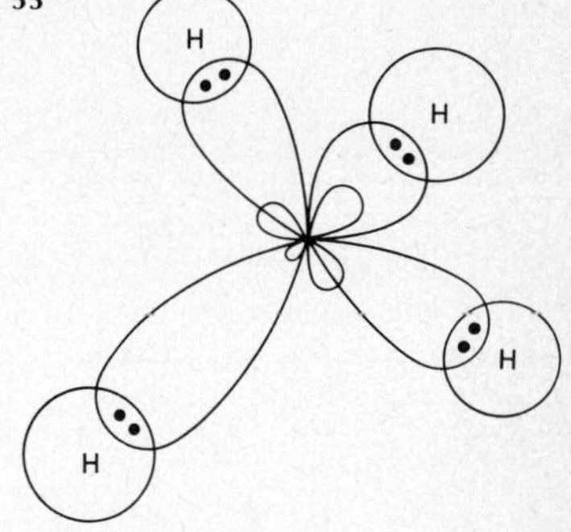

ANSWER 67 sp^2
p

ANSWER 81 sp^3 or hybrid

ITEM 13

Draw an orbital diagram for the valence electrons of sulfur.

ITEM 27

The Lewis diagram for N_2 shows that there are three covalent bonds, but it does not show the distinction between the two types of covalent bonding, that is, between __________ bonds and __________ bonds, that orbital diagrams reveal.

ITEM 41

Since boron forms three equivalent, covalent bonds in compounds such as boron trifluoride, BF_3, one of the $2s$ electrons must be promoted to the $2p_y$ or $2p_z$ orbital, thus giving the configuration __________.

ITEM 55

The Lewis diagram for ethane, C_2H_6, is __________.

ITEM 69

Draw an orbital diagram for the formaldehyde molecule.

ITEM 83

The ammonium ion, NH_4^+, is formed when H^+ is added to NH_3. The four N—H bonds in NH_4^+ are completely equivalent, so the bonding can be described on the basis of __________ orbitals of the nitrogen atom.

ANSWER 12 $1s^22s^22p^63s^23p_x^{\,2}3p_y^{\,1}3p_z^{\,1}$

$:\ddot{\mathrm{S}}\cdot$ (with a single dot below)

ANSWER 26 sigma (σ_z^b)
pi (π^b)

ANSWER 40 $1s^22s^22p_x^{\,1}$

ANSWER 54 109°28′

ANSWER 68

```
H
 \      ..
  C ═ O
 /      ..
H
```

ANSWER 82

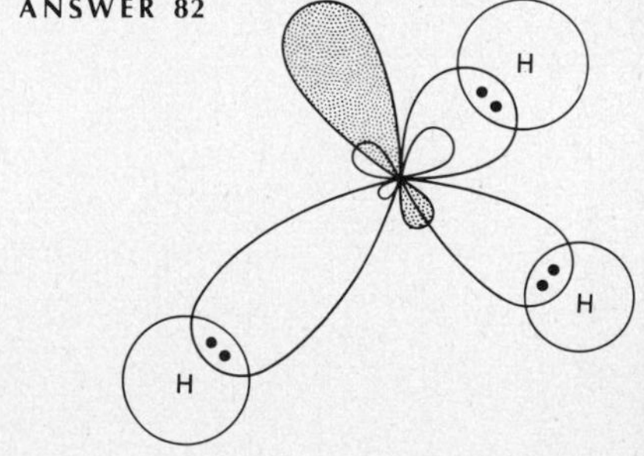

ITEM 14

Draw an orbital diagram for the molecule H_2S.

ITEM 28

The ground state electronic configuration of the beryllium atom is ________.

ITEM 42

Again, hybridization, as indicated at the right, is necessary to explain the fact that the three bonds (in BF_3, for example) are ________.

ITEM 56

Add orbitals and atom symbols to the line sketch at the right to produce an orbital diagram for ethane.

ITEM 70

If a π bond is to form as a result of the overlap of p orbitals, in the manner shown at the right, the axes of the two p orbitals that each contain one electron must lie in the same ________.

42

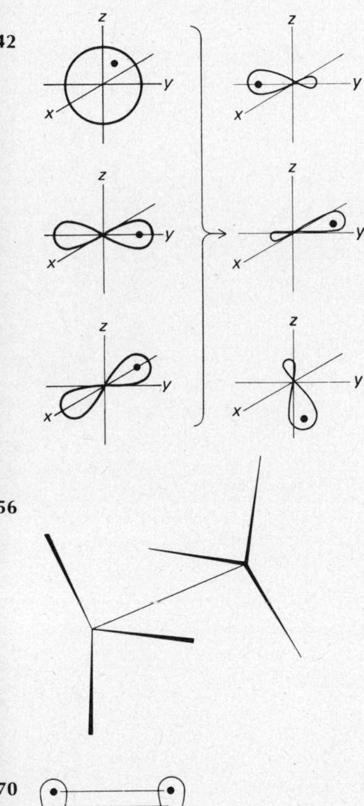

56

70

ANSWER 13

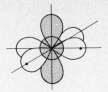

ANSWER 27 sigma
pi

ANSWER 41 $1s^2 2s^1 2p_x^1 2p_y^1$
or $1s^2 2s^1 2p_x^1 2p_z^1$

ANSWER 55

```
  H H              H   H
  ·· ··            |   |
H:C:C:H    or   H—C—C—H
  ·· ··            |   |
  H H              H   H
```

ANSWER 69

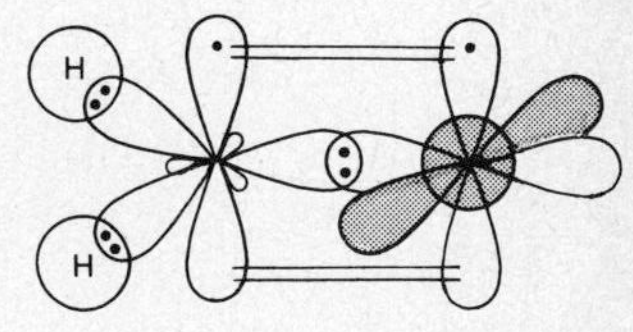

ANSWER 83 sp^3

ANSWER 14

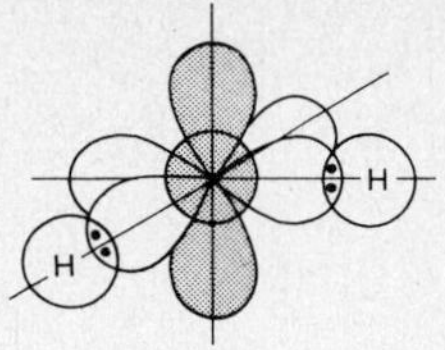

(*Item 15 is on page 526.*)

ANSWER 28 $1s^2 2s^2$

(*Item 29 is on page 526.*)

ANSWER 42 equivalent

(*Item 43 is on page 526.*)

ANSWER 56

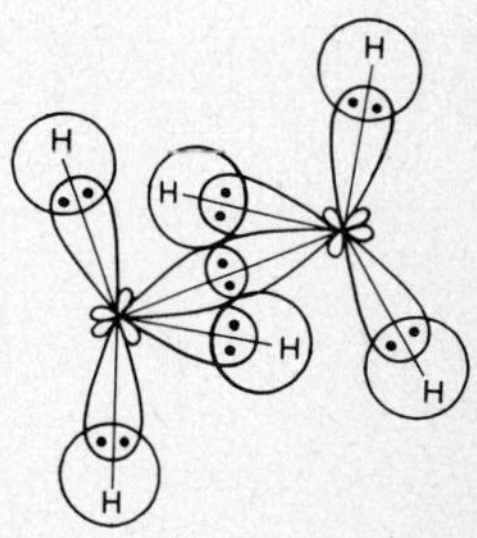

(*Item 57 is on page 526.*)

ANSWER 70 plane

(*Item 71 is on page 526.*)

REVIEW 11 *d* ORBITALS IN BONDING

Although the participation of *d* orbitals in bonding is no different in principle than the participation of *s* and *p* orbitals, it has been convenient to leave our discussion of *d* orbitals until we could use them in studying the bonding of complex ions. In this review we will consider first the use of *d* orbitals in the bonding of nonmetals and then their participation in bonding in complex ions of the transition metals. As in Review 10, the emphasis is on the geometry of molecules; therefore, we will consider only localized molecular orbital bonding as discussed in Section 11–4 of *Chemical Principles*. This review is intended as a stepping stone to your study of the ligand field treatment of complex ions.

When you have finished the review you will be able to describe the bonding of molecules or ions in terms of the hybrid orbitals of the central atom once the formula and geometry are given. And if you know the number of unpaired electrons in a complex ion, you will be able to predict whether the ion is an inner- or an outer-orbital complex.

To work through this review you must be familiar with the material of Reviews 8, 9, and 10. Furthermore, it is necessary to know the names and geometry of the five *d* orbitals. Since *p* orbitals are identical except for direction, there was little need to distinguish between them. The participation of *d* orbitals in hybridization depends entirely on their geometry. So you should remember that the d_{z^2} orbital is localized along the *z* axis, the $d_{x^2-y^2}$ orbital is localized along the *x* and *y* axes, and the d_{xy}, the d_{yz}, and the d_{xz} orbitals are localized between the *x* and *y*, *y* and *z*, and *x* and *z* axes, respectively.

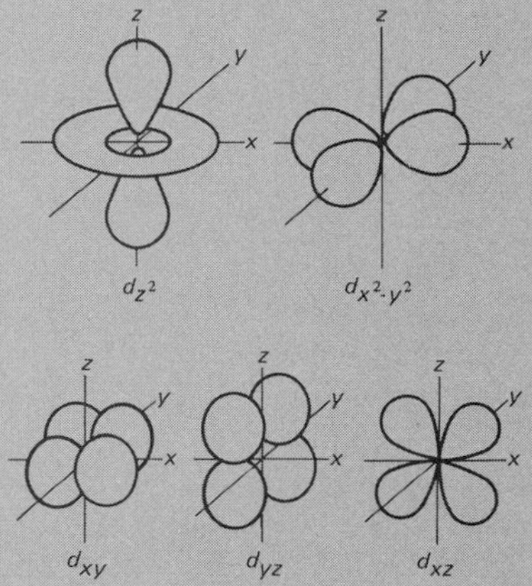

d_{z^2} $d_{x^2-y^2}$

d_{xy} d_{yz} d_{xz}

ITEM 1

Beryllium has the electronic configuration $1s^2 2s^2$. In $BeCl_2$ and other covalent compounds of beryllium, the angle between bonds is ________ degrees.

ITEM 14

These orbitals are designated sp^3d or $sp^3d_{z^2}$ because the only d orbital that has the necessary orientation in space for trigonal bipyramidal hybridization is the d_{z^2}. Draw the sp^3d orbitals in the diagram at the right. (Assume that the shape of each sp^3d orbital is the same as that of an sp orbital.)

ITEM 27

In the compound SF_4 (sulfur tetrafluoride), sulfur uses $sp^3d_{z^2}$ orbitals for bonding. The number of these orbitals is ________, and the number of sulfur electrons that must be placed in them is ________.

ITEM 40

In a Ti—F bond in TiF_6^{2-} the electrons come originally from the ________. The number of electrons in each bonding σ^b orbital is ________, and the number in each antibonding σ^* orbital is ________.

ITEM 53

The configuration of an outer-orbital complex of Cr^{2+}, including the bonding electrons of the ligands, would be ________. There will remain ________ unpaired electrons in $3d$ orbitals.

ITEM 66

The electronic configuration of an octahedral inner-orbital complex of Co^{3+} (including the bonding ligand electrons) is ________. There are ________ unpaired electrons in an inner-orbital complex of Co^{3+} and it is ________magnetic.

ITEM 79

In a complex transition metal ion the bonding electrons are supplied by the ________ molecules or ions. The d electrons are not involved in the bonding and, therefore, are described as ________.

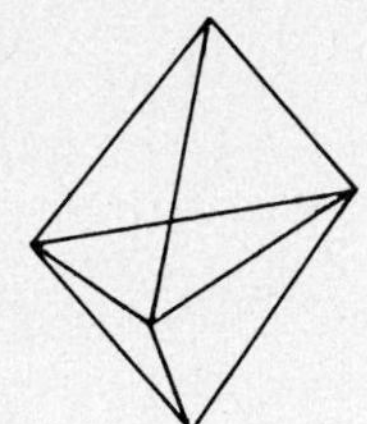

ITEM 2

We account for a bond angle of 180° by postulating *sp* hybridization of orbitals. The *sp* orbitals in beryllium result from the hybridization of __________ and __________ orbitals. Draw an orbital diagram of the molecule BeH_2.

ITEM 15

The bonding in PCl_5, then, can be described as overlap of half-filled __________ orbitals of phosphorus with the half-filled *p* orbitals of chlorine atoms.

ITEM 28

Draw an orbital diagram for SF_4 and show only the bonding orbital of each fluorine atom. (The non-bonding electrons lie in the plane that is the base of the bipyramid.)

ITEM 41

Draw an orbital diagram for *one bond* of TiF_6^{2-}.

ITEM 54

Experimental studies have shown that $Cr(H_2O)_6{}^{2+}$ has four unpaired electrons. It is therefore an __________er-orbital complex.

ITEM 67

Another type of hybrid orbital is the dsp^2. These four orbitals lie in the *xy* plane at right angles to one another. The atomic orbitals with the best orientation in space for dsp^2 hybridization are the following: $d_{x^2-y^2}$, __________, __________, __________.

ANSWER 1 180

ANSWER 14

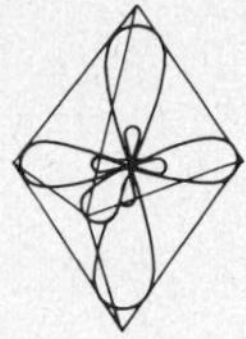

ANSWER 27 five
six

ANSWER 40 fluoride ion
two
zero

ANSWER 53

$KL\ 3s^2 3p^6 3d^4 4(sp^3d^2)^{12}$
or $KL\ 3s^2 3p^6 3d_{xy}^1 3d_{yz}^1 3d_{xz}^1 3d_{z^2}^1 4(sp^3d^2)^{12}$
four

ANSWER 66

$KL\ 3s^2 3p^6 3d^6 (d^2sp^3)^{12}$
or $KL\ 3s^2 3p^6 3d_{xy}^2 3d_{yz}^2 3d_{xz}^2 (d^2sp^3)^{12}$
no
dia

ANSWER 79 ligand
nonbonding

ITEM 3

Boron has the electronic configuration $1s^2 2s^2 2p_x^1$. The B—H bonds in BH_3 lie in a plane; the angle between any two of the bonds is ________ degrees.

ITEM 16

Draw an orbital diagram for PCl_5 (phosphorus pentachloride).

ITEM 29

When a sulfur atom is bonded to six other atoms, the geometry of the molecule is octahedral. Therefore, we assume that sulfur uses ________ orbitals for the bonding.

ITEM 42

Draw an orbital diagram for Ti^{4+} with the six ligand fluoride ions. Show only the bonding p orbital of each fluoride.

ITEM 55

The manganese(II) cation has the electronic configuration $KL\ 3s^2 3p^6 3d^5$. An octahedral inner-orbital complex of Mn^{2+} has the electronic configuration (including the bonding ligand electrons) ________.

ITEM 68

Draw a diagram showing the geometry of dsp^2 hybrid orbitals. Assume that the shape of each hybrid orbital is the same as an sp orbital.

ANSWER 2 the $2s$
one of the $2p$ (or $2p_x$)

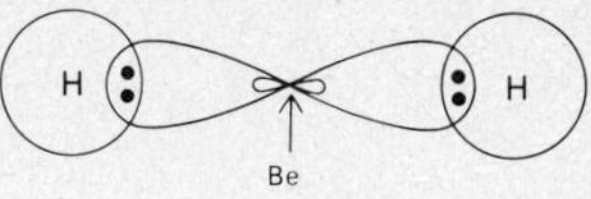

ANSWER 15 sp^3d (or $sp^3d_{z^2}$)

ANSWER 28

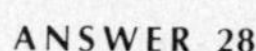

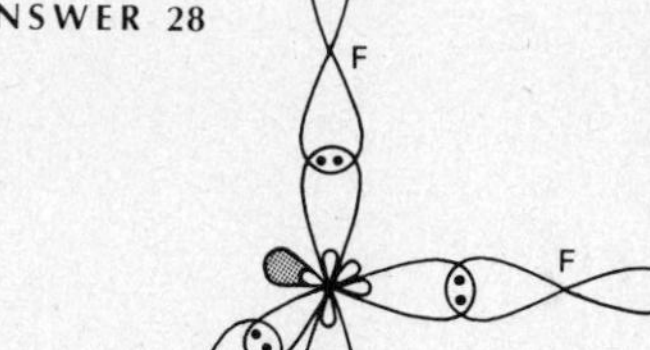

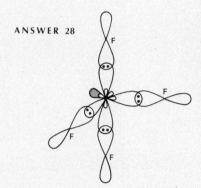

ANSWER 41

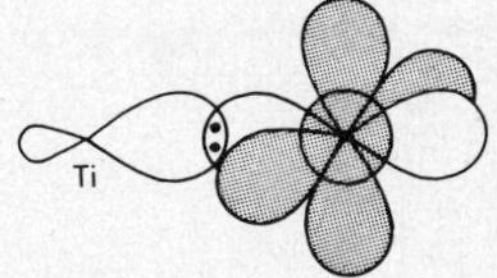

ANSWER 54 out

ANSWER 67 s
p_x
p_y

ITEM 4

The trigonal geometry of BH_3 is considered to be the result of hybridization of three atomic orbitals, $2s$, __________, and __________, to produce new ones known as __________ orbitals.

ITEM 17

The molecule PCl_5 can be considered as being produced by the addition of one phosphorus atom $\cdot\dot{\mathrm{P}}:$ to five chlorine atoms $\cdot\ddot{\underset{\cdot\cdot}{\mathrm{Cl}}}:$. Each bond is formed by the overlap of a chlorine p orbital containing __________ electron(s) with a phosphorus __________ orbital containing __________ electron(s).

ITEM 30

Draw an orbital diagram for SF_6 and show only the bonding orbital of each fluorine atom.

ITEM 43

The ion Ti^{3+} has the electronic configuration KL $3s^23p^63d^1$. Ti^{3+} reacts with water to yield the octahedral complex ion $Ti(H_2O)_6{}^{3+}$ [hexaaquotitanium(III) cation]. Since the geometry is octahedral, the titanium orbitals used for bonding are considered to be __________ hybrids that are formed from the following orbitals: __________.

ITEM 56

In an inner-orbital complex of Mn^{2+}, the number of unpaired electrons is __________.

ITEM 69

Nickel has the electronic configuration KL $3s^23p^64s^23d^8$; Ni^{2+} has the configuration __________.

$$\cdot\dot{P}: + 5\cdot\ddot{Cl}: \rightarrow PCl_5$$

(:Cl—P—Cl: with Cl above and two Cl below)

ANSWER 3 120

ANSWER 16

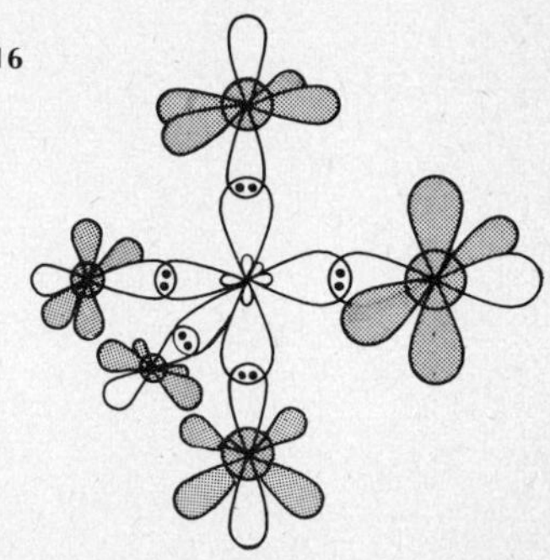

ANSWER 29 sp^3d^2

ANSWER 42

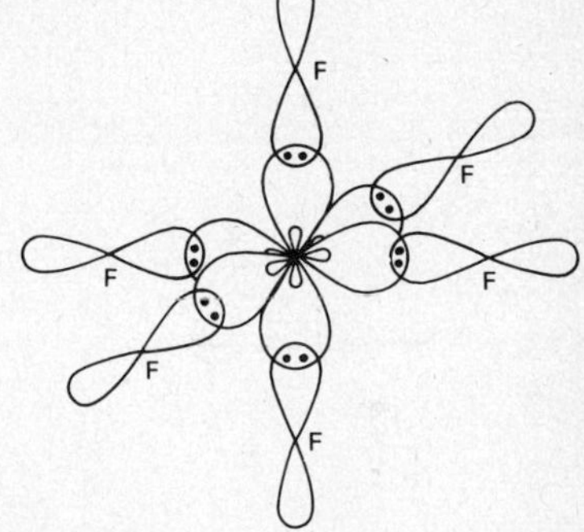

ANSWER 55

$KL\ 3s^23p^63d^5(d^2sp^3)^{12}$
or $KL\ 3s^23p^63d_{xy}^2 3d_{yz}^2 3d_{xz}^1(d^2sp^3)^{12}$

ANSWER 68

ITEM 5

The electronic configuration of carbon is $1s^2 2s^2 2p_x^1$ $2p_y^1$. The geometry of CH_4 and other compounds in which carbon is bonded to four atoms is described as __________; the angle between bonds is about __________.

ITEM 18

The ion PCl_4^+ can be considered as being formed by the bonding of one phosphorus atom, three chlorine atoms, and one chloronium ion Cl^+. Both electrons of one of the bonds in PCl_4^+ are contributed by phosphorus. Expressed in Lewis diagrams, this reaction is written __________.

ITEM 31

Bromine has the electronic configuration $1s^2 2s^2 2p^6$ $3s^2 3p^6 4s^2 3d^{10} 4p^5$. If one electron from the $4s$ and one from the $4p$ orbital are promoted to the $4d_{z^2}$ and $4d_{x^2-y^2}$ orbitals and hybridization occurs, new orbitals will be produced, __________ in number and designated __________. (Note that the $3d$ orbitals are filled and not used in hybridization.)

ITEM 44

Each bond in $Ti(H_2O)_6^{3+}$ is formed by overlap of a titanium d^2sp^3 orbital with a filled sp^3 orbital of a ligand water molecule. The number of d^2sp^3 hybrids is __________, and the total number of electrons from water molecules used in bonding is __________.

ITEM 57

An outer-orbital octahedral complex of Mn^{2+} has the electronic configuration (including the bonding electrons of the ligands) __________. There are __________ unpaired electrons.

ITEM 70

The ion Ni^{2+} reacts with cyanide ion to yield the square planar complex $Ni(CN)_4^{2-}$ [tetracyanonickelate(II)] ion, which is diamagnetic. Write the electronic configuration of the Ni^{2+} ion in the complex (including the bonding electrons from the ligand cyano groups).

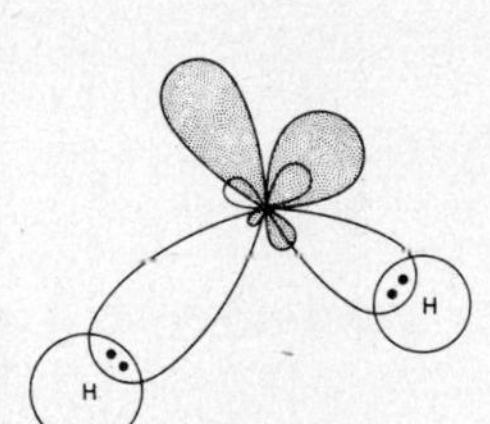

ANSWER 4 $2p_x$
$2p_y$
sp^2

ANSWER 17 one
sp^3d
one

ANSWER 30

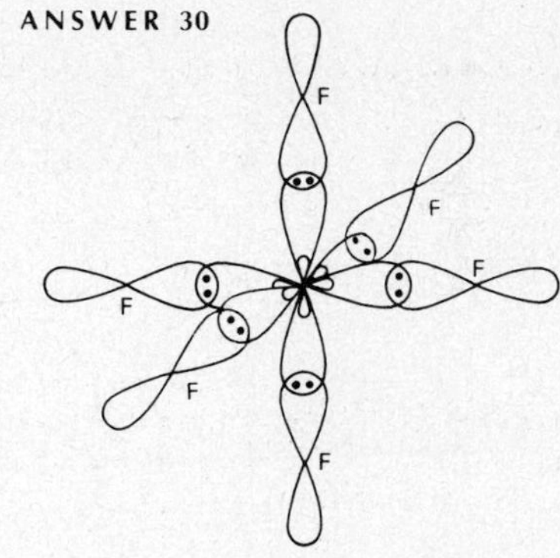

ANSWER 43 d^2sp^3
$3d_{z^2}$, $3d_{x^2-y^2}$, $4s$,
$4p_x$, $4p_y$, $4p_z$

ANSWER 56 one

ANSWER 69 $KL\ 3s^23p^63d^8$

ITEM 6

To account for the tetrahedral geometry of carbon compounds, we assume that hybridization of the __________, __________, __________, and __________ orbitals occurs. The hybrid orbitals are designated __________.

ITEM 19

The ion PCl_6^- [hexachlorophosphate(V)] can be thought of as the result of the combination of one phosphorus atom with five chlorine atoms and __________.

ITEM 32

A molecule of BrF_5 is square pyramidal, yet sp^3d^2 hybridization is used. Draw the orbital diagram and show only the bonding *p* orbitals for fluorine.

ITEM 45

Thus, the one 3*d* electron from Ti^{3+} must be in one of the *d* orbitals not used in bonding ($3d_{xy}$, $3d_{yz}$, or $3d_{xz}$) and is a *nonbonding* electron. The number of electrons in Ti^{3+} that have unpaired spins is __________. The number of unpaired electrons in $Ti(H_2O)_6{}^{3+}$ is __________.

ITEM 58

The Mn^{3+} ion has the electronic configuration *KL* $3s^23p^63d^4$. An octahedral inner-orbital complex of Mn^{3+} has the 3*d* electrons of Mn^{3+} arranged as __________. There are __________ unpaired electrons.

ITEM 71

The ion Ni^{2+} forms an octahedral complex with ammonia, the $Ni(NH_3)_6{}^{2+}$ [hexaamminenickel(II)] cation, which has two unpaired electrons. Write the electronic configuration of the Ni^{2+} ion including the bonding electrons from the NH_3 molecules.

Ti^{3+}: $KL\ 3s^2 3p^6 3d^1$

ANSWER 5 tetrahedral
109° (109°28′)

ANSWER 18

$$\cdot\ddot{P}: + 3\cdot\ddot{Cl}: + \ddot{Cl}:^{\oplus} \rightarrow :\ddot{Cl}-\underset{|\ :\ddot{Cl}:}{\overset{:\ddot{Cl}:\ |}{P^{\oplus}}}-\ddot{Cl}:$$

ANSWER 31 six
sp^3d^2

ANSWER 44 six
12

ANSWER 57

$KL\ 3s^2 3p^6 3d^5 4(sp^3d^2)^{12}$
or $KL\ 3s^2 3p^6 3d_{xy}^1 3d_{yz}^1$
$3d_{xz}^1 3d_{z^2}^1 3d_{x^2-y^2}^1 4(sp^3d^2)^{12}$

five

ANSWER 70

$KL\ 3s^2 3p^6 3d^8 (dsp^2)^8$ as
$KL\ 3s^2 3p^6 3d_{xy}^2 3d_{yz}^2 3d_{xz}^2 3d_{z^2}^2 (d_{x^2-y^2}sp^2)^8$

ITEM 7

Bonding in many compounds of nitrogen ($1s^2 2s^2 2p_x^1 2p_y^1 2p_z^1$) can be described in terms of sp^3 hybrid orbitals. The number of nitrogen electrons distributed in the four sp^3 orbitals is __________.

ITEM 20

The bonds in PCl_6^- are directed toward the apices of an octahedron. To account for this geometry, the theory assumes hybridization of the $3s$, $3p_x$, $3p_y$, $3p_z$, $3d_{z^2}$, and $3d_{x^2-y^2}$ orbitals to form sp^3d^2 orbitals. They are __________ in number and will accommodate the bonding electrons in PCl_6^-, __________ electrons from P and __________ from Cl.

ITEM 33

Although the bonding in BrF_4^- [tetrafluorobromate (III)] involves sp^3d^2 hybridization, the Br—F bonds are directed toward the corners of a square. Draw the orbital diagram and show only the p bonding orbitals for fluorine.

ITEM 46

Chromium has the electronic configuration $KL\ 3s^2 3p^6 4s^2 3d^4$. The electron configuration of Cr^{3+} is __________, in which there are __________ unpaired electrons.

ITEM 59

An octahedral outer-orbital complex of Mn^{3+} has __________ unpaired electrons.

ITEM 72

The square planar complexes of ions such as Ni(II) that contain __________ electrons in the outermost d orbitals are considered to involve __________ orbitals in formation of sigma bonds to ligands.

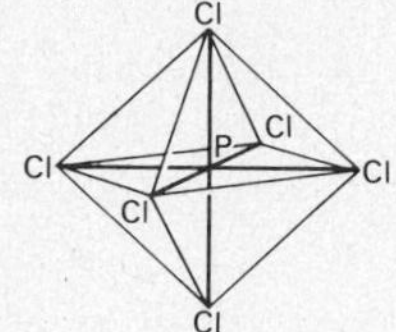

ANSWER 6 $2s$
$2p_x$
$2p_y$
$2p_z$
sp^3

ANSWER 19 one chloride ion $[:\ddot{\underset{..}{Cl}}:]^-$

ANSWER 32

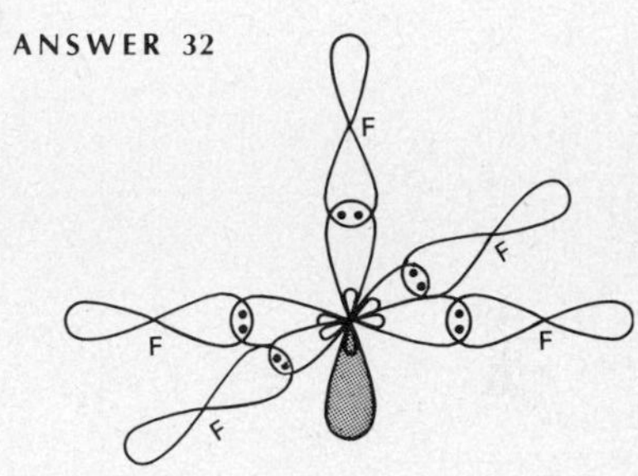

ANSWER 45 one
one

ANSWER 58 $3d_{xy}^2 3d_{yz}^1 3d_{xz}^1$
two

ANSWER 71

Ni^{2+} as the free ion has two unpaired electrons, so the configuration remains unchanged; an outer-orbital complex is formed.

$KL\ 3s^2 3p^6 3d^8 4(sp^3d^2)^{12}$

or $KL\ 3s^2 3p^6 3d_{xy}^2 3d_{yz}^2$
$3d_{xz}^2 3d_{z^2}^1 3d_{x^2-y^2}^1 4(sp^3d^2)^{12}$

ITEM 8

Draw the sp^3 orbital diagram for NH_3 and shade the orbital that contains two electrons from the nitrogen atom. Because these electrons are not involved in bond formation, they are called *nonbonding.*

ITEM 21

The sp^3d^2 hybrid orbitals are directed toward the apices of a regular octahedron. Draw the sp^3d^2 orbitals in the diagram at the right.

ITEM 34

Titanium has the electronic configuration $1s^22s^22p^6 3s^23p^64s^23d^2$ or, in abbreviated form, $KL\ 3s^23p^64s^23d^2$, in which K is a filled $n = 1$ shell and L is a filled $n = 2$ shell. The ion Ti^{4+} has the electronic configuration __________.

ITEM 47

Chromic ion, Cr^{3+}, reacts with water to yield the octahedral complex ion $Cr(H_2O)_6{}^{3+}$ [hexaaquochromium(III) cation]. As in $Ti(H_2O)_6{}^{3+}$, the bonding electrons are supplied by __________, and the bonding involves __________ orbitals of Cr^{3+} that result from hybridization of the following orbitals: __________.

ITEM 60

The $Mn(CN)_6{}^{3-}$ ion [hexacyanomanganese(III)] has two unpaired electrons. The ion is therefore an __________er-orbital complex.

ITEM 73

Many complexes of silver(I), such as $Ag(NH_3)_2{}^+$ [diamminesilver(I)], have a linear structure; thus, silver can be considered to use __________ hybrid orbitals for binding the ligands.

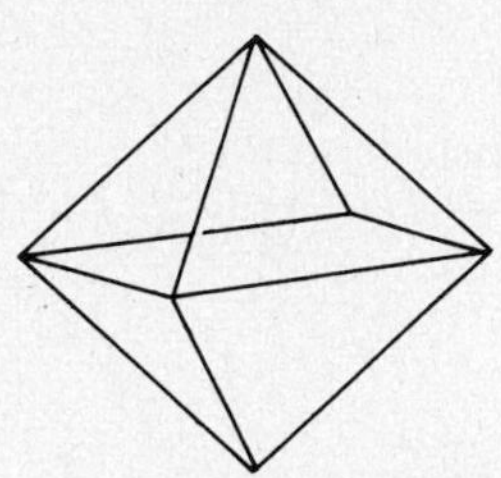

ANSWER 7 five

ANSWER 20 six
five
seven

ANSWER 33

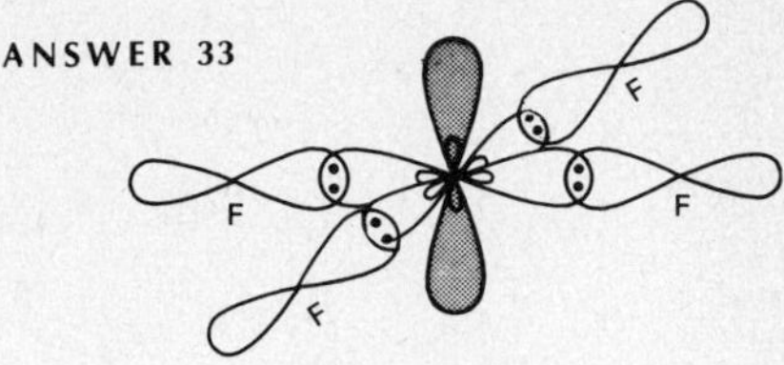

ANSWER 46 $KL\ 3s^2 3p^6 3d^3$
three

ANSWER 59 four (the four in the free Mn^{3+} ion)

ANSWER 72 eight
dsp^2

ITEM 9

If a proton, H^+, is donated to NH_3, the nonbonding sp^3 electrons become bonding by overlap of the sp^3 orbital with the __________ of the hydrogen ion. The shape of the NH_4^+ ion is __________.

ITEM 22

In bonds to four other atoms, phosphorus uses __________ hybrid orbitals and the bonds are directed toward the apices of a __________.

ITEM 35

In Ti^{4+} the $3d$, $4s$, and $4p$ orbitals are empty and can be used for bonding to atoms or ions (*ligands*) that have nonbonding electron pairs available, such as Cl^-, NH_3, and H_2O. If the $3d_{z^2}$, $3d_{x^2-y^2}$, $4s$, $4p_x$, $4p_y$, and $4p_z$ orbitals of Ti^{4+} are hybridized, six d^2sp^3 orbitals are formed. They are directed toward the apices of __________.

ITEM 48

The configuration of the Cr^{3+} ion in $Cr(H_2O)_6^{3+}$, including the electrons from water, is $KL\ 3s^23p^63d^3\ (d^2sp^3)^{12}$. It is called an inner-orbital complex because __________.

ITEM 61

The cyanide ion has the Lewis diagram $\ominus\!:C{\equiv}N:$. Draw an orbital diagram for cyanide assuming that both carbon and nitrogen use *sp* hybrids for bonding.

ITEM 74

A silver atom has the electronic configuration $KLM\ 4s^24p^65s^14d^{10}$. Write the electronic configuration for Ag^+ in $Ag(NH_3)_2^+$ (including the bonding electrons of the ammonia molecules).

ANSWER 8

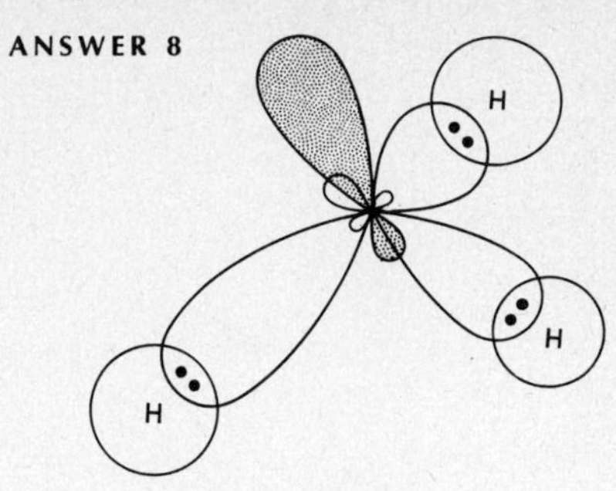

ANSWER 21

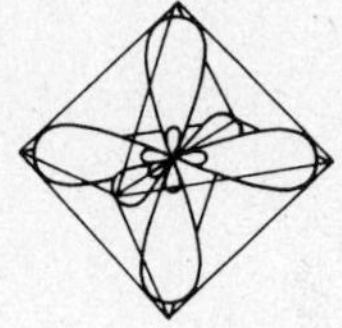

ANSWER 34 $KL\ 3s^2 3p^6$

ANSWER 47 water molecules
d^2sp^3
$3d_{z^2}$, $3d_{x^2-y^2}$, $4s$, $4p_x$, $4p_y$, $4p_z$

ANSWER 60 inn

ANSWER 73 sp

ITEM 10

Phosphorus has the electronic configuration $1s^2 2s^2 2p^6 3s^2 3p_x^1 3p_y^1 3p_z^1$. Like nitrogen, phosphorus can use __________ hybrid orbitals in bonding to four other atoms.

ITEM 23

In bonds to five other atoms, phosphorus uses __________ hybrid orbitals and the bonds are directed toward the apices of a __________.

ITEM 36

The designation d^2sp^3 indicates that the d orbitals included in hybridization have a lower principal quantum number, n, than do the s and p orbitals. They are *inner-orbital* hybrids. *Outer-orbital* hybrids of the same geometry are known as __________ orbitals.

ITEM 49

The electronic configuration of Cr^{2+} is __________. Since the number of d electrons is __________, each in a separate $3d$ orbital, the d_{z^2} and $d_{x^2-y^2}$ orbitals are no longer both available for incorporation into a d^2sp^3 hybrid that can overlap with ligand orbitals unless there is a change in the configuration.

ITEM 62

When cyanide ion is a ligand in complex ions such as $Mn(CN)_6^{3-}$, the formation of bonds involves overlap of the __________ orbitals of carbon with the __________ orbitals of Mn^{3+}.

ITEM 75

According to the localized molecular orbital view of bonding, the hybridization of bonding orbitals of the central ion of a tetrahedral complex is __________. Or when phosphorus is bonded to four other atoms in a tetrahedral molecule, the hybridization is __________.

ANSWER 9 1s orbital
tetrahedral

ANSWER 22 sp^3
tetrahedron

ANSWER 35 an octahedron

ANSWER 48 the d orbitals involved in hybridization have a lower principal quantum number than the s and p orbitals

ANSWER 61

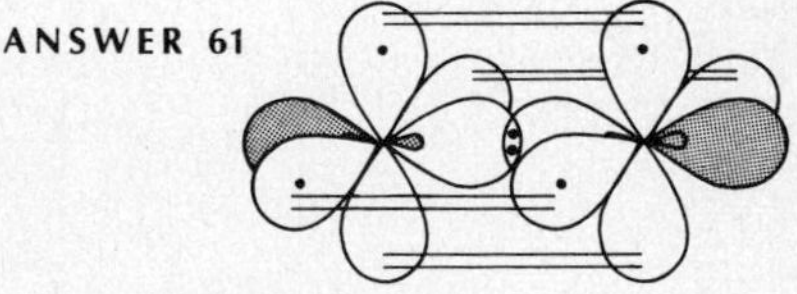

ANSWER 74 $KLM\ 4s^2 4p^6 4d^{10} 5(sp)^4$

ITEM 11

The ion PCl_4^+ [tetrachlorophosphonium or tetrachlorophosphorus(V) cation] is tetrahedral. Therefore, according to the localized molecular orbital theory, each bond is the result of overlap of a chlorine orbital with a(n) ________ orbital of phosphorus.

ITEM 24

When phosphorus is bonded to six other atoms, the ________ bonding orbitals are directed toward the apices of ________.

ITEM 37

In outer-orbital hybrids, the principal quantum number of the *d* orbitals is ________ *n* for the *s* and *p* orbitals. An example is BrF_5 (Item 32) in which the $4s$, $4p_x$, ________, ________, ________, and ________ orbitals of bromine are hybridized.

ITEM 50

If $Cr(H_2O)_6^{2+}$ is an inner-orbital complex ion, one electron originally in a $3d_{z^2}$ or $3d_{x^2-y^2}$ orbital will have to be paired with another electron in a ________, ________, or ________ orbital. These *d* electrons are nonbonding.

ITEM 63

The electronic configuration of Fe^{3+} is $KL\ 3s^2 3p^6 3d^5$. The octahedral complex ion $Fe(CN)_6^{3-}$ [hexacyanoferrate(III)] has one unpaired electron. The configuration of the Fe^{3+} in the complex (including the ligand bonding electrons) is ________.

ITEM 76

When phosphorus is bonded to five other atoms, the hybridization is known as ________. Draw a diagram showing the geometry of these orbitals.

ANSWER 10 sp^3

ANSWER 23 sp^3d (or $sp^3d_{z^2}$)
trigonal bipyramid

ANSWER 36 sp^3d^2

ANSWER 49

$KL\ 3s^23p^63d^4$
or $KL\ 3s^23p^63d_{xy}{}^13d_{yz}{}^13d_{xz}{}^13d_{z^2}{}^1$
four

ANSWER 62 sp
d^2sp^3

ANSWER 75 sp^3
sp^3

ITEM 12

Unlike nitrogen, phosphorus can be bonded to five other atoms, as in PCl_5, in which the P—Cl bonds are directed toward the apices of a triangle-base (trigonal) bipyramid. To account for five bonds, we assume participation of the ________ orbitals.

ITEM 25

Draw an orbital diagram for PCl_6^-. For the chlorine atoms, you need show only the bonding orbitals.

ITEM 38

The ion Ti^{4+} reacts with fluoride ions to form the TiF_6^{2-} ion [hexafluorotitanium(IV)]. The fluoride ions are located at the apices of an octahedron with the Ti^{4+} ion at the center. Therefore, each bond can be described as the overlap of a *p* orbital of a fluoride ion with a ________ orbital of Ti^{4+}.

ITEM 51

The configuration of an inner-orbital complex of Cr^{2+}, including the ligand bonding electrons, is ________. There are ________ unpaired electrons.

ITEM 64

Cobalt has the electronic configuration $KL\ 3s^2 3p^6 4s^2 3d^7$. The ion Co^{3+} has ________ paired 3*d* electrons and ________ unpaired electrons. An outer-orbital octahedral complex of Co^{3+} has ________ unpaired electrons and ________ nonbonding 3*d* electrons.

ITEM 77

When phosphorus is bonded to six other atoms, the hybridization is known as ________, and the shape of the molecule is ________.

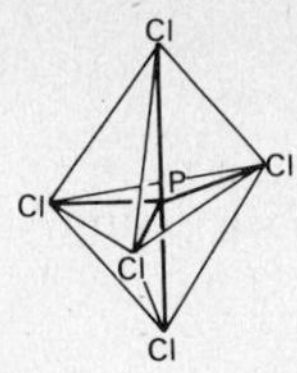

Ti^{4+}: $KL\ 3s^2 3p^6$

ANSWER 11 sp^3

ANSWER 24 sp^3d^2 or $(sp^3 d_{z^2}{}^1 d_{x^2-y^2}{}^1)$
an octahedron

ANSWER 37 the same as
$4p_y$
$4p_z$
$4d_{z^2}$
$4d_{x^2-y^2}$

ANSWER 50 $3d_{xy}$
$3d_{yz}$
$3d_{xz}$

ANSWER 63

$KL\ 3s^2 3p^6 3d^5 (d^2sp^3)^{12}$ as
$KL\ 3s^2 3p^6 3d_{xy}{}^2 3d_{yz}{}^2 3d_{xz}{}^1 (d^2sp^3)^{12}$
(The outer-orbital configuration would have five unpaired electrons, one in each $3d$ orbital.)

ANSWER 76 sp^3d (or $sp^3d_{z^2}$)

ITEM 13

Phosphorus has $3d$ orbitals that are relatively close in energy to the $3s$ and $3p$ orbitals containing the valence electrons. Calculations show that hybridization of the $3s$, $3p_x$, $3p_y$, $3p_z$, and $3d_{z^2}$ orbitals produces new orbitals, __________ in number, directed toward the apices of a trigonal bipyramid.

ITEM 26

Sulfur has the electronic configuration $1s^22s^22p^63s^2 3p_x^23p_y^13p_z^1$. The bond angle in SCl_2 indicates that the sulfur is using primarily sp^3 hybrid orbitals. Draw an orbital diagram for Cl—S—Cl and shade the orbitals filled with nonbonding electrons.

ITEM 39

This overlap results in two sigma orbitals, one bonding and the other __________.

ITEM 52

If $Cr(H_2O)_6^{2+}$ is an outer-orbital complex, the following orbitals will be incorporated into the sp^3d^2 hybrids: __________.

ITEM 65

A substance having unpaired electrons is *paramagnetic;* it is attracted into a magnetic field. A *diamagnetic* substance is repelled by a magnetic field; a diamagnetic complex ion has __________ unpaired electrons.

ITEM 78

When a transition metal is bonded octahedrally to six ligand ions or molecules, the hybridization is __________. Draw an orbital diagram for the octahedral complex of a transition metal ion with NH_3 (excluding the d orbitals).

ANSWER 12 d

ANSWER 25

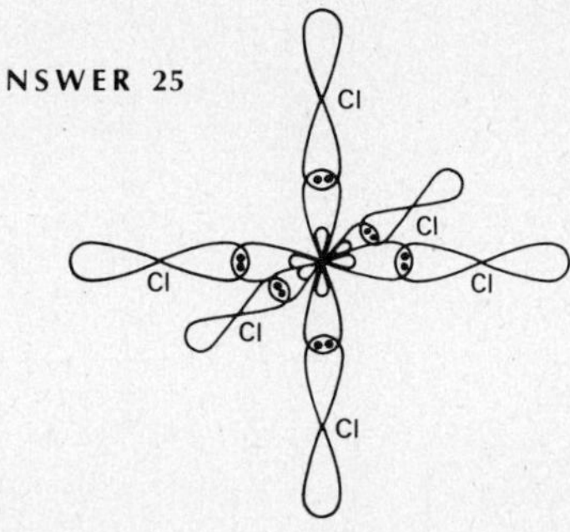

ANSWER 38 d^2sp^3

ANSWER 51

$KL\ 3s^2 3p^6 3d^4 (d^2sp^3)^{12}$

or $KL\ 3s^2 3p^6 3d_{xy}^2 3d_{yz}^1 3d_{xz}^1 (d^2sp^3)^{12}$

two

ANSWER 64

two ($KL\ 3s^2 3p^6 3d_{xy}^2 3d_{yz}^1 3d_{xz}^1 3d_{z^2}^1 3d_{x^2-y^2}^1$)

four

four

six

ANSWER 77 sp^3d^2

octahedral

ANSWER 13 five

(*Item 14 is on page 560.*)

ANSWER 26

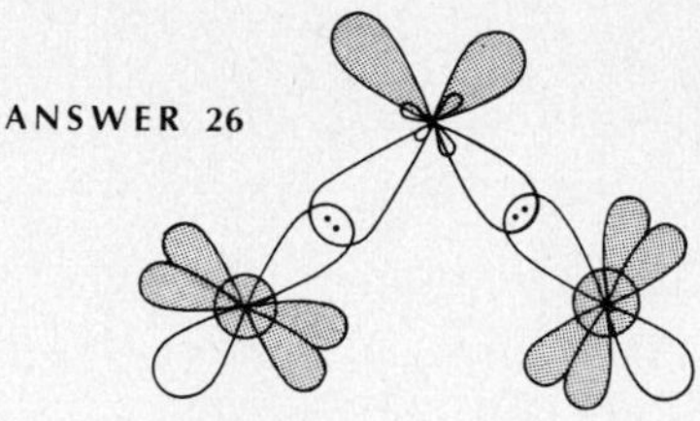

(*Item 27 is on page 560.*)

ANSWER 39 antibonding

(*Item 40 is on page 560.*)

ANSWER 52 $4s$, $4p_x$, $4p_y$, $4p_z$, $4d_{z^2}$, $4d_{x^2-y^2}$

(*Item 53 is on page 560.*)

ANSWER 65 no

(*Item 66 is on page 560.*)

ANSWER 78 d^2sp^3 or sp^3d^2

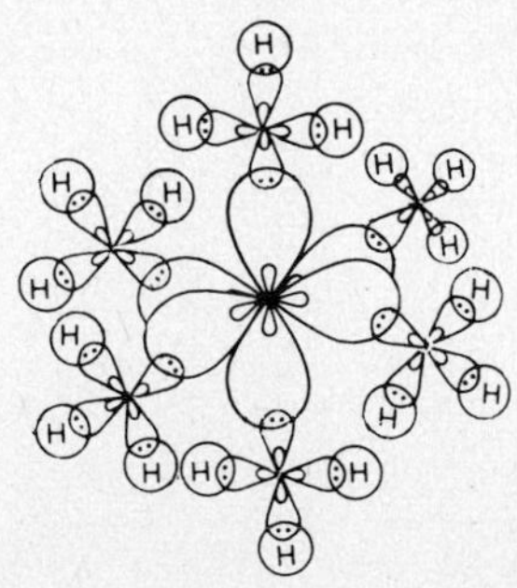

(*Item 79 is on page 560.*)

REVIEW 12 AN INTRODUCTION TO ORGANIC COMPOUNDS

Organic chemistry is the chemistry of carbon compounds. To introduce you to the marvelous number and variety of them, this review will work through the structure and names of the simplest, most common organic compounds. You need not learn all the names, but you should be able to write the structure of most of them, once you are given the name.

The names presented in this review are generally those assigned in accordance with the rules agreed upon by the International Union of Pure and Applied Chemistry (IUPAC). They are not as interesting as the common names, but they have the advantage of being logical and systematic. To interpret systematic names, you really only need to know how to count (preferably in Greek as well as in English): The name will indicate the structure. The common name is usually no help in deducing the structure. For example, the name tricyclo [3.3.2.$0^{2,8}$] deca-3,6,9-triene, once you have deciphered it, clearly says the molecule is

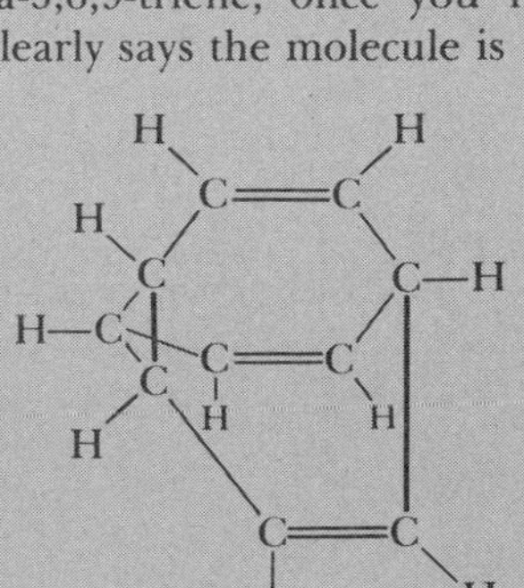

The name by which it is actually known to organic chemists, bullvalene, provides food for thought, but little clue to the structure of the molecule.

12-1 HYDROCARBONS

This section is devoted to hydrocarbons of all kinds, but mainly small ones, and is a review of parts of Sections 12–2, 12–4, and 12–5 of *Chemical Principles.* Before you try this section, you should know the material of Reviews 6 and 10; in particular, you should be able to draw Lewis diagrams. When an item asks for a Lewis diagram, you should include all the valence electrons, either as lines if they are bonded or as dots if nonbonded. If an item asks for a diagram or a structural formula, the nonbonded electrons or single bonds need not be shown if you understand that they are present. Double bonds and triple bonds always should be shown.

To complete items involving double bonds, you should know that if two groups of atoms (such as the CHBrCl groups in the following example) are connected by a single bond, the two groups can rotate relative to one another. In two-dimensional representation, the structural formulas below are equally valid representations of the *same molecule*.

```
     H   H                 H   Cl
     |   |                 |   |
 Cl—C—C—Cl           Cl—C—C—Br
     |   |                 |   |
     Br  Br                Br  H

     H   Br
     |   |
 Cl—C—C—H
     |   |
     Br  Cl
```

However, if the groups (such as CHCl in the following example) are connected by a double bond, no rotation around that bond can occur. Therefore, the formulas that follow represent two distinctly different compounds.

```
H       H     H       Cl
 \     /       \     /
  C=C           C=C
 /     \       /     \
Cl      Cl    Cl      H
```

The formula on the left shows the two hydrogens in what is called a cis conformation: they are on the same "side" of the double bond (as distinct from being on the same carbon). In the formula on the right, the two hydrogen atoms are trans, that is, on different sides of the double bond.

Remember that your answer need not be exactly the same as the one given on the answer page, but you should satisfy yourself that your answer is equivalent. If you can figuratively turn your diagram around or rotate a single bond to obtain our diagram, the answers are equivalent. For example, the formulas below represent the same molecule.

```
    H                            H
    |                            |
H—C—H      H     H         H—C—H
     \    /       \        /
      C=C          C=C
     /    \       /    \
    H      H     H      H

H       H
 \     /
  C=C
 /     \
H     H—C—H
         |
         H
```

ITEM 1

Hydrocarbons are compounds containing only hydrogen and __________. The simplest has the formula CH and the name *methyne.* Draw a Lewis diagram for the electrically neutral methyne molecule.

ITEM 15

Usually, the suffix "ene" is indicative of a compound with a double bond, an *olefin.* In the systematic nomenclature this is always true, as in the name *ethene.* The names *ethylene* and *ethene* refer to the same molecule, for which the Lewis diagram is __________.

ITEM 29

Give the systematic names and structural formulas of the six saturated and unsaturated three-carbon hydrocarbons.

ITEM 43

The common names, systematic names, and structural formulas of the two isomeric butanes are __________.

ITEM 57

If six hydrogen atoms are removed from butane, from carbons 1, 2, 3, and 4, a *triene* could be produced. The Lewis diagram of *butatriene* is __________.

ITEM 71

Double bonds that have the structure C═C—C═C are called *conjugated.* An example is 1,3-butadiene. Draw Lewis diagrams for the *cis-* and *trans-*1,3-pentadienes, which also have conjugated double bonds.

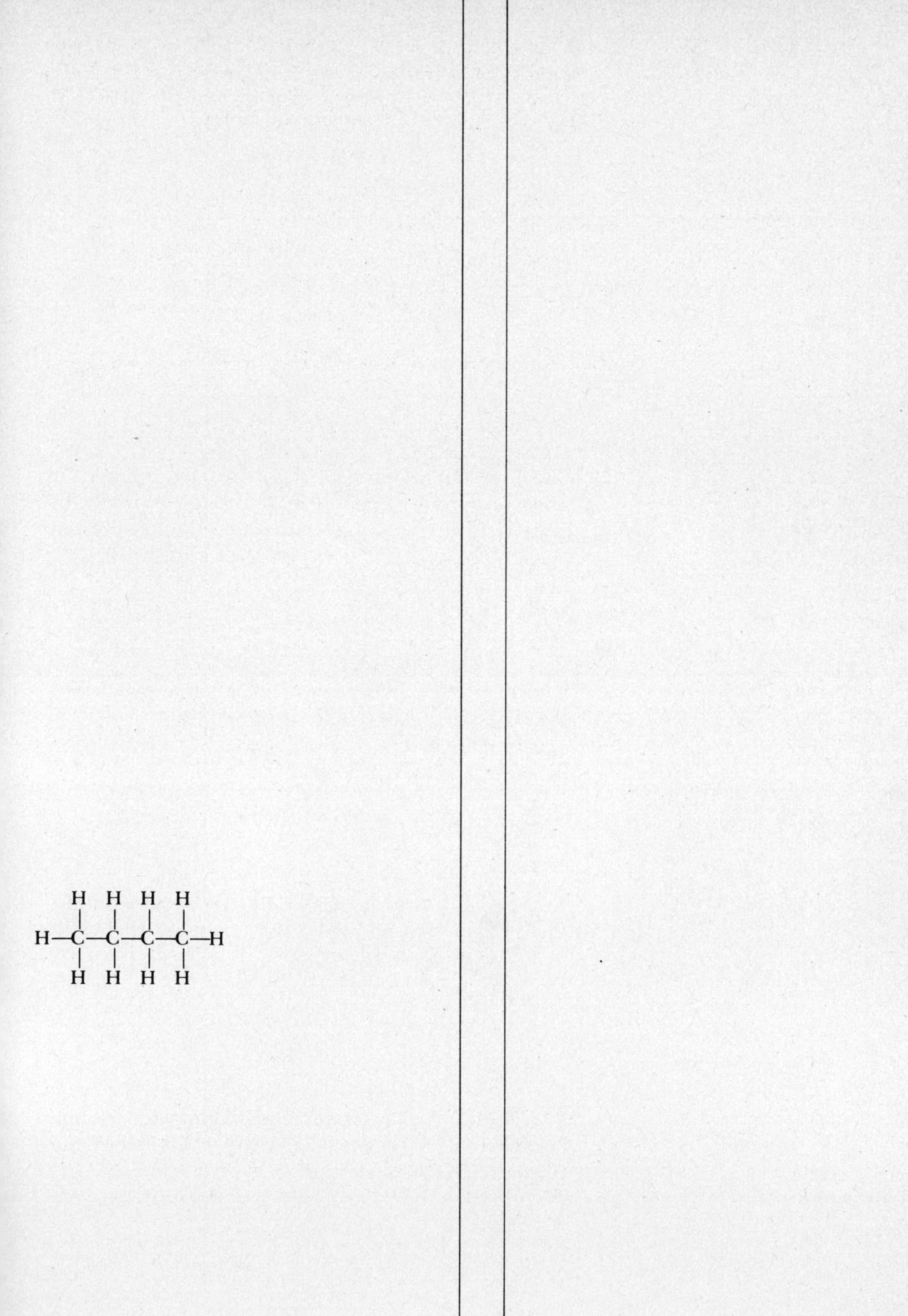
H H H H
H—C—C—C—C—H
H H H H

ITEM 2

You will never find CH in a bottle—it is far too reactive. But it does have a name, which is ______.

ITEM 16

Draw an orbital diagram for ethylene.

ITEM 30

Consider the removal of one hydrogen atom from a propane molecule. The six hydrogen atoms at the ends of the chain are all equivalent. Removal of any one of them would leave a *propyl* radical with the Lewis diagram ______.

ITEM 44

There are two isomeric four-carbon cycloalkanes, too. The Lewis diagram for cyclobutane is ______, and the Lewis diagram for methylcyclopropane is ______.

ITEM 58

Isomers of the butadienes are the two possible *butynes.* Write the names and Lewis diagrams of the two butyne molecules.

ITEM 72

Two other hydrocarbons with conjugated double bonds are 1,3-cyclohexadiene and 1,3,5-hexatriene. Draw their Lewis diagrams.

```
   H  H  H
   |  |  |
H—C—C—C—H
   |  |  |
   H  H  H
```

ANSWER 1 carbon

$\dot{\underset{\cdot}{C}}$—H

ANSWER 15

```
H       H
 \     /
  C=C
 /     \
H       H
```

ANSWER 29

$CH_3CH_2CH_3$	$H_2C{=}CHCH_3$
propane	propene
H_2C——CH_2 \/ CH_2	$HC{=}CH$ \/ CH_2
cyclopropane	cyclopropene
$H_2C{=}C{=}CH_2$	$HC{\equiv}CCH_3$
propadiene	propyne

ANSWER 43

n-butane, butane

```
   H  H  H  H
   |  |  |  |
H—C—C—C—C—H
   |  |  |  |
   H  H  H  H
```

isobutane
2-methylpropane

```
        H
        |
     H—C—H
        |
   H    |    H
   |    |    |
H—C—C—C—H
   |    |    |
   H    H    H
```

ANSWER 57

```
H              H
 \            /
  C=C=C=C
 /            \
H              H
```

ANSWER 71

```
H        H
 \      /
  C=C         H
 /      \    /
H        C=C
        /    \
       H      C—H
             /|
            H H
```

trans

```
H        H       H
 \      /        |
  C=C        H—C—H
 /      \        /
H        C=C
        /    \
       H      H
```

cis

ITEM 3

The hydrocarbon CH_2 is also extremely reactive. It is known as *methylene*, and the neutral molecule has the Lewis diagram ______.

ITEM 17

A double bond is one kind of *unsaturation;* a triple bond is another. Triple bonds are denoted by the suffix "yne" in the name of a compound. Ethyne has the Lewis diagram ______.

ITEM 31

Removal of a hydrogen atom from the central carbon of the propane chain yields another radical, different from the propyl radical. It has the name *isopropyl* and the Lewis diagram ______.

ITEM 45

Four carbons in a molecule also allow the existence of another cyclic alkane, *bicyclobutane*, in which there are two rings. The Lewis diagram is shown at the right. Is it an isomer of cyclobutane?

ITEM 59

The Lewis diagram for 3-*buten*-1-*yne* is ______ and for *butadiyne* is ______.

ITEM 73

An extreme example—so extreme that it belongs to another class of compounds—is 1,3,5-cyclohexatriene, better known as *benzene*. A Lewis diagram for benzene is ______.

```
   H  H  H
   |  |  |
H—C—C—C—H
   |  |  |
   H  H  H
```

```
   H      H
   |     /
H—C—C
   | / |
   C  C  H
  /    |
 H     H
```

ANSWER 2 methyne

ANSWER 16

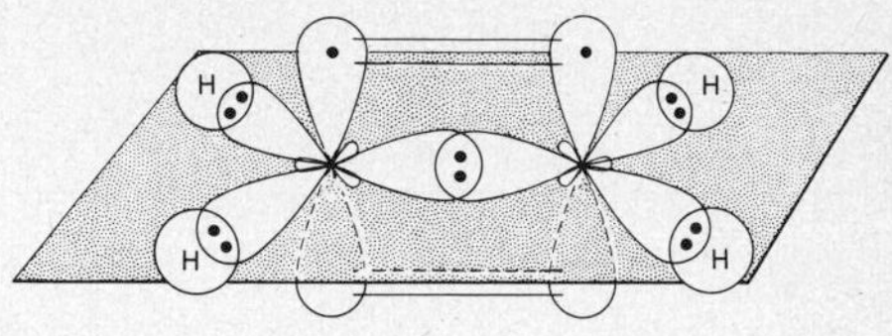

ANSWER 30

```
   H  H  H         H  H  H
   |  |  |         |  |  |
H—C—C—C    or   C—C—C—H   etc.
   |  |  |         |  |  |
   H  H  H         H  H  H
```

(They are equivalent because you need only to turn one to have the other.)

ANSWER 44

```
                        H
                        |
   H  H              H—C—H  H
   |  |                 |      |
H—C—C—H           H—C———C—H
   |  |                   \  /
H—C—C—H                   C
   |  |                    /  \
   H  H                   H    H
```

(The methyl group can be attached to any of the three equivalent carbons.)

ANSWER 58

```
            H  H              H          H
            |  |              |          |
H—C≡C—C—C—H             H—C—C≡C—C—H
            |  |              |          |
            H  H              H          H
```

1-butyne (or ethylacetylene) 2-butyne (or dimethylacetylene)

ANSWER 72

```
      H
      |
   H  C
   | / \\
H—C     C—H
   |     |
H—C     C—H
   | \ //
   H  C
      |
      H
```

```
                H          H
                  \       /
        H          C=C
          \       /     \
H          C=C          H
  \       /     \
   C=C          H
  /     \
H        H
      trans
```

```
            H      H
              \   /
H             C=C               H
  \          /     \           /
   C=C               C=C
  /     \            /     \
H        H       H         H
             cis
```

ITEM 4

Neutral CH_3 is called the *methyl radical;* a radical is an atom or molecule with one unpaired electron. The Lewis diagram for the methyl radical is ________.

ITEM 18

You may realize already that the common name for H—C≡C—H is *acetylene,* which is quite different from the systematic name, ________.

ITEM 32

Draw Lewis diagrams for the isopropyl carbonium ion and the isopropyl carbanion.

ITEM 46

Two different radicals can be formed by the removal of one hydrogen atom from a *n*-butane molecule. The Lewis diagrams are ________.

ITEM 60

And finally, there are two cyclic dienes. *Methylenecyclopropene* has the Lewis diagram ________, and 1,3-*cyclobutadiene* has the Lewis diagram ________.

ITEM 74

Neither of the two Lewis diagrams given in Answer 73 is a good representation of benzene because the electrons in a benzene molecule are distributed equally among the carbons of the ring. One way to show this in a diagram is at the right. Draw a similar diagram for 1,2-dimethylbenzene.

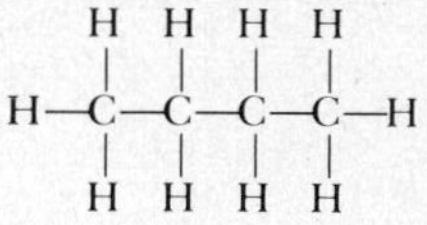

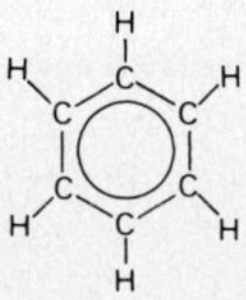

ANSWER 3

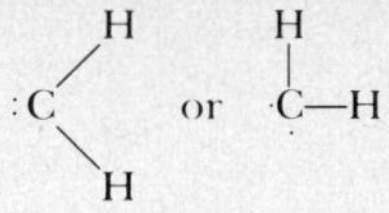

ANSWER 17 H—C≡C—H

ANSWER 31

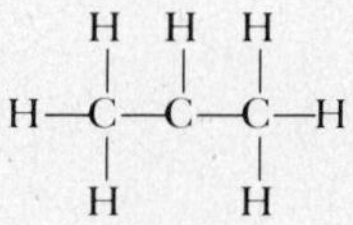

ANSWER 45 No, because bicyclobutane has two fewer hydrogen atoms than does cyclobutane.

ANSWER 59

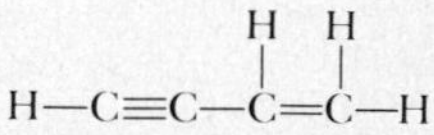

H—C≡C—C≡C—H

ANSWER 73

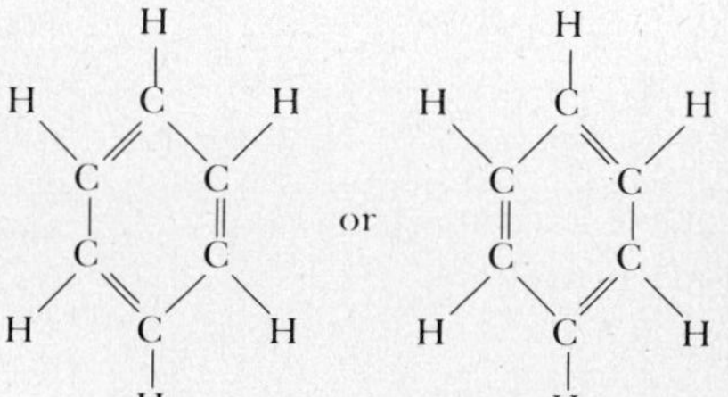

ITEM 5

The methyl radical is also very reactive. If it reacts with a hydrogen atom, a molecule having the Lewis diagram ________ will be formed. You would expect its name to begin with the letters ________.

ITEM 19

Draw an orbital diagram for acetylene.

ITEM 33

Removal of any hydrogen atom from the cyclopropane molecule yields one radical. It has the name ________ and the Lewis diagram ________.

ITEM 47

Similarly, two different radicals can result from the removal of a hydrogen atom from isobutane. The Lewis diagrams are ________.

ITEM 61

1,2-Cyclobutadiene is impossible because the C═C═C structure is always linear (180°) and cannot form a square or nearly square molecule such as cyclobutane, in which you would expect the C—C—C bond angle to be about ________ degrees.

ITEM 75

The compound 1,2-dimethylbenzene is often named *ortho*-dimethylbenzene, usually written *o*-dimethylbenzene. *meta*-Dimethylbenzene is the 1,3-dimethylbenzene isomer. Draw a diagram representing *m*-dimethylbenzene.

ANSWER 4

$$\begin{array}{l} H \\ | \\ C{-}H \\ | \\ H \end{array}$$

ANSWER 18 ethyne

ANSWER 32

$$\begin{array}{ccccccc} & & H & & H & & H \\ & & | & & | & & | \\ H & - & C & - & \underset{\oplus}{C} & - & C & - & H \\ & & | & & & & | \\ & & H & & & & H \end{array}$$

$$\begin{array}{ccccccc} & & H & & \ominus & & H \\ & & | & & \ddot{} & & | \\ H & - & C & - & C & - & C & - & H \\ & & | & & | & & | \\ & & H & & H & & H \end{array}$$

ANSWER 46

$$\begin{array}{ccccccccc} & & H & & H & & H & & H \\ & & | & & | & & | & & | \\ H & - & C & - & C & - & C & - & C\cdot \\ & & | & & | & & | & & | \\ & & H & & H & & H & & H \end{array}$$

(The electron can be on either end carbon, for *butyl* or *n-butyl*).

$$\begin{array}{ccccccccccc} & & H & & H & & H & & H \\ & & | & & | & & | & & | \\ H & - & C & - & \underset{\cdot}{C} & - & C & - & C & - & H \\ & & | & & & & | & & | \\ & & H & & & & H & & H \end{array}$$

(The electron can be on either central carbon for *sec-butyl,* said "secondary-butyl.")

ANSWER 60

ANSWER 74

HHH

$$\begin{array}{ccccccc} & & H & & H & & H \\ & & | & & | & & | \\ H & - & C & - & C & - & C & - & H \\ & & | & & | & & | \\ & & H & & | & & H \\ & & H & - & C & - & H \\ & & & & | \\ & & & & H \end{array}$$

ITEM 6

CH_4 has the name *methane*. The names and formulas, then, of the four possible neutral hydrocarbons containing one carbon are as follows: __________.

ITEM 20

A saturated hydrocarbon has an IUPAC name ending in __________, an olefin has a name ending in __________, and a compound with a triple bond has a name ending in __________. The suffix indicating a radical is __________.

ITEM 34

A general name (other than paraffin) for a saturated hydrocarbon is *alkane*. The radical derived from an alkane is named *alkyl*. By analogy with the names you have learned, you know that an *alkene* has a __________ and an *alkyne* has a __________.

ITEM 48

Consider the removal of two hydrogen atoms from butane (at the right) to form a double bond. If hydrogens are removed, one each from carbons 1 and 2 (or 3 and 4), an alkene named *1-butene* would result. (The carbons that are doubly bonded are given the lowest possible numbers.) The Lewis diagram is __________.

ITEM 62

Alkanes containing five carbons are known collectively as *pentanes*. Draw Lewis diagrams for the three noncyclic pentanes. (Remember that, regardless of how the molecule is drawn on paper, branching, by definition, does not occur on an end carbon.)

ITEM 76

The 1,4-substituted benzene compounds are known as *para*-disubstituted compounds. For example, *p*-diethylbenzene can be drawn as __________.

```
    H   H   H   H
    |   |   |   |
H—C—C—C—C—H
    |   |   |   |
    H   H   H   H
    1   2   3   4
```

ANSWER 5

```
    H
    |
H—C—H
    |
    H
```

meth

ANSWER 19

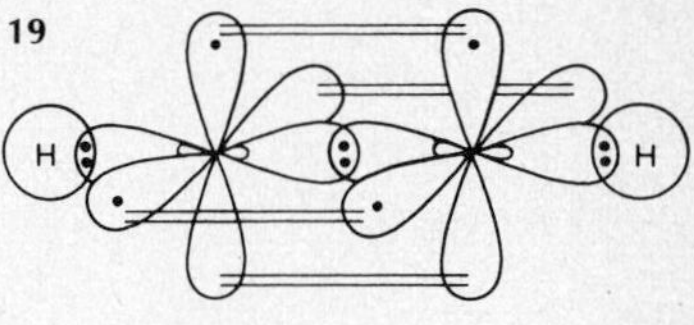

ANSWER 33 cyclopropyl

```
  H     H
    \  /
     C
    / \
H—C———C—H
  ·      |
         H
```

(The electron can be on any of the carbons.)

ANSWER 47

```
    H   H
    |   |   ·
H—C—C—C—H
    |   |   |
    H   |   H
        |
    H—C—H
        |
        H
```

(The hydrogen can be taken from any CH_3 group for *iso-butyl.*)

```
    H       H
    |   ·   |
H—C—C—C—H
    |   |   |
    H   |   H
        |
    H—C—H
        |
        H
```

(*t-butyl,* said "tertiary-butyl")

ANSWER 61 90

ANSWER 75

```
        H       H
        |       |
   H    C       C—H
    \ //  \    /|
     C      C   H
     |  ( ) |
     C      C
    /  \  //  \
   H    C      H
        |
     H—C—H
        |
        H
```

ITEM 7

If a hydrogen atom, H·, is lost from a methane molecule, the resulting radical has the Lewis diagram __________ and the name __________. If a hydrogen ion, H^+, is removed from methane, the resulting ion has the name *methide* and the Lewis diagram ________.

ITEM 21

The names and structural formulas (written in the form H_3CCH_3) of the hydrocarbon molecules containing two carbons (including the one important radical) are __________.

ITEM 35

The only *cycloalkane* mentioned so far has been __________, and the only *cycloalkene* mentioned has been __________.

ITEM 49

If two hydrogen atoms are removed, one each from carbons 2 and 3, 2-*butene* is produced, but two isomers are possible. In one, the hydrogen atoms are on the same side of the double bond and the methyl groups on the other. This is *cis*-2-*butene,* for which the Lewis diagram is __________.

ITEM 63

The systematic name for the compound at the right is __________.

ITEM 77

Let *R* stand for any alkyl radical and draw diagrams for *o*-, *m*-, and *p*-dialkylbenzenes. Label them as *ortho, meta,* or *para.*

```
    H        H
    |        |
H—C—H  H—C—H
     \      /
   H—C—C—H
      |   |
      H   H
1    2   3   4
```

```
            H
            |
         H—C—H
  H   H     |
  |   |     |
H—C—C—C—H
  |   |   |
  |   H   H
H—C—H
  |
  H
```

ANSWER 6 methyne, CH
methylene, CH_2
methyl, CH_3
methane, CH_4

ANSWER 20 ane
ene
yne
yl

ANSWER 34 double bond
triple bond

ANSWER 48

```
H         H
  \     /
   C═C
  /     \
H      H—C—H
          |
       H—C—H
          |
          H
```

ANSWER 62

```
   H  H  H  H  H
   |  |  |  |  |
H—C—C—C—C—C—H
   |  |  |  |  |
   H  H  H  H  H
```

(*n*-pentane)

```
   H  H  H  H
   |  |  |  |
H—C—C—C—C—H
   |  |  |  |
   H  |  H  H
      |
   H—C—H
      |
      H
```

(isopentane)

```
          H
          |
       H—C—H
          |
   H      |      H
   |      |      |
H—C——C——C—H
   |      |      |
   H      |      H
          |
       H—C—H
          |
          H
```

(neopentane)

ANSWER 76

```
          H     H  H
     H    |     |  |
       \  C  —C—C—H
        C    C  |  |
 H  H   |    |  H  H
 |  |   C    C
H—C—C—  C  \
 |  |   |    H
 H  H   H
```

ITEM 8

If a hydride ion, H:$^-$, is removed from methane, the resulting ion is called the *methyl carbonium ion.* It has the Lewis diagram ________.

ITEM 22

Ethyl carbonium ion has the Lewis diagram ________. Ethide, $CH_3CH_2^-$, is the corresponding carbanion, which is defined as ________.

ITEM 36

The letters indicating a four-carbon compound are "but." The Lewis diagram for the straight-chain *butane* molecule is ________. It is commonly known as *n-butane* (normal-butane).

ITEM 50

If two hydrogen atoms are removed from carbons 2 and 3 so the hydrogen atoms are on opposite sides of the double bond, the resulting molecule is known as *trans-2-butene.* Draw the Lewis diagram.

ITEM 64

The systematic name for the isomer shown at the right is 2,2-*dimethylpropane,* in which the 2 refers to the number of the carbon atom of the chain and "dimethyl" refers to the two methyl groups. The systematic name for

```
CH3—CH—CH2CH3
    |
    CH3
```

is ________.

ITEM 78

If a hydrogen atom is removed from a benzene molecule, the resulting radical is called *phenyl.* Draw the Lewis diagram for diphenylmethane.

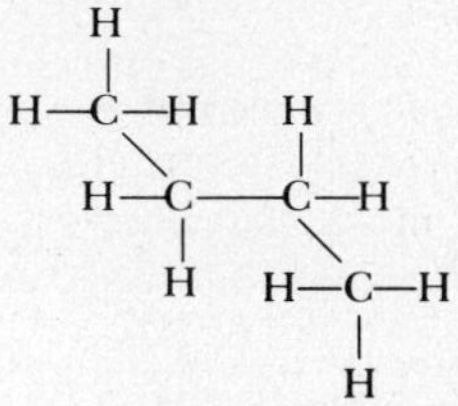

$$\begin{array}{c} H \\ | \\ H-C-H \quad H \\ \diagdown \quad\quad | \\ H-C——C-H \\ | \quad\quad \diagdown \\ H \quad H-C-H \\ | \\ H \end{array}$$

1 2 3 4

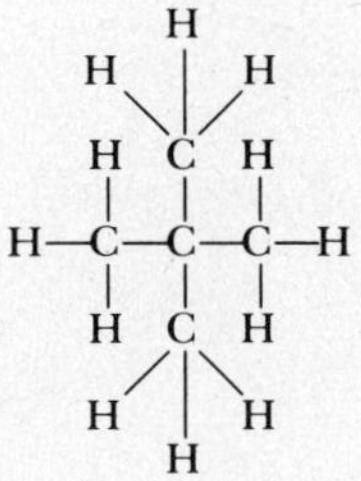

$$\begin{array}{c} H \\ H \diagdown | \diagup H \\ H \quad C \quad H \\ | \quad | \quad | \\ H-C-C-C-H \\ | \quad | \quad | \\ H \quad C \quad H \\ H \diagup | \diagdown H \\ H \end{array}$$

ANSWER 7

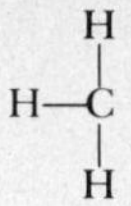

$$\begin{array}{c} H \\ | \\ H-C \\ | \\ H \end{array}$$

methyl

$$\begin{array}{c} H \\ | \\ H-C:^{-} \\ | \\ H \end{array}$$

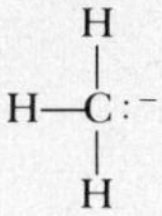

ANSWER 21

ethane, H_3CCH_3
ethene or ethylene, $H_2C{=}CH_2$
ethyne or acetylene, $H{-}C{\equiv}C{-}H$
ethyl, H_3CCH_2

ANSWER 35 cyclopropane
cyclopropene

ANSWER 49

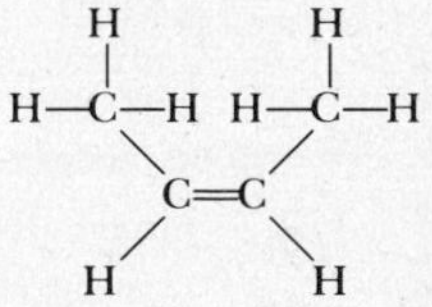

$$\begin{array}{c} H \quad\quad\quad H \\ | \quad\quad\quad | \\ H-C-H \quad H-C-H \\ \diagdown \quad \diagup \\ C{=}C \\ \diagup \quad \diagdown \\ H \quad\quad\quad H \end{array}$$

ANSWER 63 pentane

ANSWER 77

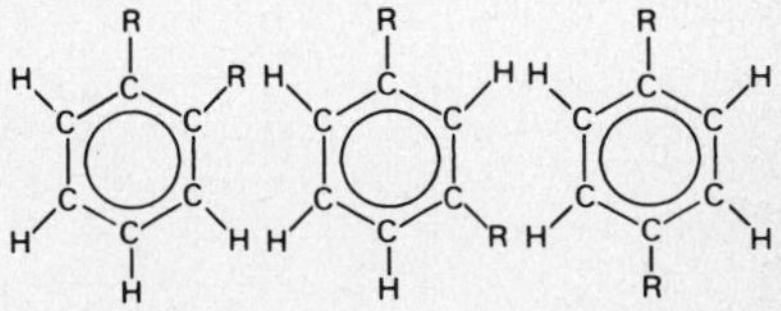

ortho *meta* *para*

ITEM 9

The name *carbonium ion* indicates that a carbon atom of the ion carries a positive charge. The name *carbanion* indicates that a carbon atom has a __________, as in the methide ion.

ITEM 23

The three-carbon compounds have names that begin with the letters "prop." The Lewis diagram for propane, C_3H_8, is __________.

ITEM 37

Another four-carbon alkane is possible, one in which the carbons are not linked in a single chain but in which the carbons are branched. Draw the Lewis diagram for *isobutane,* the branched-chain compound.

ITEM 51

Draw the Lewis diagram for cyclobutene.

ITEM 65

Draw the Lewis diagrams for the two 2-*pentenes* and label them cis and trans.

ITEM 79

Draw a diagram for diphenylacetylene.

ANSWER 8

$$\begin{array}{c} \quad\ \ \mathrm{H} \\ \quad\ \ | \\ \mathrm{H}-\mathrm{C}^{+} \\ \quad\ \ | \\ \quad\ \ \mathrm{H} \end{array}$$

ANSWER 22

$$\begin{array}{ccccc} & & \mathrm{H} & & \mathrm{H} \\ & & | & & | \\ \mathrm{H} & - & \mathrm{C} & - & \mathrm{C}^{+} \\ & & | & & | \\ & & \mathrm{H} & & \mathrm{H} \end{array}$$

an ion having a negative charge on a carbon atom

ANSWER 36

$$\begin{array}{ccccccccccc} & & \mathrm{H} & & \mathrm{H} & & \mathrm{H} & & \mathrm{H} & & \\ & & | & & | & & | & & | & & \\ \mathrm{H} & - & \mathrm{C} & - & \mathrm{C} & - & \mathrm{C} & - & \mathrm{C} & - & \mathrm{H} \\ & & | & & | & & | & & | & & \\ & & \mathrm{H} & & \mathrm{H} & & \mathrm{H} & & \mathrm{H} & & \end{array}$$

ANSWER 50

$$\begin{array}{c} \mathrm{H_3C} \qquad \mathrm{H} \\ \diagdown \quad \diagup \\ \mathrm{C}=\mathrm{C} \\ \diagup \quad \diagdown \\ \mathrm{H} \qquad \mathrm{CH_3} \end{array}$$

ANSWER 64 2-methylbutane

ANSWER 78

$$\mathrm{C_6H_5}-\mathrm{CH_2}-\mathrm{C_6H_5}$$

ITEM 10

Methyl carbonium ion has the Lewis diagram __________. Methide, the methyl carbanion, has the Lewis diagram __________.

ITEM 24

Propane is a "straight-chain" paraffin, but three carbons also can be arranged in a ring. This cyclic molecule is known as *cyclopropane.* It has the Lewis diagram __________.

ITEM 38

The two butanes are *structural isomers* of one another. Structural isomers are molecules having the same number and kind of atoms but different arrangements of atoms. Each of the isomeric butanes has __________ carbon atoms and __________ hydrogen atoms, but they are arranged in different ways.

ITEM 52

The olefin *methylenecyclopropane* is named as a cyclopropane in which two hydrogen atoms on the same carbon are replaced by CH_2, which is called the __________ group.

ITEM 66

The six-carbon chains are *hexanes,* the seven carbon chains are *heptanes.* Draw the structural formula for the alkane 3,3,5-trimethylheptane.

ITEM 80

Benzene and certain other planar, cyclic, conjugated compounds have characteristic chemical properties that distinguish them from other compounds. These benzene or benzenoid compounds are called *aromatic.* 1,3,5-Hexatriene (is, is not) aromatic.

```
  H     H
   \   /
    C
    ‖
    C
   / \
  C———C
 /|   |\
H |   | H
  H   H
```

ANSWER 9 negative charge

ANSWER 23

```
   H  H  H
   |  |  |
H—C—C—C—H
   |  |  |
   H  H  H
```

ANSWER 37

```
   H  H  H
   |  |  |
H—C—C—C—H
   |  |  |
   H  C  H
     /|\
    H | H
      H
```

ANSWER 51

```
H      H
 \     |
  C—C—H
  ‖  |
  C—C—H
 /     |
H      H
```

ANSWER 65

```
   H          H  H
   |          |  |
H—C—H      C—C—H
     \     / |  |
      C═C   H  H
     /    \
    H      H
```

cis-2-pentene

```
   H
   |
H—C—H      H
     \     /
      C═C   H  H
     /    \  |  |
    H      C—C—H
            |  |
            H  H
```

trans-2-pentene

ANSWER 79

```
  H       H           H       H
   \     /             \     /
    C—C                 C—C
   /    \               /    \
H—C  ( )  C—C ≡ C—C  ( )  C—H
   \    /               \    /
    C—C                 C—C
   /     \             /     \
  H       H           H       H
```

ITEM 11

The suffix "ane" indicates a *paraffin,* or *saturated* hydrocarbon: a hydrocarbon that has no double or triple bonds or other electrons available for bonding. Every carbon atom in a saturated hydrocarbon is bonded to __________ other atoms.

ITEM 25

If one double bond is present in a three-carbon, straight-chain hydrocarbon, the compound is named *propene* or *propylene.* It has the Lewis diagram ________.

ITEM 39

The butane isomers are clearly different compounds, differing in boiling point, melting point, and all other properties. The Lewis diagram of *n*-butane is __________ and of its isomer is __________.

ITEM 53

Draw the Lewis diagrams for the two isomeric methylcyclopropenes. (Two carbon atoms in cyclopropene are equivalent, but different from the third.)

ITEM 67

Write the structural formulas of the compounds (a) 3-ethyl-2-hexene and (b) 3-methyl-1,5-hexadiene.

ITEM 81

Compounds such as the ones in Items 1–72 are called *aliphatic,* as distinguished from aromatic. Aliphatic compounds do not contain __________.

	n	*iso*
bp	0°C	−12°C
mp	−138°C	−159°C

ANSWER 10

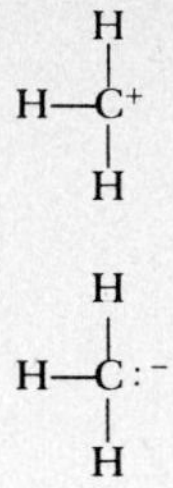

```
    H
    |
 H—C+
    |
    H

    H
    |
 H—C:⁻
    |
    H
```

ANSWER 24

```
    H         H
    |         |
 H—C—————C—H
      \   /
        C
      /   \
    H       H
```

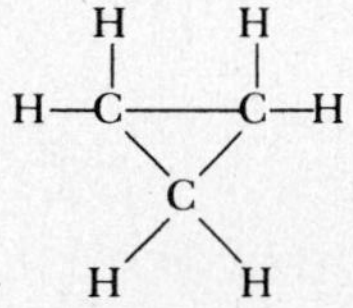

ANSWER 38 four
ten

ANSWER 52 methylene

ANSWER 66

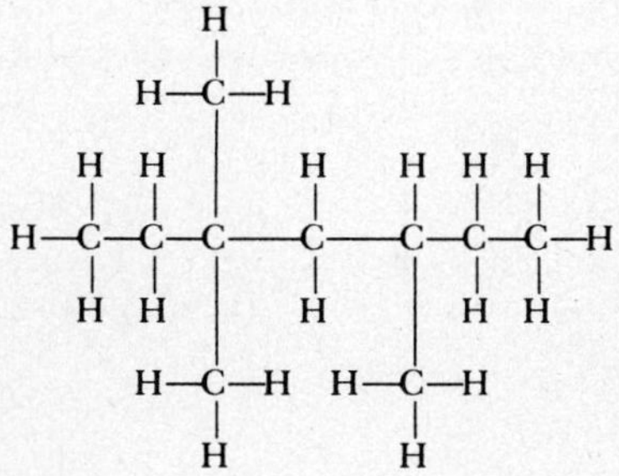

```
          H
          |
       H—C—H
          |
  H  H    |     H     H  H  H
  |  |    |     |     |  |  |
H—C—C—C———C———C—C—C—H
  |  |    |     |     |  |  |
  H  H    |     H     |  H  H
          |           |
       H—C—H     H—C—H
          |           |
          H           H
```

ANSWER 80 is not
(It is not cyclic.)

ITEM 12

In methane the H—C—H bond angle is 109°28′, and the carbon atom uses __________ orbitals for bonding. Two-dimensional drawings cannot show the actual bond angles in saturated hydrocarbons, but we know that nearly all the bond angles (C—C—C, C—C—H, and H—C—H) will be about __________.

ITEM 26

The cyclic olefin, *cyclopropene,* has the Lewis diagram __________.

ITEM 40

The branched-chain butane has the common name *isobutane,* but the systematic name is 2-*methylpropane.* Thus it is considered a propane in which the central carbon is bonded to a __________.

ITEM 54

Consider now the removal of four hydrogen atoms from butane to produce two double bonds. Can this be done to isobutane without changing the carbon atom arrangement?

ITEM 68

Eight carbon chains are *octanes,* nine carbon chains are *nonanes,* and ten carbon chains are *decanes.* Name the hexamethyl compound at the right. Number the chain to give the methyl groups the lowest possible numbers.

ITEM 82

Which of the following compounds are aromatic: cyclohexene, butylbenzene, diphenylmethane, naphthalene (shown at the right).

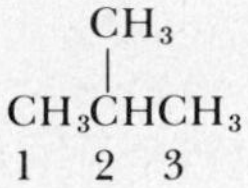

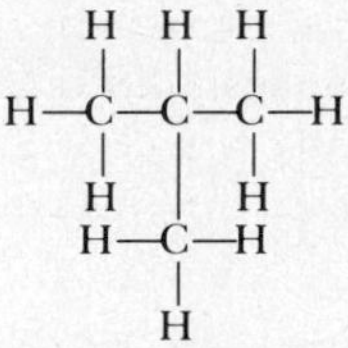

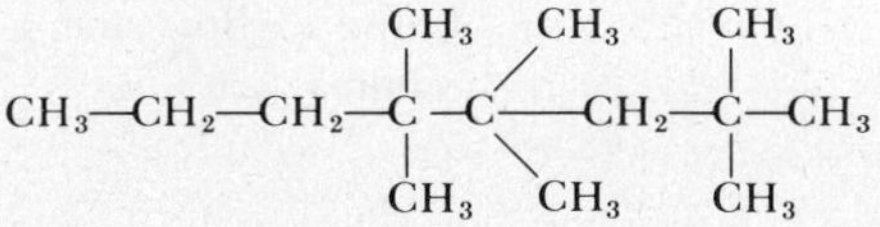

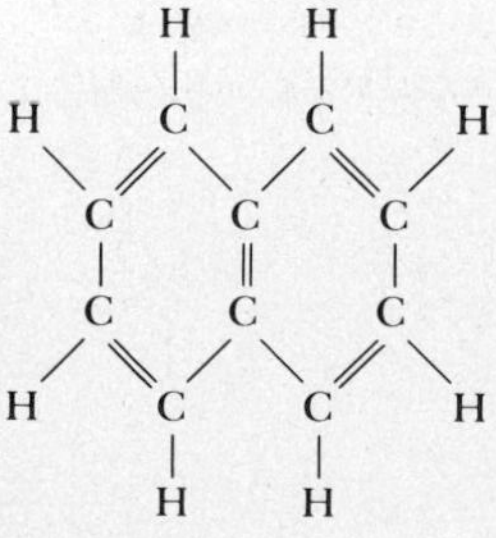

$$\begin{array}{l} \quad CH_3 \\ \quad | \\ CH_3CHCH_3 \\ 1 \quad 2 \quad 3 \end{array}$$

$$\begin{array}{ccccc} & H & H & H & \\ & | & | & | & \\ H- & C- & C- & C & -H \\ & | & | & | & \\ & H & | & H & \\ & H- & C & -H & \\ & & | & & \\ & & H & & \end{array}$$

$$CH_3-CH_2-CH_2-C(CH_3)_2-C(CH_3)_2-CH_2-C(CH_3)_2-CH_3$$

ANSWER 11 four

ANSWER 25

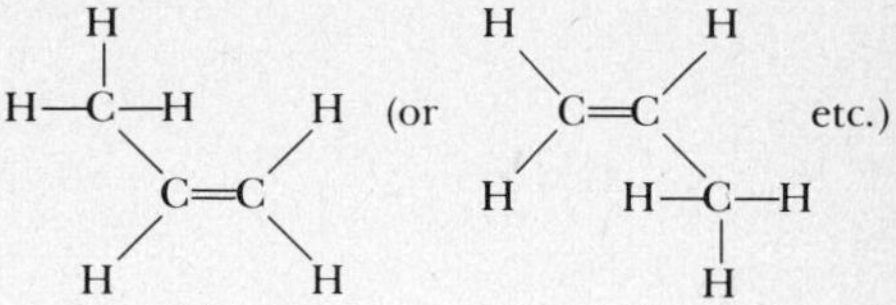

$$CH_3-CH=CH_2$$

(or $H_2C=CH-CH_3$ etc.)

ANSWER 39

$$CH_3-CH_2-CH_2-CH_3 \qquad CH_3-CH(CH_3)-CH_3$$

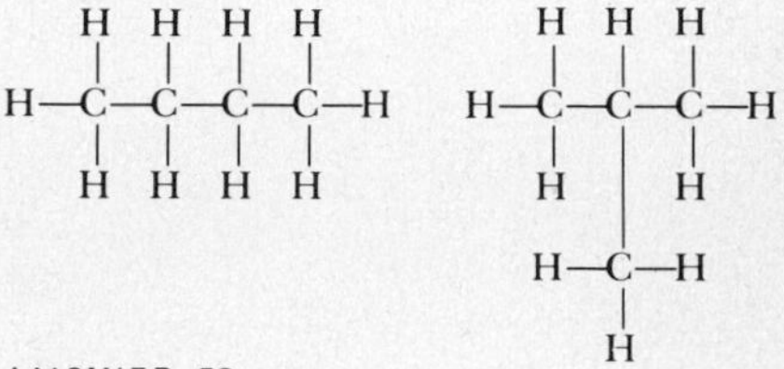

ANSWER 53

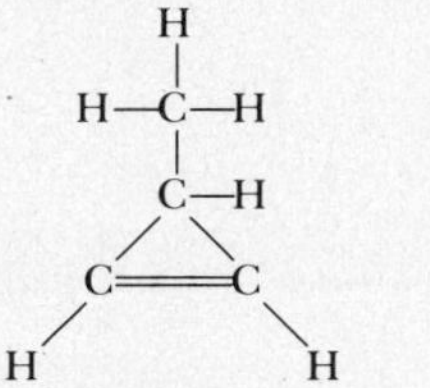

3-methylcyclopropene 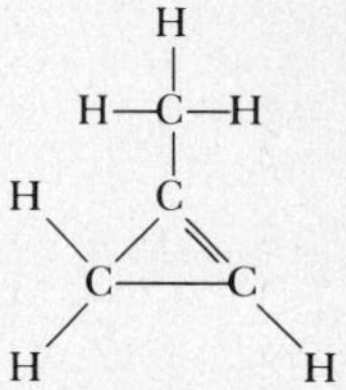 1-methylcyclopropene

ANSWER 67

(a) 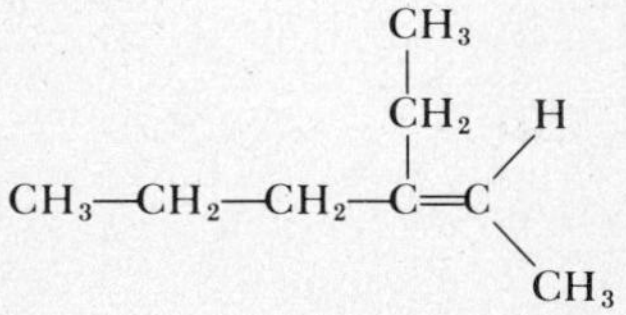 $CH_3-CH_2-CH_2-C(CH_2CH_3)=CH-CH_3$

or $CH_3-CH_2-CH_2-C(CH_2CH_3)=C(CH_3)H$

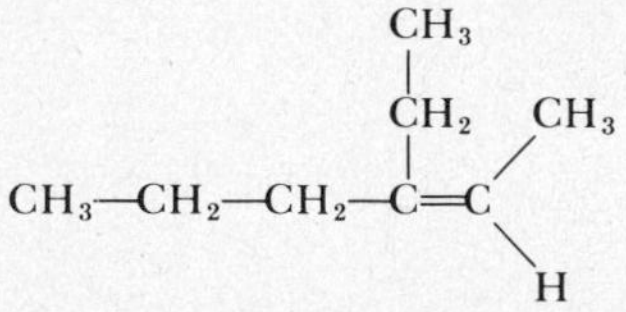

(b) 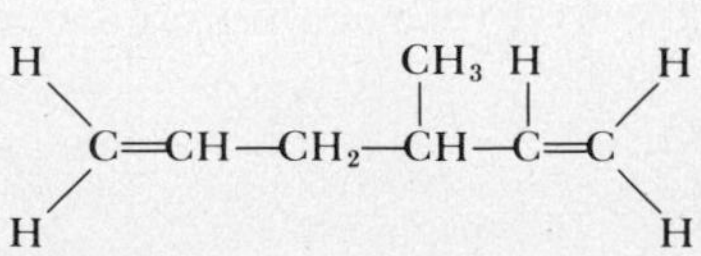 $H_2C=CH-CH_2-CH(CH_3)-CH=CH_2$

ANSWER 81 benzene
or benzenoid rings

ITEM 13

The saturated hydrocarbon containing two carbon atoms is named *ethane.* Draw the Lewis diagram for ethane.

ITEM 27

If two double bonds are present in a compound, it is called a *diene* or *diolefin.* The Lewis diagram for propadiene is ________.

ITEM 41

In the IUPAC system an alkane is given the name corresponding to the longest single carbon chain, and the branches are identified by name and by the number of the carbon to which they are attached. The central carbon of propane is numbered ________, no matter which end of the chain is numbered 1.

ITEM 55

From *n*-butane, however, four hydrogens can be removed, one each from carbons 1 and 3, and two from carbon 2. The resulting 1,2-*butadiene* has the Lewis diagram ________.

ITEM 69

The structural formula of 1,4-decadiyne is ________ and of cyclooctene is ________.

```
    H  H  H  H
    |  |  |  |
H—C—C—C—C—H
    |  |  |  |
    H  H  H  H
```

ANSWER 12 sp^3
109°

ANSWER 26

```
H—C═══C—H
   \   /
    C
   / \
  H   H
```

ANSWER 40 methyl radical
(or methyl group)

ANSWER 54 No.

ANSWER 68 2,2,4,4,5,5-hexamethyloctane

ANSWER 82 butylbenzene,
diphenylmethane,
naphthalene

ITEM 14

The suffix "yl" designates a radical. If one hydrogen atom is removed from ethane, the resulting radical will be named ________ and will have the Lewis diagram ________.

ITEM 28

The systematic name for the three-carbon hydrocarbon with a triple bond is ________. It also can be named as an acetylene in which one hydrogen is replaced by a methyl group: The common name is *methylacetylene.* Draw the Lewis diagram for this compound.

ITEM 42

In the compound at the right, the longest single chain contains ________ carbon atoms. The name therefore *cannot* correspond to any propane. The systematic name must be ________.

ITEM 56

Or, if hydrogens are removed, one each from carbons 1, 2, 3, and 4, the resulting 1,3-*butadiene* has the Lewis diagram ________.

ITEM 70

The straight-chain radicals having five, six, seven, eight, nine, and ten carbon atoms are known as ________, ________, ________, ________, ________, and ________.

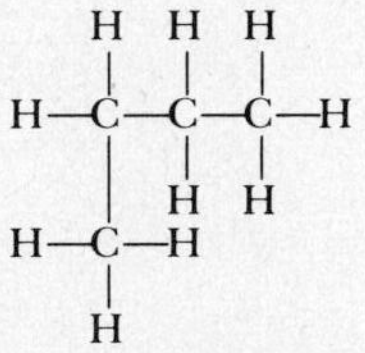

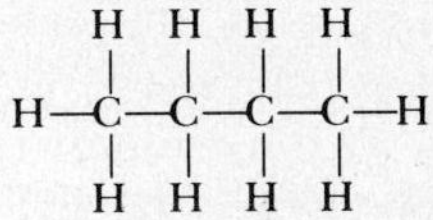

```
    H   H   H
    |   |   |
H — C — C — C — H
    |   |   |
    |   H   H
H — C — H
    |
    H
```

```
    H   H   H   H
    |   |   |   |
H — C — C — C — C — H
    |   |   |   |
    H   H   H   H
```

ANSWER 13

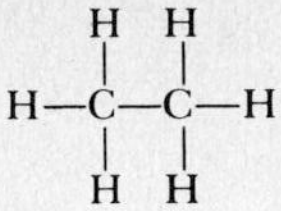

```
    H   H
    |   |
H — C — C — H
    |   |
    H   H
```

ANSWER 27

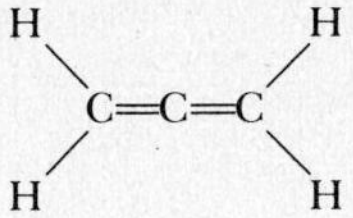

```
H             H
 \           /
  C == C == C
 /           \
H             H
```

ANSWER 41 2

ANSWER 55

```
              H
              |
H             C — H
 \           / |
  C == C == C  H
 /           \
H             H
```

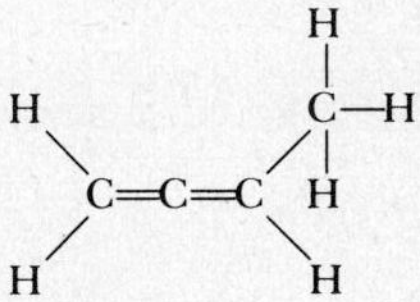

ANSWER 69

HC≡C—CH$_2$—C≡C—CH$_2$CH$_2$CH$_2$CH$_2$CH$_3$

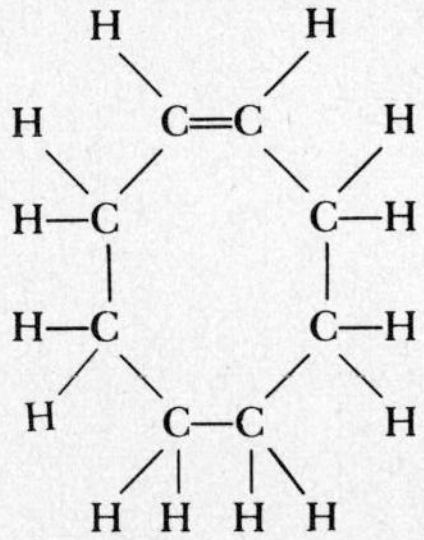

```
          H       H
           \     /
   H        C = C        H
    \      /     \      /
 H — C                 C — H
     |                 |
 H — C                 C — H
    /   \           /   \
   H     C   —   C       H
        / |      | \
       H  H      H  H
```

ANSWER 14 ethyl

```
    H  H
    |  |
 H—C—C·
    |  |
    H  H
```

(Any H can be removed because they are all equivalent.)

(*Item 15 is on page 592.*)

ANSWER 28 propyne

```
    H
    |
 H—C—C≡C—H
    |
    H
```

(*Item 29 is on page 592.*)

ANSWER 42 four
butane

(*Item 43 is on page 592.*)

ANSWER 56

```
           H         H
            \       /
 H           C═C
  \         /    \
   C═C            H
  /    \
 H      H
```

(*Item 57 is on page 592.*)

ANSWER 70 pentyl, hexyl, heptyl, octyl, nonyl, and decyl

(*Item 71 is on page 592.*)

12-2 THE FUNCTIONAL GROUPS

Alkanes are unreactive compounds: Drastic conditions are required to break the sigma bonds between the carbon and hydrogen atoms. The pi bonds of alkenes and alkynes are broken more easily, so these compounds are more reactive; addition of atoms to double and triple bonds is a common reaction. However, the presence of halogen, oxygen, nitrogen, sulfur, or phosphorus atoms in a molecule greatly extends the possibilities for reaction. In part, this is because of their nonbonded valence electrons.

In this section, which is a review of parts of Section 12-3 of *Chemical Principles,* again you will draw Lewis diagrams to emphasize the presence of nonbonded electrons. In doing so, you will learn the names and structures of simple organic compounds containing the most common reactive atoms or groups of atoms, the *functional groups.* When you have finished this section, you will be able to recognize names and draw structural diagrams for halides, alcohols, aldehydes, ketones, acids, ethers, esters, and amines.

ITEM 1

Although alkanes are extremely unreactive compounds, they will react with chlorine gas in sunlight to produce *chloroalkanes.* For example, from methane, one product is *chloromethane* (also called monochloromethane and methyl chloride); it has the Lewis diagram ________.

ITEM 16

Alcohols containing two *hydroxyl* (OH) groups are called *diols* or *glycols.* Draw the Lewis diagram for 1,2-ethanediol, better known as ethylene glycol.

ITEM 31

The compound at the right is another example of an enol. Its systematic name is ________. Like its isomer, $CH_3CH{=}CHOH$, this enol rapidly rearranges to a carbonyl compound.

ITEM 46

The simplest carboxylic acid is *formic acid.* Since its systematic name is *methanoic acid,* it must contain ________ carbon atom(s). The Lewis diagram is ________.

ITEM 61

The reaction of a carboxylic acid with an alcohol in the presence of an inorganic acid leads to the formation of an *ester.* For example,

$$\begin{array}{c} \mathrm{H} \\ | \\ \mathrm{H{-}\overset{}{C}{-}O{-}H} \\ | \\ \mathrm{H} \end{array} + \mathrm{H{-}O{-}\overset{\overset{\displaystyle O}{\|}}{C}{-}H}$$

$$\rightarrow \begin{array}{c} \mathrm{H} \\ | \\ \mathrm{H{-}C{-}O{-}\overset{\overset{\displaystyle O}{\|}}{C}{-}H} \\ | \\ \mathrm{H} \end{array} + \underline{\qquad\qquad}.$$

ITEM 76

Draw structural formulas for the following organic compounds: 2-methyl-4-chloro-1-butene, methyl alcohol, and acetone.

$$H_3C-C(OH)=CH_2$$

$$-C(=\ddot{O}:)-\ddot{O}-H$$

Carboxyl group

ITEM 2

The other products are *dichloromethane* (methylene chloride), *trichloromethane* (chloroform), and *tetrachloromethane* (carbon tetrachloride). Draw Lewis diagrams for these compounds.

ITEM 17

Draw the structural formulas for 1,4-butanediol and 1-pentene-3,5-diol.

ITEM 32

The proton transfer reaction shown at the right leads to a *ketone*, a general name for a compound containing a carbonyl as a nonterminal carbon rather than at the end of the chain, as it is in a(n) __________.

ITEM 47

Acetic acid has the systematic name *ethanoic* acid, but its IUPAC name is rarely used. The Lewis diagram of acetic acid is __________.

ITEM 62

All esters contain the partial structure

$$\mathrm{C{-}O{-}\overset{\overset{\displaystyle O}{\|}}{C}}$$

of which the C=O group is derived from an acid and the other carbon (and the other oxygen) is derived from an __________.

LTEM 77

Draw structural formulas for the following: phenol, *p*-dibromobenzene, formaldehyde.

$$\begin{array}{c} H \\ | \\ H-C-H \quad\; H \\ \diagdown \quad\;\; \diagup \\ C{=}C \\ \diagup \quad\;\; \diagdown \\ OH \qquad\quad H \end{array} \rightarrow H-\underset{H}{\overset{H}{C}}-\underset{O}{\overset{}{C}}-\underset{H}{\overset{H}{C}}-H$$

ANSWER 1

$$H-\underset{:\ddot{Cl}:}{\overset{H}{C}}-H$$

ANSWER 16

$$H-\underset{\underset{H}{:\ddot{O}:}}{\overset{H}{C}}-\underset{\underset{H}{:\ddot{O}:}}{\overset{H}{C}}-H \quad \text{or} \quad H-\underset{\underset{H}{:\ddot{O}:}}{\overset{H}{C}}-\underset{H}{\overset{H}{C}}-\ddot{\underset{..}{O}}-H \text{ etc.}$$

ANSWER 31 1-propen-2-ol

ANSWER 46 one

$$\begin{array}{l} H \\ \quad\diagdown \\ \quad\;\; C-\ddot{\underset{..}{O}}-H \\ \quad\;\; \| \\ \quad\; :\ddot{O}: \end{array}$$

ANSWER 61 H_2O

ANSWER 76

$$H-\overset{H}{C}{=}\underset{\underset{\underset{H}{H-C-H}}{|}}{C}-\underset{H}{\overset{H}{C}}-\underset{H}{\overset{Cl}{C}}-H$$

$$H-\underset{H}{\overset{H}{C}}-O-H$$

$$H-\underset{H}{\overset{H}{C}}-\overset{O}{\overset{\|}{C}}-\underset{H}{\overset{H}{C}}-H$$

ITEM 3

Organic halogen compounds are named systematically by a prefix designating the number and positions of the halogen atoms followed by the appropriate hydrocarbon name. As an example, the compound shown at the right is named 1,2,2,4-*tetrachlorobutane.* Draw a similar diagram for 1,1,2-trifluoropropane.

ITEM 18

The term "alcohol" is used for aliphatic hydroxyl compounds. The term for compounds containing a hydroxyl group directly bonded to a benzene ring is *phenol.* The parent compound is *phenol* itself, which is shown at the right. 2-Methylphenol can be drawn as __________.

ITEM 33

Since a ketone carbonyl group cannot be at the end of a chain, the simplest ketone is *propanone,* usually called *acetone* (shown at the right). The carbonyl group is bonded to two __________.

ITEM 48

The systematic names of acids are those of the corresponding hydrocarbons with the final "e" replaced by "oic" and the word "acid" added. The carboxyl carbon is counted as number 1. Thus, the three-carbon acid at the right is named __________.

ITEM 63

An ester is named after the alcohol and acid from which it is derived. The ester at the right is named methyl formate (or methyl methanoate) and is derived from __________ alcohol and __________ acid.

ITEM 78

Draw structural formulas for the following: 2-pentanone, acetic acid, 3-bromocyclohexanone.

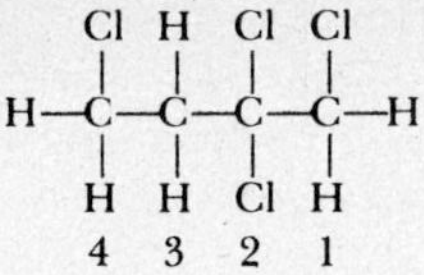

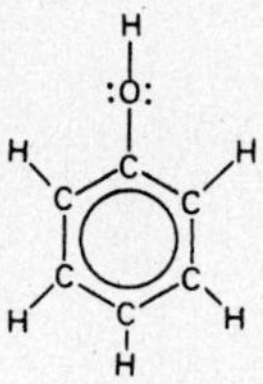

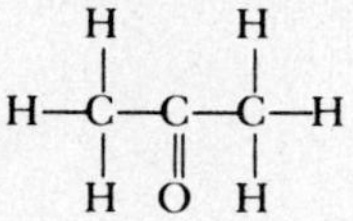

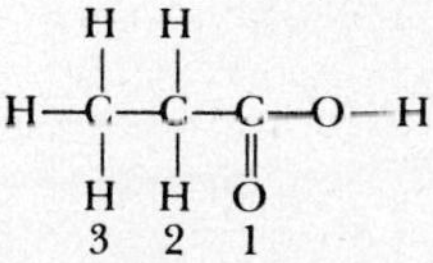

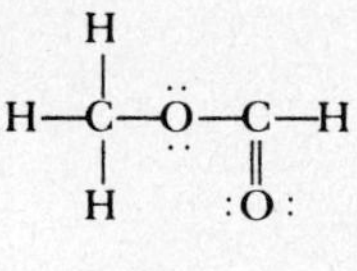

```
  Cl  H  Cl  Cl
  |   |   |   |
H—C———C———C———C—H
  |   |   |   |
  H   H   Cl  H
  4   3   2   1
```

```
        H
        |
       :O:
 H      |      H
   \    C    /
      C   C
      C   C
   /    C    \
 H      |      H
        H
```

```
  H       H
  |       |
H—C———C———C—H
  |   ‖   |
  H   O   H
```

```
  H   H
  |   |
H—C———C———C—O—H
  |   |   ‖
  H   H   O
  3   2   1
```

```
  H
  |     ..
H—C———O———C—H
  |   ..  ‖
  H      :O:
```

ANSWER 2

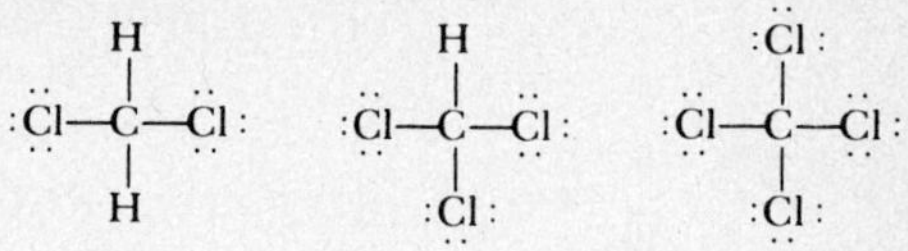

```
     H              H              ..
     |              |             :Cl:
 ..  |  ..      ..  |  ..      ..  |  ..
:Cl—C—Cl:      :Cl—C—Cl:      :Cl—C—Cl:
 ..  |  ..      ..  |  ..      ..  |  ..
     H             :Cl:           :Cl:
                    ..             ..
```

ANSWER 17

```
      H   H   H   H
      |   |   |   |
H—O———C———C———C———C—O—H
      |   |   |   |
      H   H   H   H
```

```
          H   H   H
          |   |   |
H—C═C———C———C———C—O—H
  |  |    |   |   |
  H  H    O   H   H
          |
          H
```

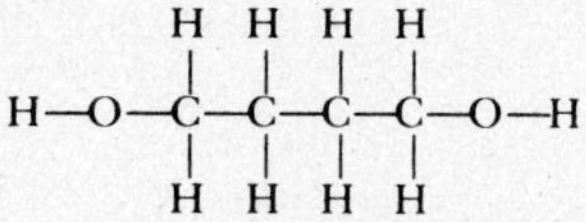

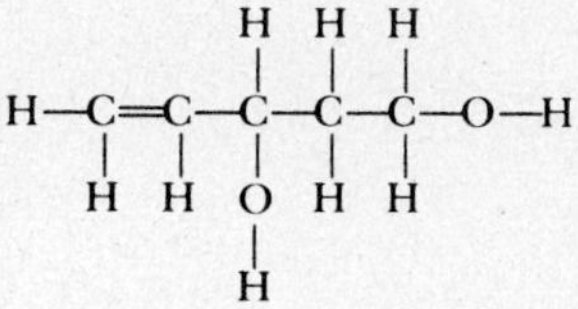

ANSWER 32 aldehyde

ANSWER 47

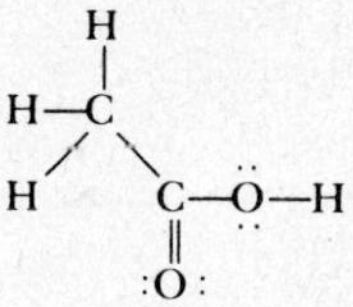

```
   H
   |
H—C
  /  \      ..
H      C—O—H
       ‖  ..
      :O:
```

ANSWER 62 alcohol

ANSWER 77

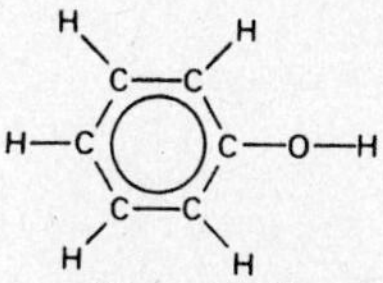

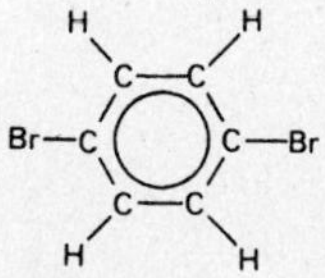

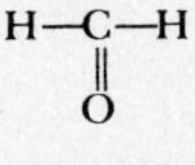

```
   H       H
    \     /
     C—C
H—C       C—O—H
     C—C
    /     \
   H       H
```

```
   H       H
    \     /
     C—C
Br—C       C—Br
     C—C
    /     \
   H       H
```

```
H—C—H
  ‖
  O
```

ITEM 4 Some common names consist of the hydrocarbon radical name followed by the halide name with a prefix "di" or "tri" indicating two or three halide atoms. The compound at the right is named *n-hexyl fluoride.* The systematic name is ________.

ITEM 19 Draw diagrams for 2-ethylphenol and 2-phenylethanol. (When a number is not given, it is assumed to be 1, as in 2-phenyl-1-ethanol.)

ITEM 34 A ketone is named systematically by considering the longest chain carrying the ketone group to be the parent hydrocarbon. The carbonyl carbon is given the lowest possible number and the "e" of the hydrocarbon name is replaced by ________.

ITEM 49 The acid that has the systematic name propenoic acid has the Lewis diagram ________.

ITEM 64 The name of the ester thus includes the name of an acid in which the suffix "ic" of the acid name is replaced by "ate" in the ester name. The Lewis diagram of ethyl formate is ________.

ITEM 79 Draw structural formulas for the following: benzoic acid, ethyl acetate, trimethylamine.

$$\begin{array}{cccccccc} & H & H & H & H & H & H & \\ & | & | & | & | & | & | & \\ H- & C- & C- & C- & C- & C- & C- & F \\ & | & | & | & | & | & | & \\ & H & H & H & H & H & H & \end{array}$$

$$\begin{array}{c} CH_3CH_2\underset{\underset{\displaystyle O}{\|}}{C}CH_2CH_3 \end{array}$$

3-pentanone

ANSWER 3

$$\begin{array}{ccccc} & H & F & F & \\ & | & | & | & \\ H- & C- & C- & C- & H \\ & | & | & | & \\ & H & H & F & \end{array}$$

ANSWER 18

ANSWER 33 methyl groups

ANSWER 48 propanoic acid
(The common name is propionic.)

ANSWER 63 methyl (methanol)
formic (methanoic)

ANSWER 78

$$\begin{array}{cccccc} & H & & H & H & H \\ & | & & | & | & | \\ H- & C- & C- & C- & C- & C-H \\ & | & \| & | & | & | \\ & H & O & H & H & H \end{array}$$

$$\begin{array}{cccc} & H & & \\ & | & & \\ H- & C- & C- & O-H \\ & | & \| & \\ & H & O & \end{array}$$

ITEM 5

The haloalkane $CH_3CH_2CH_2Br$ has the systematic name ________ and the common name ________.

ITEM 20

Write names and structural formulas for the one-carbon and two-carbon alcohols and diols.

ITEM 35

In the case of propanone, the number is redundant. The common name of propanone is ________.

ITEM 50

Many times the formula for an acid is written without showing the double bond. For example, CH_3COOH is equivalent to the formula shown at the right. The common name of this acid is ________, and the systematic name is ________.

ITEM 65

The Lewis diagrams for methyl formate and for methyl acetate are ________.

ITEM 80

Draw structural formulas for 3,3-difluoroheptanal, ether (i.e., diethyl ether), isopropyl formate, and 3-phenylpropanoic acid.

CH_3CCH_3

```
CH3CCH3
   ‖
   O
```

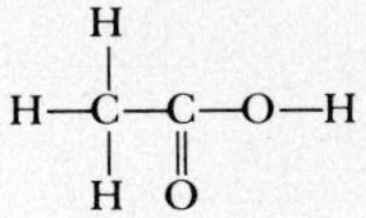

```
   H
   |
H—C—C—O—H
   |  ‖
   H  O
```

ANSWER 4 1-fluorohexane

ANSWER 19

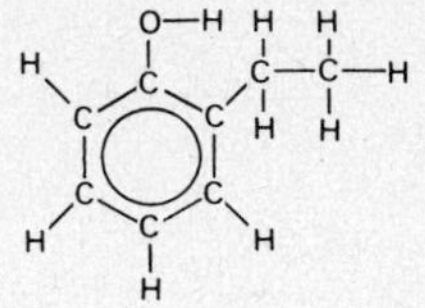

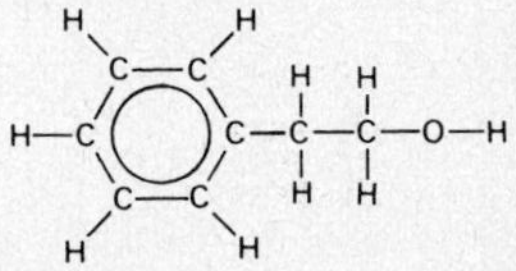

```
        O—H  H   H
 H      |    |   |
   \    C    C—C—H
     C     C   |   |
     |  ○  |   H   H
     C     C
   /    C    \
 H      |     H
        H
```

```
    H       H
     \     /
      C—C        H   H
     /   \       |   |
H—C   ○   C—C—C—O—H
     \   /       |   |
      C—C        H   H
     /     \
    H       H
```

ANSWER 34 one

ANSWER 49

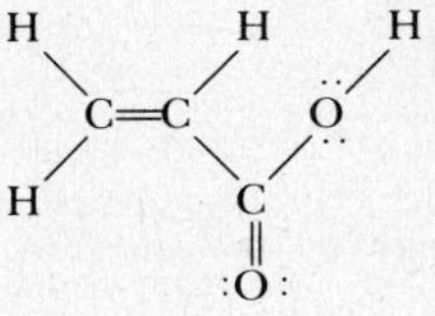

```
H          H           H
  \       /           /
    C═C          Ö
  /       \     /  ··
H          C
           ‖
          :O:
```

ANSWER 64

```
     H   H
     |   |    ··
H—C—C—O—C—H
     |   |    ··  ‖
     H   H       :O:
```

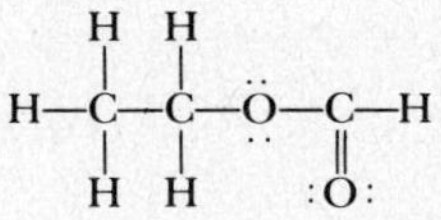

ANSWER 79

```
     HO   O
       \ //
        C
 H      |      H
   \    C    /
     C     C
     |  ○  |
     C     C
   /    C    \
 H      |      H
        H
```

```
  H  H         H              H        H
  |  |         |              |        |
H—C—C—O—C—C—H      H—C—N—C—H
  |  |     ‖   |              |   |    |
  H  H     O   H              H   |    H
                                H—C—H
                                  |
                                  H
```

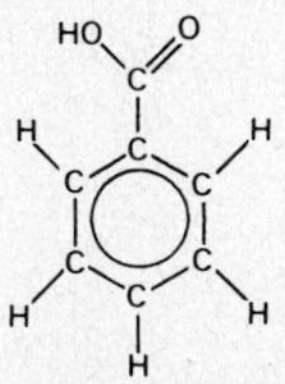

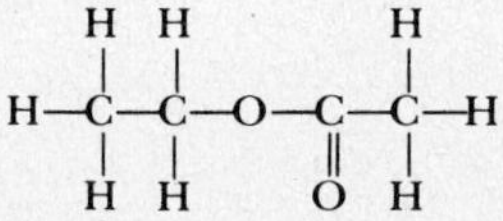

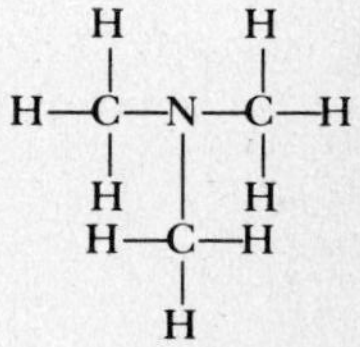

ITEM 6

There are two structural isomers of monoiodopropane. Write the common names and formulas for both.

ITEM 21

Alcohols do not ionize in water: They do not have the basic properties of NaOH in water nor the acidic properties of HOCl in water. However, phenols are acidic in water and, therefore, capabie of donating a ________ to ________.

ITEM 36

The Lewis diagram of 2-pentanone is ________. The common name is methyl *n*-propyl ketone.

ITEM 51

Acids in which the carboxyl group is bonded directly to a benzene ring are known as *benzoic acids.* The parent acid, benzoic acid, has the structural formula ________.

ITEM 66

Draw diagrams for methyl benzoate and phenyl acetate.

ANSWER 5 1-bromopropane
n-propyl bromide

ANSWER 20

methanol, methyl alcohol H_3COH
methanediol H_2COH with OH on the C (structure: $H_2C(OH)OH$)

ethanol, ethyl alcohol CH_3CH_2OH
1,2-ethanediol, ethylene glycol
$HOCH_2CH_2OH$
1,1-ethanediol $HOCHCH_3$ with OH on the CH (structure: $HOCH(OH)CH_3$)

ANSWER 35 acetone

ANSWER 50 acetic
ethanoic

ANSWER 65

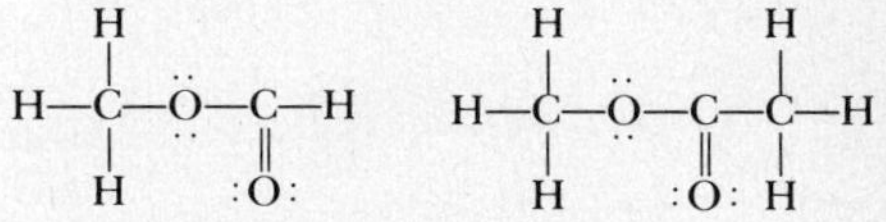

```
   H                          H           H
   |    ..                    |     ..    |
H—C—O—C—H                H—C—O—C—C—H
   |    ..  ||                |     ..  ||  |
   H       :O:                H        :O: H
```

ANSWER 80

```
      H  F  H  H  H  H
      |  |  |  |  |  |
H—C—C—C—C—C—C—C—H
   ||  |  |  |  |  |  |
   O  H  F  H  H  H  H
```

```
   H  H     H  H
   |  |     |  |
H—C—C—O—C—C—H
   |  |     |  |
   H  H     H  H
```

```
   H  H
   |  |
H—C—C—O—C—H
   |  |     ||
   H  C—H   O
     /|
    H  H
```

```
   H       H
    \     /
     C==C        H   H
    /    \       |   |
H—C  ( )  C—C—C—C—OH
    \    /       |   |   ||
     C==C        H   H   O
    /     \
   H       H
```

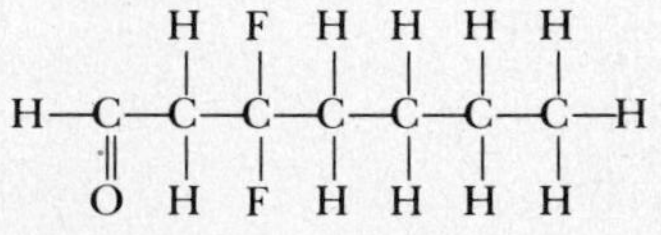

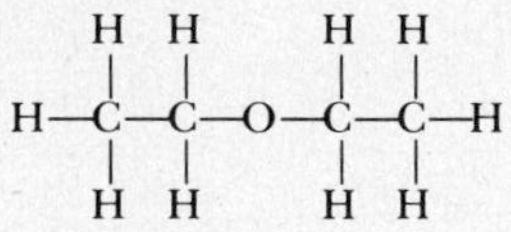

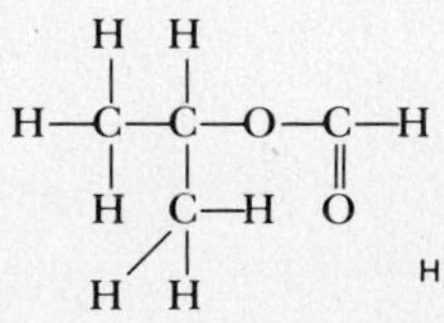

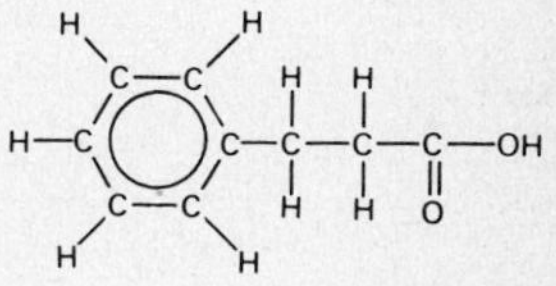

(*Turn to page 656.*)

ITEM 7 How many isomers are possible for dibromopropane? Give the systematic names.

ITEM 22 The characteristic structure of a phenol is __________. Draw a diagram for *meta*-chlorophenol.

ITEM 37 The structural formula of 2,4-dimethyl-3-pentanone is __________. The common name is diisopropyl ketone.

ITEM 52 Draw a diagram to represent 2,4,6-tribromobenzoic acid.

ITEM 67 The ester derived from isopropyl alcohol and *o*-chlorobenzoic acid has the name *isopropyl o-chlorobenzoate* and the structural formula __________.

ANSWER 6 *n*-propyl iodide

```
   H  H  H
   |  |  |
H—C—C—C—I
   |  |  |
   H  H  H
```

iso-propyl iodide

```
   H  H  H
   |  |  |
H—C—C—C—H
   |  |  |
   H  I  H
```

ANSWER 21 proton (or hydrogen ion)
H_2O

ANSWER 36

```
   H        H  H  H
   |        |  |  |
H—C——C——C——C——C—H
   |   ||   |  |  |
   H  :O:   H  H  H
```

ANSWER 51

```
        O
         \\
          C—O—H                     COOH
 H        |        H          H      |      H
   \      C      /              \    C    /
     C  /   \  C                  C /  \ C
     | (  O  ) |        or        | ( O ) |
     C  \   /  C                  C \  / C
   /      C      \              /    C    \
 H        |        H          H      |      H
          H                          H
```

ANSWER 66

```
                         H           H
                           \        /
       H                    C —— C
       |                  /   ◯    \
  H—C—Ö—C—C               C—H
       |        ||         \        /
       H       :O:          C —— C
                           /        \
                         H           H

         H        H
           \     /
            C—C                     H
          /  ◯  \                    |
     H—C         C—Ö—C—C—H
          \     /     ..   ||   |
            C—C            :O:  H
           /     \
         H        H
```

ITEM 8

Draw diagrams to represent the compounds 1,1-dichloro-2-methyl-1-propene and *cis*-2,3-dibromo-2-butene.

ITEM 23

Consider the diagram at the right. The name of this compound would be ________. This kind of alcohol, in which the OH group is attached to a doubly bonded carbon, is known as an *enol.*

ITEM 38

The common names of ketones, as you have seen, usually indicate structure. For example, ethyl phenyl ketone has the structural formula at the right. Methyl *o*-bromophenyl ketone has the structural formula ________.

ITEM 53

Write structural formulas for phenol, benzaldehyde, and benzoic acid.

ITEM 68

Write partial structures characteristic of the following types of compounds: alcohols, ________; aldehydes, ________; ketones, ________; esters, ________; acids, ________.

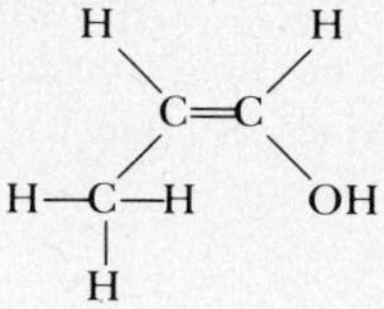

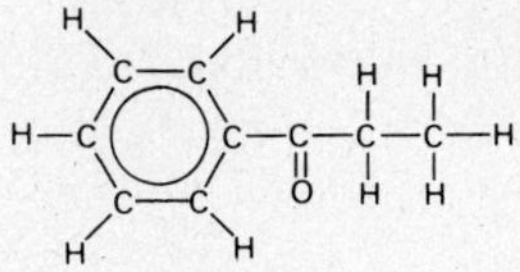

ANSWER 7

four

1,1-dibromopropane ($CH_3CH_2CHBr_2$)
1,2-dibromopropane ($CH_3CHBrCH_2Br$)
1,3-dibromopropane ($CH_2BrCH_2CH_2Br$)
2,2-dibromopropane ($CH_3CBr_2CH_3$)

ANSWER 22

a hydroxyl group bonded directly to a benzene ring

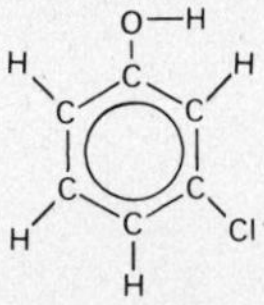

ANSWER 37

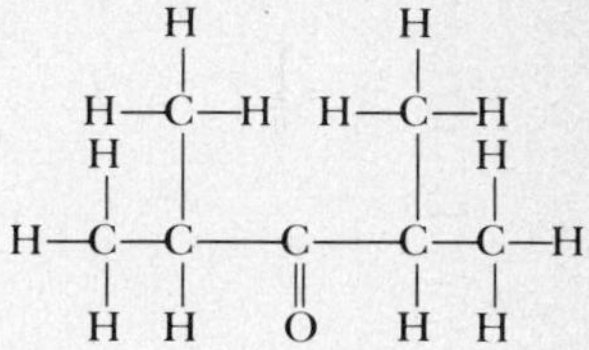

ANSWER 52

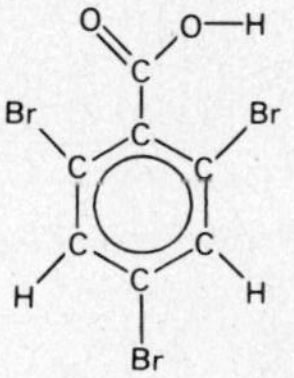

ANSWER 67

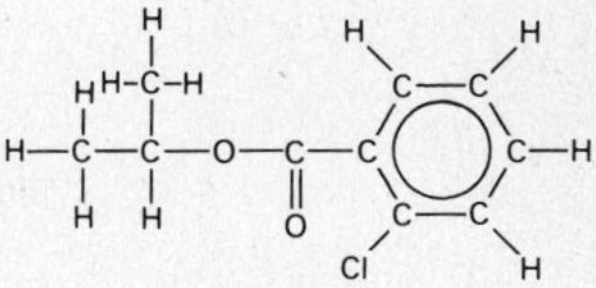

ITEM 9

Draw a diagram for 3,4-diiodo-1-pentyne.

ITEM 24

Enols are rare species because they are extremely unstable. They rearrange to give a *carbonyl* compound (containing a C═O group), as shown at the right. A proton is transferred from _________ of the enol to carbon atom 2, which was doubly bonded in the enol.

ITEM 39

The Lewis diagrams of *cis-* and *trans*-3-penten-2-one are _________.

ITEM 54

Write structural formulas for methyl alcohol, formaldehyde, and formic acid.

ITEM 69

An important class of organic nitrogen compounds is *amines:* These are derivatives of ammonia in which one or more hydrogen atoms of NH_3 are replaced by aliphatic or aromatic radicals. *Methylamine* is shown at the right. The Lewis diagram for *ethylamine* is _________.

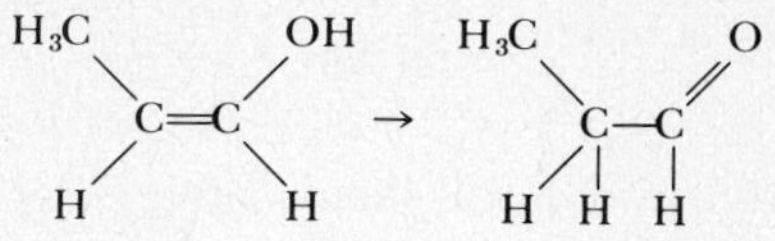

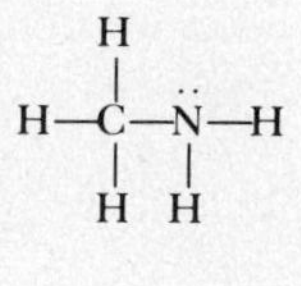

$$H_3C-CH=CH-OH \rightarrow H_3C-CH_2-CH=O$$

$$H_3C-\ddot{N}H_2$$

ANSWER 8

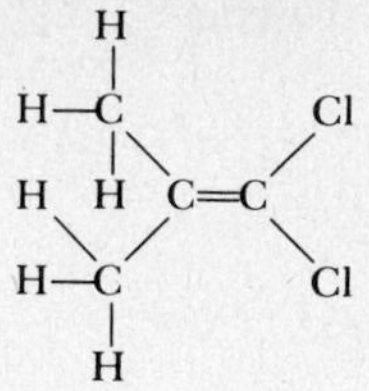

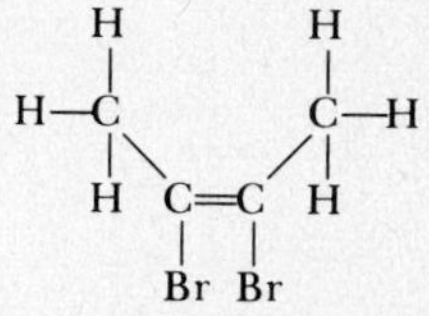

$(CH_3)_2C=CCl_2$

$CH_3-C(Br)=C(Br)-CH_3$

ANSWER 23 *cis*-1-propenol
or *cis*-1-propen-1-ol

ANSWER 38

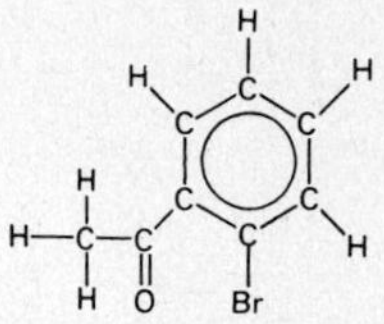

$CH_3-C(=O)-C_6H_4Br$

ANSWER 53

C_6H_5OH $\quad C_6H_5CHO$ $\quad C_6H_5COOH$

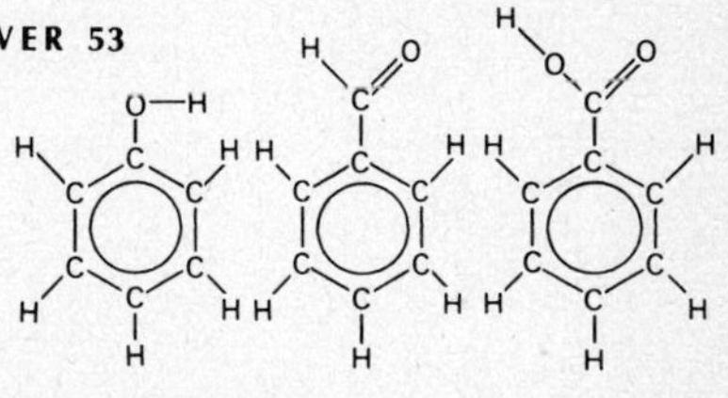

ANSWER 68

C—O—H
H—C=O
C—C(=O)—C

C—O—C(=O)
H—O—C=O

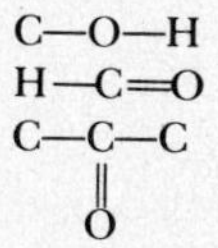

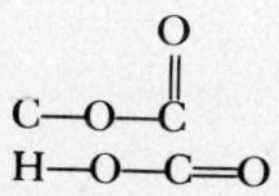

ITEM 10

Under proper conditions, 2-bromopropane (isopropyl bromide) can be converted to an alcohol that has the Lewis diagram shown at the right and the common name *isopropyl alcohol.* The reaction is a displacement of Br by ________.

ITEM 25

This proton transfer reaction is instantaneous, so any enol made in a reaction will be transformed almost completely to a compound containing a C—O group, known as the ________ group. (Phenols are special cases and do not undergo this reaction.)

ITEM 40

The structural formula of acetone is ________.

ITEM 55

Write structural formulas for ethanol, acetaldehyde, and acetic acid.

ITEM 70

Propylamine has the Lewis diagram ________. (If isopropyl is not specified, the *n*-propyl derivative is meant.)

```
    H  H  H
    |  |  |
 H—C—C—C—H
    |  |  |
    H :O: H
       |
       H
```

ANSWER 9

```
            H  H  H
            |  |  |
 H—C≡C—C—C—C—H
            |  |  |
            I  I  H
```

ANSWER 24 the OH group

ANSWER 39

```
                 H
                 |
  H           H—C—H
  |              |
H—C              C
  |   \        /  \\
  H    C==C        :O:
      /    \
     H      H
```

cis

```
                 H
                 |
              H—C—H
                 |
     H           C
       \       /  \\
 H      C==C        :O:
   \   /     \
    C—H       H
   /
  H
```

trans

ANSWER 54

```
   H
   |
H—C—O—H
   |
   H
```

```
H     H
  \  /
   C
   ||
   O
```

```
H     OH
  \  /
   C
   ||
   O
```

ANSWER 69

```
   H  H
   |  |  ..
H—C—C—N—H
   |  |  |
   H  H  H
```

ITEM 11

The systematic name of isopropyl alcohol is 2-*propanol,* in which the 2 designates __________ and the suffix "ol" designates the __________ group.

ITEM 26

All *aldehydes* contain the partial structure H—C═O, a hydrogen bonded to a __________ group. The simplest aldehyde (in which the rest of the structure is only a H atom) has the Lewis diagram __________.

ITEM 41

Both ketones and aldehydes contain a __________. The difference between ketones and aldehydes is __________.

ITEM 56

The reaction of two alcohol molecules in concentrated sulfuric acid produces an *ether.* For example, from two molecules of ethyl alcohol, one molecule of *diethyl ether* and one molecule of __________ are produced, as shown at the right.

ITEM 71

Dimethylamine is a derivative of NH_3 in which two hydrogen atoms are replaced by __________. The Lewis diagram is __________.

$$
\begin{array}{ccccccccccccccc}
 & & \mathrm{H} & & \mathrm{H} & & & & & & & \mathrm{H} & & \mathrm{H} & \\
 & & | & & | & & & & & & & | & & | & \\
\mathrm{H} & — & \mathrm{C} & — & \mathrm{C} & — & \mathrm{O} & — & \mathrm{H} + \mathrm{H} & — & \mathrm{O} — & \mathrm{C} & — & \mathrm{C} & — \mathrm{H} \quad \rightarrow \\
 & & | & & | & & & & & & & | & & | & \\
 & & \mathrm{H} & & \mathrm{H} & & & & & & & \mathrm{H} & & \mathrm{H} &
\end{array}
$$

$$
\begin{array}{cccccccccc}
 & & \mathrm{H} & & \mathrm{H} & & \mathrm{H} & & \mathrm{H} & \\
 & & | & & | & & | & & | & \\
\mathrm{H} & — & \mathrm{C} & — & \mathrm{C} & — \mathrm{O} — & \mathrm{C} & — & \mathrm{C} & — \mathrm{H} + ? \\
 & & | & & | & & | & & | & \\
 & & \mathrm{H} & & \mathrm{H} & & \mathrm{H} & & \mathrm{H} &
\end{array}
$$

ANSWER 10 OH

ANSWER 25 carbonyl

ANSWER 40

$$
\begin{array}{ccccccccc}
 & & \mathrm{H} & & & & \mathrm{H} & & \\
 & & | & & & & | & & \\
\mathrm{H} & — & \mathrm{C} & — & \mathrm{C} & — & \mathrm{C} & — & \mathrm{H} \\
 & & | & & \| & & | & & \\
 & & \mathrm{H} & & \mathrm{O} & & \mathrm{H} & &
\end{array}
$$

ANSWER 55

$$
\begin{array}{ccccccccc}
 & & \mathrm{H} & & \mathrm{H} & & & & \\
 & & | & & | & & & & \\
\mathrm{H} & — & \mathrm{C} & — & \mathrm{C} & — & \mathrm{O} & — & \mathrm{H} \\
 & & | & & | & & & & \\
 & & \mathrm{H} & & \mathrm{H} & & & &
\end{array}
$$

$$
\begin{array}{cccccc}
 & & \mathrm{H} & & & \mathrm{H} \\
 & & | & & \diagup & \\
\mathrm{H} & — & \mathrm{C} & — & \mathrm{C} & \\
 & & | & & & \diagdown\!\!\diagdown \\
 & & \mathrm{H} & & & \mathrm{O}
\end{array}
$$

$$
\begin{array}{ccccccc}
 & & \mathrm{H} & & & & \\
 & & | & & & & \\
\mathrm{H} & — & \mathrm{C} & — & \mathrm{C} & — & \mathrm{OH} \\
 & & | & & \| & & \\
 & & \mathrm{H} & & \mathrm{O} & &
\end{array}
$$

ANSWER 70

$$
\begin{array}{ccccccccc}
 & & \mathrm{H} & & \mathrm{H} & & \mathrm{H} & & \\
 & & | & & | & & | & & \\
\mathrm{H} & — & \mathrm{C} & — & \mathrm{C} & — & \mathrm{C} & — & \ddot{\mathrm{N}} — \mathrm{H} \\
 & & | & & | & & | & & | \\
 & & \mathrm{H} & & \mathrm{H} & & \mathrm{H} & & \mathrm{H}
\end{array}
$$

ITEM 12

Methanol, or *methyl alcohol* as it is commonly known, has the Lewis diagram ________, and *ethanol,* or *ethyl alcohol,* has the Lewis diagram ________.

ITEM 27

The one-carbon aldehyde is usually called *formaldehyde,* although the systematic name is *methanal. Acetaldehyde,* or *ethanal,* has one carbonyl group and a methyl group. Draw the Lewis diagram.

ITEM 42

Draw structural formulas for cyclohexanone and cyclohexanol.

ITEM 57

Ether molecules contain an oxygen atom bonded to two carbon atoms. Diethyl ether has the Lewis diagram shown at the right. Dimethyl ether has the Lewis diagram ________.

ITEM 72

Triethylamine has the Lewis diagram ________.

```
H—C—H
  ||
 :O:
```
Formaldehyde

ANSWER 11 the number of the carbon to which a group is attached OH (hydroxyl)

ANSWER 26 carbonyl

```
H—C—H
  ||
 :O:
```

ANSWER 41 carbonyl group (C═O) in the position of the carbonyl. In aldehydes the carbonyl carbon is at the end of the chain; in ketones it is not.

```
  H H     H H
  | |  ..  | |
H—C—C—O—C—C—H
  | |  ..  | |
  H H     H H
```

ANSWER 56 H_2O

ANSWER 71 Methyl groups (CH_3)

```
  H     H
  |  ..  |
H—C—N—C—H
  |  |  |
  H  H  H
```

ITEM 13

2-Pentanol has the structural formula __________, and 3-heptanol has the structural formula __________.

ITEM 28

An aldehyde carbon must always be at the end of a chain, and it is understood to be carbon 1. The name is then that of the corresponding hydrocarbon with the final "e" replaced by "al." The compound at the right is named __________.

ITEM 43

Draw structural formulas for formaldehyde and acetaldehyde.

ITEM 58

The names of simple ethers consist of the names of the two radical groups attached to the oxygen followed by the word "ether." The Lewis diagram for methyl ethyl ether is __________.

ITEM 73

Draw a diagram for cyclopentylamine.

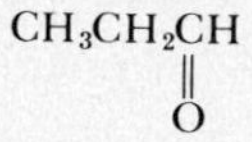

$$\underset{\displaystyle \text{O}}{\text{CH}_3\text{CH}_2\overset{}{\text{C}}\text{H}}$$

(CH₃CH₂CH with =O double-bonded below the terminal C)

ANSWER 12

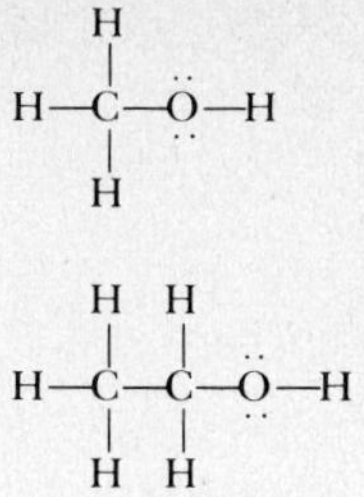

```
     H
     |   ..
  H—C—O—H
     |   ..
     H
```

```
     H  H
     |  |   ..
  H—C—C—O—H
     |  |   ..
     H  H
```

ANSWER 27

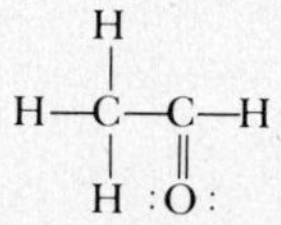

```
     H
     |
  H—C—C—H
     |  ‖
     H :O:
```

ANSWER 42

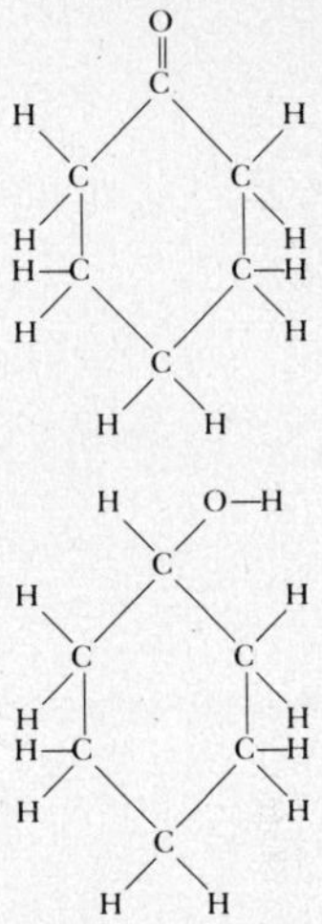

```
            O
            ‖
            C
  H       /   \       H
    \   /       \   /
     C             C
  H /|             |\ H
  H—C              C—H
  H /               \ H
       \         /
            C
          /   \
         H     H
```

```
         H    O—H
           \ /
            C
  H       /   \       H
    \   /       \   /
     C             C
  H /|             |\ H
  H—C              C—H
  H /               \ H
       \         /
            C
          /   \
         H     H
```

ANSWER 57

```
     H         H
     |  ..     |
  H—C—O—C—H
     |  ..     |
     H         H
```

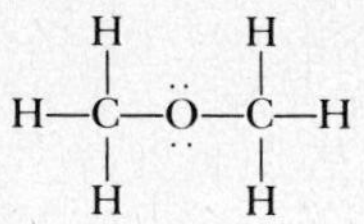

ANSWER 72

```
     H  H      H  H
     |  |  ..  |  |
  H—C—C—N—C—C—H
     |  |  |   |  |
     H  H  C   H  H
          /|\
         H C H
          /|\
         H H H
```

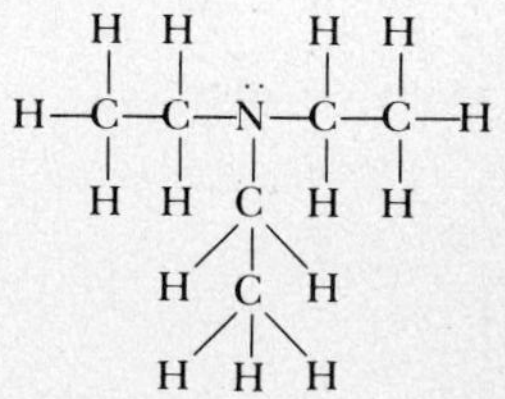

ITEM 14 Draw the Lewis diagrams for *trans*-2-buten-1-ol and *cis*-2-buten-1-ol.

ITEM 29 The Lewis diagram for butanal is __________, and the Lewis diagram for 3-butenal is __________.

ITEM 44 Aldehydes in which the HC═O group is bonded directly to a benzene ring are known as *benzaldehydes.* Benzaldehyde, the parent compound, is shown at the right. Draw the structural formula for *para*-methylbenzaldehyde.

ITEM 59 Write structural formulas for diphenyl ether and *n*-propyl cyclobutyl ether.

ITEM 74 Amines in which the nitrogen atom is bonded directly to a benzene ring are called *anilines.* The parent compound is shown at the right. Draw a diagram for *N*-methylaniline; *N*-methyl indicates that the methyl is bonded to the nitrogen atom.

```
     H  O..
      \ //
       C
       |
  H    C    H
   \ /   \ /
    C     C
    |     |
    C     C
   / \   / \
  H    C    H
       |
       H
```

```
    H—N̈—H
      |
  H   C   H
   \ / \ /
    C   C
    |   |
    C   C
   / \ / \
  H   C   H
      |
      H
```

ANSWER 13

```
    H   H   H   H   H
    |   |   |   |   |
H — C — C — C — C — C — H
    |   |   |   |   |
    H   O   H   H   H
        |
        H
```

```
            H
            |
    H   H   O   H   H   H   H
    |   |   |   |   |   |   |
H — C — C — C — C — C — C — C — H
    |   |   |   |   |   |   |
    H   H   H   H   H   H   H
```

ANSWER 28 propanal

(1-Propanal is redundant. The common name is propionaldehyde.)

ANSWER 43

```
    O
    ‖
H — C — H
```

```
    H       H
    |      /
H — C — C
    |      \\
    H       O
```

ANSWER 58

```
    H         H   H
    |   ..    |   |
H — C — O — C — C — H
    |   ..    |   |
    H         H   H
```

ANSWER 73

```
 H      H      H
   \     \    /
 H — C —— C        H
     |      \     /   ..
     |        C — N — H
     |      /       |
 H — C —— C         H
   /     /    \
 H      H      H
```

ITEM 15

Draw Lewis diagrams for 3-chloro-1-octanol and 2-methyl-2-propanol (usually known as *t*-butyl alcohol).

ITEM 30

The Lewis diagram for 3,4-dibromopentanal is ________.

ITEM 45

Aldehydes react with oxygen to form *carboxylic acids* that contain the partial structure H—O—C=O. As the name suggests, most carboxylic acids can donate ________ to water molecules.

ITEM 60

All ketones contain the partial structure

```
     O
     ‖
C—C—C
```

and all aldehydes contain the partial structure ________. All ethers contain the partial structure ________.

ITEM 75

The prefix "amino" indicates an NH_2 group, as in *p*-aminophenol. Draw a diagram for this phenol.

ANSWER 14

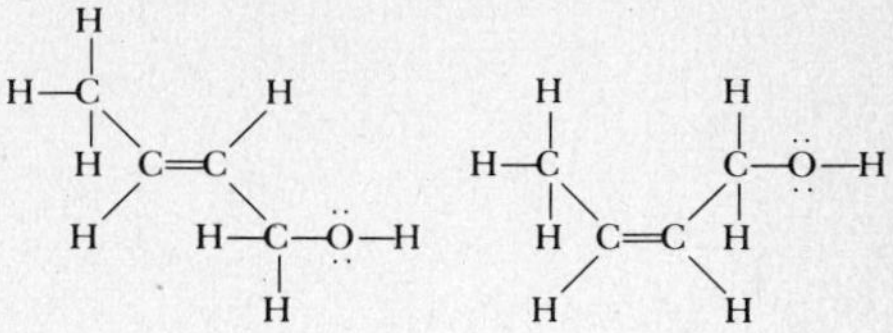

trans

cis

ANSWER 29

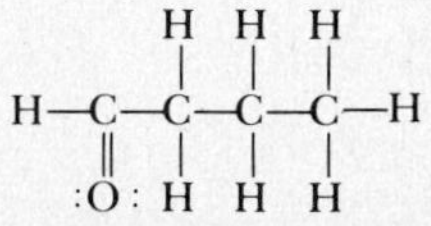

(The common name is butyraldehyde.)

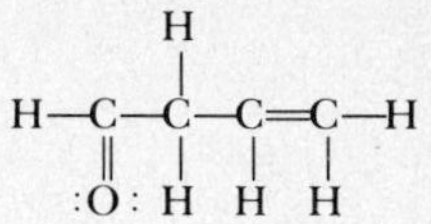

ANSWER 44

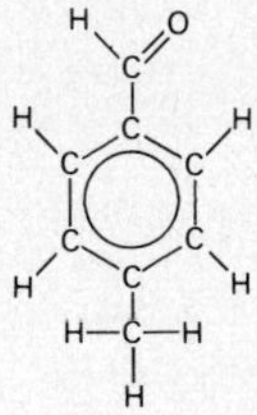

ANSWER 59

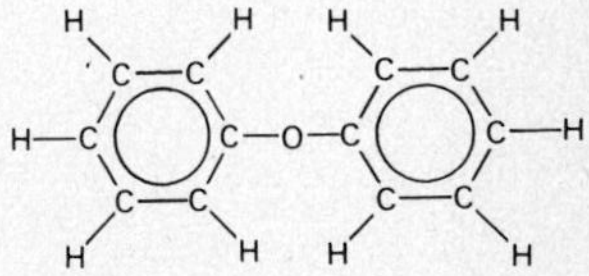

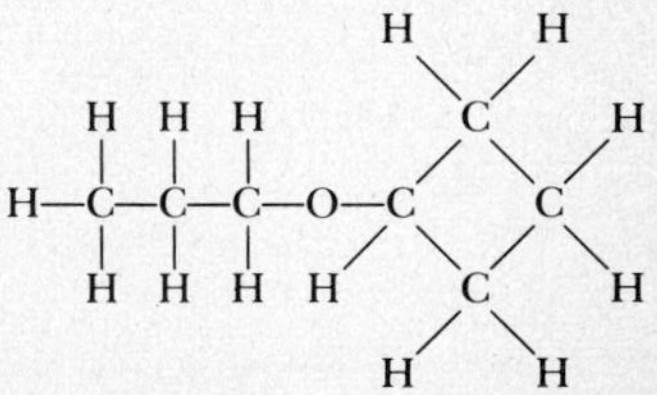

ANSWER 74

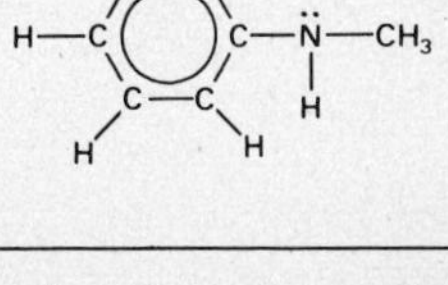

ANSWER 15

```
            H   H  :Cl: H   H   H   H   H
      ..    |   |   |   |   |   |   |   |
   H—O—C—C—C—C—C—C—C—C—H
      ..    |   |   |   |   |   |   |   |
            H   H   H   H   H   H   H   H
```

```
          H
          |
       H—C—H
    H     |
    |     |   ..
 H—C———C—O—H
    |     |   ..
    H     |
       H—C—H
          |
          H
```

(*Item 16 is on page 624.*)

ANSWER 30

```
                      ..
     H   H       :Br:     H
     |   |        |       |
 H—C—C————C————C—C—H
     |   |        |       |   ‖
     H  :Br:      H       H  :O:
          ..
```

(*Item 31 is on page 624.*)

ANSWER 45 protons (or hydrogen ions)

(*Item 46 is on page 624.*)

ANSWER 60 H—C═O
C—O—C

(*Item 61 is on page 624.*)

ANSWER 75

```
           OH
           |
   H       C       H
     \   /   \   /
       C       C
       |   ○   |
       C       C
     /   \   /   \
   H       C       H
           |
           N—H
           |
           H
```

(*Item 76 is on page 624.*)

Beyond the bare-bones knowledge of organic compounds that you now have, lie the fascinating questions of what compounds and reactions are already known, which remain to be discovered, how reactions occur, and ultimately, why do they occur?

REVIEW 13 COLLIGATIVE PROPERTIES OF SOLUTIONS

The study of solutions is important because most chemical reactions occur in solution and because many methods of purification of compounds depend on the properties of their solutions. This review will present calculation methods involving the colligative properties of solutions, the properties that depend on *how many* particles of solute are present in solution, but not *what* these solute particles are. The material treated in this review corresponds to Section 15–10 of *Chemical Principles*.

The four colligative properties we will examine are the lowering of the vapor pressure (Section 13–2), elevation of the boiling point (Section 13–3), depression of the freezing point (Section 13–4), and osmotic pressure (Section 13–5) that accompany solution formation. Although the calculations in this review are concerned primarily with the determination of molecular weights, this knowledge of solutions will serve as a base for further exploration of their properties.

13-1 MOLE FRACTION AND MOLALITY

To do calculations involving the colligative properties of solutions, you should be able to describe the relative amounts of solute and solvent in terms not only of molarity but also of mole fraction and molality. In this section you will learn the definitions of these terms and how to perform calculations involving them. Then you will be in a position to study colligative properties of solutions, the subject of the following sections.

If you can do correctly Items 35 through 42 at the end of this section, you are already prepared for Section 13-2.

ITEM 1

The total number of moles in a solution containing 2.0 moles of alcohol and 3.0 moles of water is __________.

ITEM 8

Note that in Item 7 the sum of the mole fractions is one. Prove that this is true for any binary (two-component) solution containing n_A moles of A and n_B moles of B.

ITEM 15

In a solution containing 0.50 mole of water and 0.030 mole of alcohol, we would consider __________ to be the solute; its mole fraction would be $X_A =$ __________.

ITEM 22

A 0.36-*molal* KCl solution has __________ mole(s) of KCl in 1000 g of solvent.

ITEM 29

To perform calculations involving molality, you must remember, and make use of, its definition, which is __________.

ITEM 36

A sample of solution containing 0.037 mole of solute in 1000 g of water has the molality __________. The number of moles of water present is __________; the mole fraction of solute, X_A, is __________.

ITEM 2

What fraction of the total moles in the solution at the right is alcohol?

ITEM 9

In a solution of A and B, if $X_A = 0.12$, then X_B must be 1.00 − ________ = ________.

ITEM 16

If 0.0010 mole of H_2SO_4 is dissolved in 55 moles of water, the mole fraction of H_2SO_4 is X_A=________.

ITEM 23

In a solution consisting of 0.062 mole of HCl and 450 g of alcohol, the number of moles of solute per gram of alcohol is ________; the molality, *m*, is ________.

ITEM 30

To make mole fraction calculations you must remember that

$$X_{solute} = \frac{\text{moles of solute present}}{\text{total number of moles present}}$$

and that, for very dilute solutions, this becomes X_{solute} = ________.

ITEM 37

What is the mole fraction of solute in a solution containing 0.00241 mole of solute and 1.96 moles of solvent?

2.0 moles alcohol
+
3.0 moles water

0.00241 mole of solute
1.96 moles of solvent

ANSWER 1 5.0

ANSWER 8

$$X_A + X_B = \frac{n_A}{n_A + n_B} + \frac{n_B}{n_A + n_B}$$

$$= \frac{n_A + n_B}{n_A + n_B} = 1$$

ANSWER 15 alcohol

$$\frac{0.030}{0.53} = 0.057$$

ANSWER 22 0.36

ANSWER 29 the number of moles of solute per 1000 g of solvent

ANSWER 36 $m = 0.037$

$$\frac{1000 \text{ g}}{18 \text{ g mole}^{-1}} = 55.6 \text{ moles}$$

$$\frac{0.037 \text{ mole}}{55.6 \text{ moles}} = 0.00067$$

ITEM 3

What fraction of the total number of moles in the solution at the right is water?

ITEM 10

Calculate the mole fractions of the three components in a solution containing 4.73 moles of water, 1.40 moles of alcohol, and 3.21 moles of acetic acid.

ITEM 17

For very dilute solutions, such as that in Item 16, n_A is much smaller than __________. In such a case, the expression for X_A can be simplified to $X_A \simeq$ __________.

ITEM 24

Molality is defined as __________.

ITEM 31

Calculate the mole fraction and the molality of KBr in a solution made by dissolving 1.0×10^{-3} mole of KBr in 200 g of water.

ITEM 38

To calculate the molality of the solution in Item 37, we need to know the molecular weight of the solvent ($MW_B = 98$). Then the weight of solvent present is __________; the moles of solute per 1000 g solvent is __________.

0.12 mole water
+
0.79 mole alcohol

$$X_A = \frac{n_A}{n_A + n_B}$$

0.00241 mole of solute
1.96 moles of solvent
(*MW* of solvent = 98)

ANSWER 2 $\frac{2.0}{5.0} = 0.40$

ANSWER 9 0.12
0.88

ANSWER 16 $\frac{0.0010}{55} = 1.8 \times 10^{-5}$

ANSWER 23 $\frac{0.062 \text{ mole}}{450 \text{ g}}$

$= 1.4 \times 10^{-4} \text{ mole g}^{-1}$

$10^3 \times 1.4 \times 10^{-4} = 0.14$

ANSWER 30 $\frac{\text{moles of solute}}{\text{moles of solvent}}$

ANSWER 37

$$X_A = \frac{0.00241 \text{ mole}}{1.96 \text{ moles}} = 0.00123$$

ITEM 4

In this solution the fraction of moles contributed by water is 0.12/0.91 = 0.13. We say that the *mole fraction* of water is 0.13. The mole fraction of alcohol in the same solution is ________.

ITEM 11

The sum of the mole fractions in Answer 10 is ________. The sum of the mole fractions of all substances in any solution must equal ________.

ITEM 18

Calculate X_A for a solution of 0.020 g of sulfuric acid, H_2SO_4, in 50 g of benzene, C_6H_6. (See the periodic table inside the front cover for atomic weights.)

ITEM 25

Note the difference between "molality" and the term used in solution stoichiometry, that is, "molarity." The latter is defined as ________.

ITEM 32

The number of moles of solute is often designated by n_A. If we consider solutions that contain 1000 g of solvent, we see from the definition of molality that the symbol n_A and the symbol ________ will both represent the number of moles of ________ present.

ITEM 39

For calculations of both m and X_A, it is necessary to know the number of ________ of solute.

ANSWER 3 $\frac{0.12}{0.91} = 0.13$

ANSWER 10

$$X_{\text{water}} = \frac{4.73}{9.34} = 0.506$$

$$X_{\text{alcohol}} = \frac{1.40}{9.34} = 0.150$$

$$X_{\text{acetic acid}} = \frac{3.21}{9.34} = 0.344$$

ANSWER 17 n_B

$$\frac{n_A}{n_B}$$

ANSWER 24 the number of moles of solute per 1000 g of solvent

ANSWER 31

$$X_{\text{KBr}} = \frac{1.0 \times 10^{-3} \text{ mole}}{200 \text{ g}/(18 \text{ g mole}^{-1})} = 9.0 \times 10^{-5}$$

$$m_{\text{KBr}} = \left(\frac{1.0 \times 10^{-3} \text{ mole}}{200 \text{ g}}\right)1000$$

$$= 5.0 \times 10^{-3} \text{ mole } (1000 \text{ g})^{-1}$$

ANSWER 38

$$(1.96 \text{ moles})(98 \text{ g mole}^{-1}) = 192 \text{ g}$$

$$m = \left(\frac{0.00241 \text{ mole}}{192 \text{ g}}\right)(1000)$$

$$= 0.0126$$

ITEM 5

The mole fraction of a substance in any mixture is defined as the number of moles of that substance divided by the total ________ in the mixture.

ITEM 12

In a solution containing a small amount of substance A and a large amount of substance B, the first (A) is often called the *solute* and the second (B) is called the ________. Such a solution is a *dilute* solution of substance A in substance B.

ITEM 19

The number of moles of water in the solution at the right is ________; the mole fraction of KCl is ________.

ITEM 26

What would you do to prepare a 0.15-molal aqueous solution of acetic acid from a 0.15-mole sample of acetic acid?

ITEM 33

For dilute solutions $X_A = n_A/n_B$. We shall now see that X_A can be simply related to m, the molality. Let MW_B be the molecular weight of the solvent. For a solution containing 1000 g of solvent, $n_B =$ ________.

ITEM 40

Moreover, for calculations of m we also need to know the weight of ________ and, for calculations of X_A, the ________ of solvent.

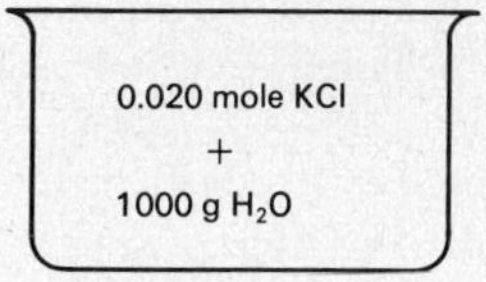

ANSWER 4 $\frac{0.79}{0.91} = 0.87$

ANSWER 11 1.000
one

ANSWER 18

$$n_B = \frac{50 \text{ g}}{78 \text{ g mole}^{-1}} = 0.64 \text{ mole}$$

$$n_A = \frac{0.020 \text{ g}}{98 \text{ g mole}^{-1}} = 0.00020 \text{ mole}$$

$$X_A = \frac{0.00020 \text{ mole}}{0.64 \text{ mole}} = 0.00031$$

ANSWER 25 the number of moles of solute per liter of solution

ANSWER 32 m
solute

ANSWER 39 moles

ITEM 6

If n_A moles of substance A and n_B moles of substance B form a solution, the mole fractions X_A and X_B are given by the relationships $X_A =$ ________ and $X_B =$ ________.

ITEM 13

If n_B represents the number of moles of solvent and n_A the number of moles of solute in a solution, then the mole fraction of solvent is $X_B =$ ________; the mole fraction of solute is $X_A =$ ________.

ITEM 20

The relative amounts of solute and solvent can also be expressed by the *molality, m,* which is defined as the number of moles of solute per 1000 g of solvent. In a solution containing 0.020 mole of KCl and 1000 g of H_2O, the molal concentration of KCl is ________.

ITEM 27

What weight of ethyl alcohol, C_2H_5OH, would you add to 1000 g of water to prepare a 0.50-molal solution?

ITEM 34

If a dilute solution is assumed, substitution for n_A and n_B in terms of the molality and the molecular weight of solvent gives $X_A = n_A/n_B =$ ________.

ITEM 41

Calculate the mole fraction and the molality of HBr in a solution of 0.49 g of HBr in 60 g of acetic acid, CH_3COOH.

ANSWER 5 number of moles

ANSWER 12 solvent

ANSWER 19

$$\frac{1000\ \text{g}}{18\ \text{g mole}^{-1}} = 55.6\ \text{moles}$$

$$X_A = \frac{0.020\ \text{mole}}{55.6\ \text{mole}} = 3.6 \times 10^{-4}$$

ANSWER 26 Dissolve the acetic acid in 1000 g of water.

ANSWER 33 $1000/MW_B$

ANSWER 40 solvent
number of moles

ITEM 7

The mole fractions of a solution containing 1.34 moles of A and 0.260 mole of B are $X_A =$ __________ and $X_B =$ __________.

ITEM 14

The colligative properties of solutions are usually dealt with in terms of the amount of solute relative to the total amount of solution. Calculations involving these properties, therefore, use the variable X_A, which is the __________ of __________.

ITEM 21

In a solution containing 0.27 mole of solute and 325 g of solvent, the number of moles of solute per gram of solvent is __________; the molality, moles per 1000 g, is $m =$ __________.

ITEM 28

To make a 0.50-molal solution of ethyl alcohol by adding alcohol to 250 g of water, you would not add 23 g, as in Item 27, but only __________ × 23 g = __________ g.

ITEM 35

According to the relation $X_A = (MW_B/1000)m$, which is valid for __________ solutions, the molality of the solution and the mole fraction of solute are proportional to one another; the proportionality factor depends only on the molecular weight of the __________.

ITEM 42

Show again that, for dilute solutions, the values of m and X_A are proportional to one another (a proportionality that will be used in the following section): $X_A = n_A/n_B =$ __________.

ANSWER 6

$$\frac{n_A}{n_A + n_B}$$

$$\frac{n_B}{n_A + n_B}$$

ANSWER 13

$$\frac{n_B}{n_A + n_B}$$

$$\frac{n_A}{n_A + n_B}$$

ANSWER 20 0.020

ANSWER 27

0.50 mole (46.1 g mole^{-1}) = 23 g

ANSWER 34

$$\frac{m}{1000/MW_B} = \left(\frac{MW_B}{1000}\right)m$$

ANSWER 41

$$\text{moles HBr} = \frac{0.49 \text{ g}}{81 \text{ g mole}^{-1}} = 0.0060$$

$$\text{moles acetic acid} = \frac{60 \text{ g}}{60 \text{ g mole}^{-1}} = 1.0$$

$$X_{HBr} = \left(\frac{0.0060 \text{ mole}}{1.0 \text{ mole}}\right) = 0.0060$$

$$m_{HBr} = \left(\frac{0.0060 \text{ mole}}{60 \text{ g}}\right)1000 = 0.10$$

ANSWER 7 $\frac{1.34}{1.60} = 0.838$

$\frac{0.260}{1.60} = 0.162$

(*Item 8 is on page 660.*)

ANSWER 14 mole fraction
solute

(*Item 15 is on page 660.*)

ANSWER 21 $\frac{0.27 \text{ mole}}{325 \text{ g}} = 0.00083$

$(1000)(0.00083) = 0.83$

(*Item 22 is on page 660.*)

ANSWER 28 $\left(\frac{250}{1000}\right)$

5.8 g

(*Item 29 is on page 660.*)

ANSWER 35 dilute
solvent

(*Item 36 is on page 660.*)

ANSWER 42 $X_A = n_A/n_B$

$$= \frac{m}{1000/MW_B} = \left(\frac{MW_B}{1000}\right)m$$

13-2 VAPOR PRESSURE LOWERING

Now that the background material relating the mole fraction of solute to molality has been developed, we can consider four properties of solutions that depend on these quantities. The treatment here is limited, first, to binary solutions (i.e., to a single solute dissolved in a solvent). Second, we will examine only binary solutions in which the solute is relatively nonvolatile and the solute molecules show little tendency to associate, or dissociate, when dissolved in the solvent.

The principal application of the theory developed in this review is the determination of the molecular weight of the solute. When you have finished this section, you will be able to calculate the molecular weight of a compound from data on its concentration in a solution and the decrease in vapor pressure of the solution from that of the pure solvent. Items 21 and 22 will test your skill in doing this.

It is assumed in this review that you are familiar with the concept of vapor pressure and its relationship to the boiling point and freezing point of a pure liquid. We will begin by considering the effect that the addition of solute to a solvent has on the vapor pressure of the solvent. You will find that the resulting decrease in the vapor pressure of the solvent is one of the colligative properties of the solution.

ITEM 1

First let us examine a solution in which the solute is nonvolatile (i.e., the solute has a negligible vapor pressure). As the solute is added to the solvent, the solvent vapor pressure, P_B, decreases, as shown at the right. Because this relationship is linear, with $P_B = 0$ when $X_B = 0$, we can write $P_B \propto$ ________ or $P_B =$ (const) ________.

ITEM 5

For dilute solutions it is usually more convenient to consider the amount of solute present and to have an expression involving X_A, the ________ of solute.

ITEM 9

The relationship at the right was shown in Section 13–1 to be a good approximation for dilute solutions. Thus, $\Delta P = P_B{}^0 X_A$ can be written $\Delta P =$ (________) m.

ITEM 13

When 1.76 g of a solute is dissolved in 1000 g of CCl_4, the lowering of the vapor pressure is 0.58 torr at 20°C. The solution must have a molality of ________; therefore, 1.76 g of solute must be equivalent to ________ mole(s).

ITEM 17

Since 0.165 mole of the solute weighs 13.7 g, the weight per mole (i.e., the ________ weight) is ________ g.

ITEM 21

Calculate the molecular weight of compound Y if a 6.9-g sample dissolved in 150 g of benzene lowers the vapor pressure of benzene at 50°C by 9.8 torr.

ITEM 25

The value of the constant is best obtained experimentally from measurements of ΔP for solutions of known ________, but the value of the constant can also be calculated from the ________ and ________ of the pure solvent.

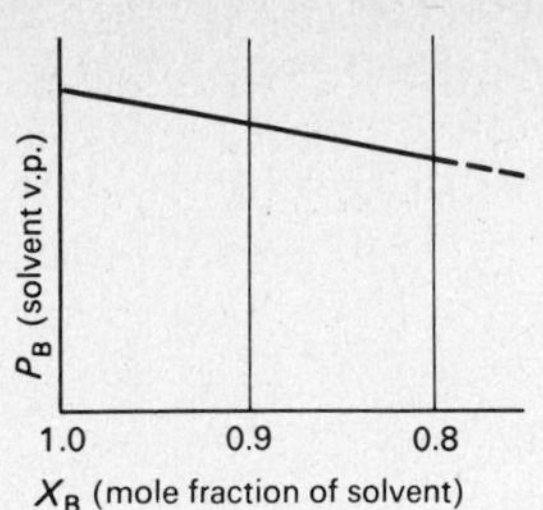

$$X_A \simeq \frac{n_A}{n_B} = \frac{m}{1000/MW_B}$$

$$\simeq \left(\frac{MW_B}{1000}\right)m$$

For CCl_4:
$\Delta P = (14.0)m$ at 20°C

For C_6H_6 at 50°C: $\Delta P = 21.2m$
(for ΔP in units of torr)

ITEM 2

A pure solvent (containing no solute) has a value of X_B equal to ________ and a vapor pressure denoted by P_B^0. Application of the equation $P_B = (\text{const})X_B$ to the pure solvent shows that (const) = ________.

ITEM 6

Since the mole fractions X_A and X_B are related by $X_A + X_B = 1$, then $X_B =$ ________. If this relation is substituted in $P_B = P_B^0 X_B$, we obtain $P_B =$ ________.

ITEM 10

You can see from the expression

$$\Delta P = (MW_B P_B^0/1000)m$$

that ΔP is ________ to m and that the proportionality factor depends only on properties of the ________.

ITEM 14

Since 1.76 g of solute in Item 13 is 0.041 mole, there are ________ g per mole.

ITEM 18

The value of the proportionality constant ($MW_B P_B^0$/1000) for the solvent acetone at 20°C is 10.7 torr for a 1-molal solution. Calculate the molecular weight of compound D if a 3.54-g sample lowers the vapor pressure of 100 g of acetone by 2.83 torr.

ITEM 22

At 50°C, the vapor pressure of water is 92.5 torr. Calculate the molecular weight of compound Z if a 7.84-g sample dissolved in 50.0 g of water lowers the vapor pressure to 89.6 torr.

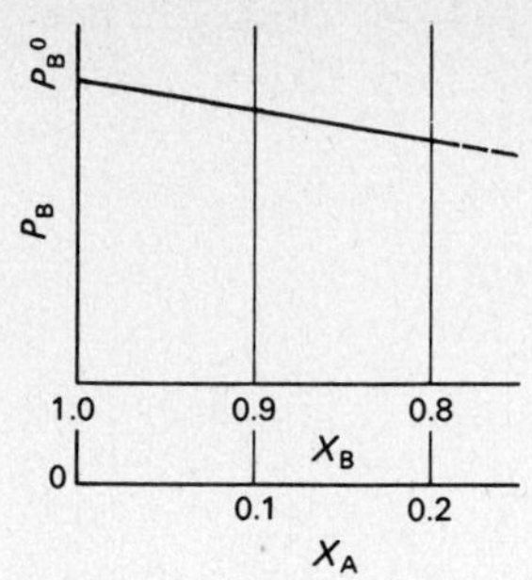

ANSWER 1 X_B
X_B

ANSWER 5 mole fraction

ANSWER 9 $\dfrac{MW_B P_B^0}{1000}$

ANSWER 13

$$m = \frac{0.58 \text{ torr}}{14.0 \text{ torr for 1-molal solution}} = 0.041$$

0.041

ANSWER 17 molecular

$$\frac{13.7 \text{ g}}{0.165 \text{ mole}} = 83$$

ANSWER 21

6.9 g in 150 g benzene = 46 g in 1000 g
$\Delta P = 21.2m$

$$m = \frac{9.8 \text{ torr}}{21.2 \text{ torr per 1-molal solution}} = 0.46$$

$$MW_Y = \frac{46 \text{ g}}{0.46 \text{ mole}} = 100 \text{ g mole}^{-1}$$

ANSWER 25 molality
molecular weight
vapor pressure

ITEM 3

Solutions whose vapor pressure behavior conforms to the graph at the right (i.e., to the equation __________) obey Raoult's law.

ITEM 7

Addition of solute lowers the vapor pressure of the solvent by the amount $\Delta P = P_B^0 - P_B$. Since $P_B = (P_B^0 - P_B^0 X_A)$, we have $\Delta P = P_B^0 -$ __________ $=$ __________.

ITEM 11

For carbon tetrachloride at 20°C, for example, the value of $(MW_B P_B^0/1000)$ is 14.0 torr per 1.00-molal solution. It follows that the vapor pressure lowering when a 1.00-molal solution is made by using CCl_4 as solvent with any __________ will be __________.

ITEM 15

In the solution at the right there is __________ g of solute per gram of CCl_4, or __________ g of solute per 1000 g of CCl_4.

ITEM 19

The proportionality constant between ΔP and m can be determined experimentally by measuring ΔP for a solution of known molality. The value of the constant can be calculated also if the molecular weight and __________ of the pure solvent are known.

ITEM 23

To obtain the molecular weight of a nonvolatile material you need (1) the __________ of solute in a known weight of solvent and (2) the number of __________ of solute in the same weight of solvent.

3

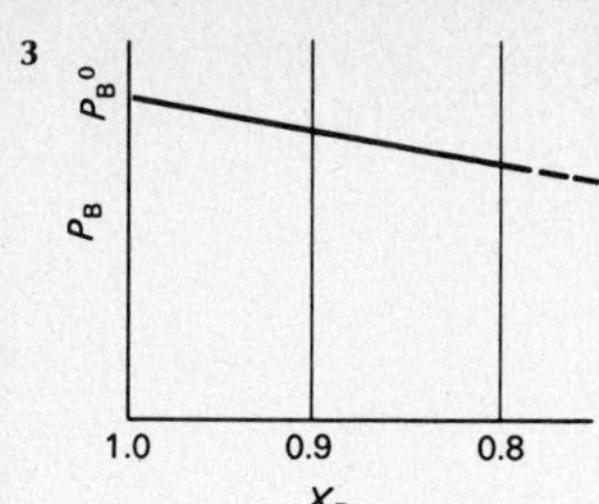

7

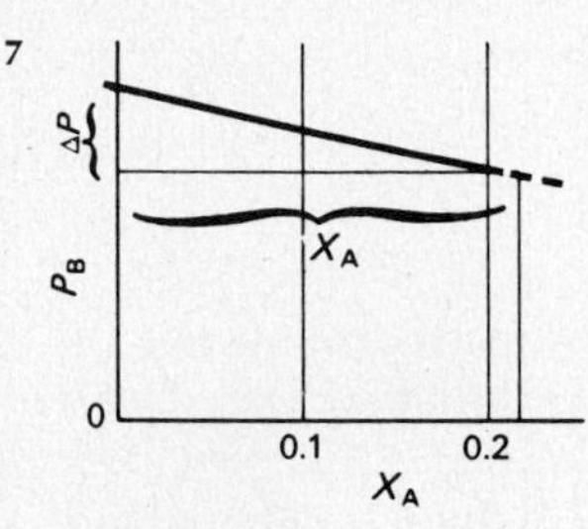

11 $\Delta P = \left(\frac{MW_B P_B^0}{1000}\right) m$

15

1.04 g solute
+
76.0 g CCl_4 (at 20°C)

$\Delta P_{CCl_4} = (14.0)m$ at 20°C

19 $\Delta P = \left(\frac{MW_B P_B^0}{1000}\right) m$

ANSWER 2 one (1.0)
P_B^0

ANSWER 6

$1 - X_A$
$P_B = P_B^0(1 - X_A) = P_B^0 - P_B^0 X_A$

ANSWER 10 directly proportional
solvent

ANSWER 14 $\frac{1.76 \text{ g}}{0.041 \text{ mole}} = 43 \text{ g mole}^{-1}$

ANSWER 18

$3.54 \text{ g}(100 \text{ g of solvent})^{-1}$
$= 35.4 \text{ g}(1000 \text{ g solvent})^{-1}$

$$m = \frac{\Delta P}{(MW_B P_B^0/1000)}$$

$$= \frac{2.83 \text{ torr}}{10.7 \text{ torr per 1-molal solution}}$$

$$= 0.264$$

$$MW_D = \frac{35.4 \text{ g}}{0.264 \text{ mole}} = 134 \text{ g mole}^{-1}$$

ANSWER 22

7.84 g in 50.0 g H_2O = 157 g in 1000 g

$$m = \frac{\Delta P}{MW_B P_B^0/1000} = \frac{1000 \text{ g} \times 2.9 \text{ torr}}{18.0 \text{ g mole}^{-1} \times 92.5 \text{ torr}}$$

$$= 1.74$$

$$MW_Z = \frac{157 \text{ g}}{1.74 \text{ mole}} = 90 \text{ g mole}^{-1}$$

ITEM 4

Raoult's law, which is a reasonably good approximation for dilute solutions, states that the vapor pressure of the solvent varies directly with the __________ of the solvent.

ITEM 8

For dilute solutions obeying __________ law, the lowering of the vapor pressure of the solvent is proportional to the mole fraction of __________.

ITEM 12

Similarly, any solution consisting of a solute in CCl_4 at 20°C that gives a vapor pressure lowering of 14.0 torr must have m equal to __________; that is, it contains __________ mole of solute for every 1000 g of CCl_4.

ITEM 16

For the solution of Item 15, $\Delta P = 2.31$ torr. Thus, there is __________ mole of solute per 1000 g CCl_4; that is, 13.7 g solute is __________ mole.

ITEM 20

The vapor pressure of pure benzene, C_6H_6, at 50°C is 271.3 torr. What is the value of the proportionality constant between ΔP and m at 50°C?

ITEM 24

The number of moles of solute in 1000 g of solvent (i.e., the __________) can be obtained from the relation $\Delta P = (\text{const})m$ once the value of the constant is known.

ANSWER 3 $P_B = (\text{const})X_B$ or $P_B = P_B{}^0X_B$

$P_B = P_B{}^0X_A$

ANSWER 7 $(P_B{}^0 - P_B{}^0X_A)$
$P_B{}^0X_A$

$\Delta P = 14.0m$
(for ΔP in units of torr)

ANSWER 11 solute
14.0 torr

13.7 g solute
per 1000 g CCl_4
(at 20°C)

$\Delta P_{CCl_4} = (14.0)m$ (at 20°C)

ANSWER 15 $\frac{1.04\text{ g}}{76.0\text{ g}} = 0.0137$

$0.0137(1000) = 13.7$

ANSWER 19 vapor pressure

ANSWER 23 weight
moles

ANSWER 4 mole fraction

(*Item 5 is on page 678.*)

ANSWER 8 Raoult's
solute

(*Item 9 is on page 678.*)

ANSWER 12 1.00
1.00

(*Item 13 is on page 678.*)

ANSWER 16

per 1000 g solvent: $\dfrac{2.31 \text{ torr}}{14.0 \text{ torr}} = 0.165$

0.165

(*Item 17 is on page 678.*)

ANSWER 20

$$\frac{MW_B P_B^0}{1000} = \frac{78.1 \text{ g mole}^{-1} \times 271.3 \text{ torr}}{1000 \text{ g}}$$

$$= 21.2 \text{ torr per 1-molal solution}$$

(*Item 21 is on page 678.*)

ANSWER 24 molality

(*Item 25 is on page 678.*)

13-3 BOILING POINT ELEVATION

We have noted that the lowering of the vapor pressure, ΔP, of a particular solvent at a given temperature depends on the relative numbers of moles of solute and solvent present, hence, on the molality of the solution. You have seen how this colligative property of a solution can be used to determine the molecular weight of the solute.

Other colligative properties of solutions often can be measured more conveniently, and these also can be used to determine molecular weights. The relationships of these properties to the molality of the solutions can be deduced from the effect of the solute on the vapor pressure of the solvent. The next colligative property to be studied is the raising, or elevation, of the boiling point that accompanies the addition of a nonvolatile solute to a solvent.

Typical of the problems you should be able to solve at the end of the review is this: "A solution of 46.8 g of a solute in 1000 g water has a ΔT_b value of 0.38°C. The freezing point depression constant for water is 0.51. Calculate the molecular weight of the solute."

ITEM 1

The vapor pressure of a liquid, as shown at the right, always __________ as the temperature rises.

ITEM 4

The two vapor pressure curves in the accompanying figure represent pure solvent and dilute solution containing a nonvolatile solute. Label the curves *solvent* and *solution.*

ITEM 7

The difference in temperature between the boiling point of the solution and the boiling point of the pure __________ is known as the *boiling point elevation.*

ITEM 10

Since, for dilute solutions, ΔT_b and ΔP are proportional to one another, we can write $\Delta T_b \propto$ __________ or __________ = __________.

ITEM 13

As was true of the proportionality constant between ΔP and m, the constant k_b depends on the nature of the __________ but is not influenced by the nature of the __________.

ITEM 16

The k_b values for various solvents at the right can be regarded as indicating the boiling point __________ for 1-molal solutions of nonvolatile solutes in these solvents. (Theoretically, k_b is $RT^2MW_B/1000\Delta H_{vap}$, in which R is the gas constant, T is the boiling point, MW_B is the molecular weight of the solvent, and ΔH_{vap} is the molar heat of vaporization. The constant k_b is nearly always determined experimentally.)

ITEM 19

The molal boiling point elevation constant for carbon tetrachloride is 5.03. If a solution containing 0.929 g of compound L dissolved in 89.3 g of carbon tetrachloride has a boiling point elevation of 0.460 deg, what is the molecular weight of L?

1

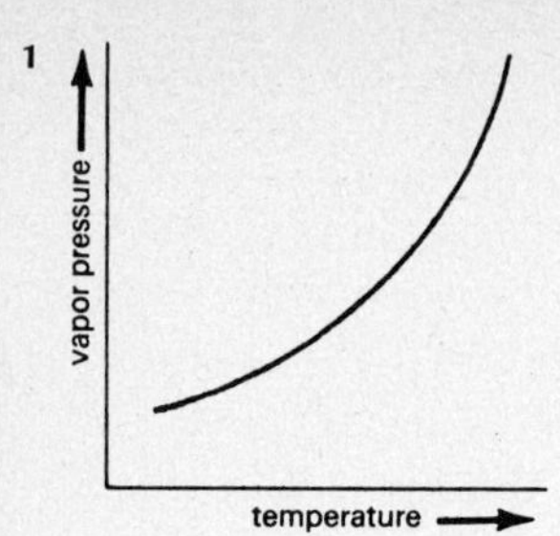

4

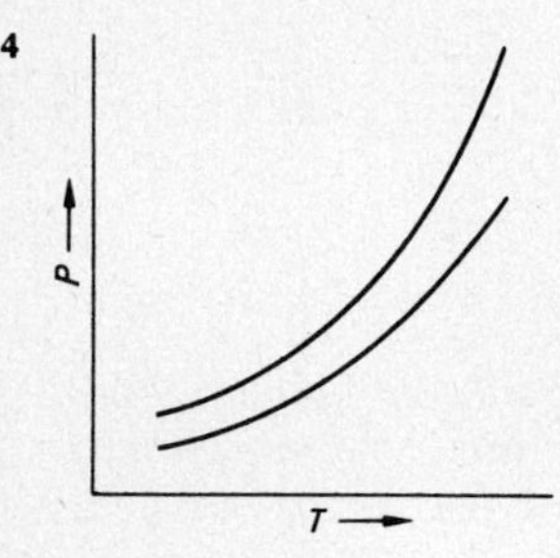

16

Solvent	k_b
Water	0.5
Benzene	2.5
Chloroform	3.7
Ethyl alcohol	1.2

ITEM 2 The boiling point of a liquid is the temperature at which the vapor pressure is equal to the external pressure. Thus, the boiling point can be located in the figure at the right as the temperature at which the dashed line, representing the external pressure, and the solid curve, representing the ________ of the liquid, intersect.

ITEM 5 Redraw these two curves, label the axes, and locate on the temperature axis the boiling points of the pure solvent and of the solution.

ITEM 8 The vapor pressure lowering and the boiling point elevation of dilute solutions are relatively small. Therefore, for comparisons of the vapor pressure curves for solvents and dilute solutions, the curves near the boiling point can be approximated, as shown, by straight parallel lines. Label these lines and the boiling points on the diagram at the right.

ITEM 11 Furthermore, in Section 13–2 we saw that ΔP is proportional to *m*, the ________. From this and the preceding discussion (Items 9 and 10), it follows that $\Delta T_b \propto$ ________.

ITEM 14 The constant k_b can be determined for a solvent by measuring ΔT_b for a solution of known ________.

ITEM 17 Since k_b for water is 0.51, an aqueous solution that boils 1.37°C higher than pure water must have a molality of ________.

ITEM 20 Chloroform, $CHCl_3$, has a boiling point of 61.3°C. A solution of 1.68 g of methylcyclohexane, C_7H_{14}, in 30.0 g of $CHCl_3$ has a boiling point of 63.4°C. What is the boiling point elevation constant of chloroform?

2

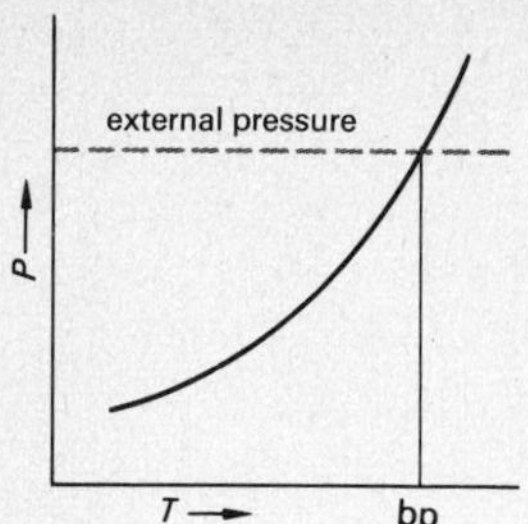

5

8

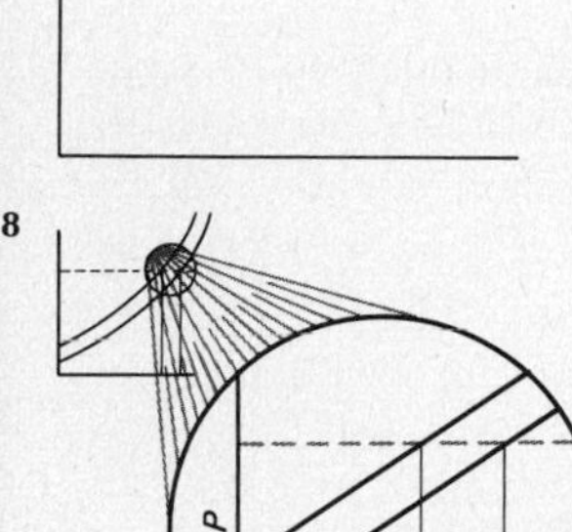

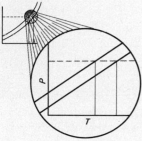

14 $\Delta T_b = k_b m$

17 $\Delta T_b = 0.51\, m$
for aqueous solutions

ANSWER 1 increases

ANSWER 4

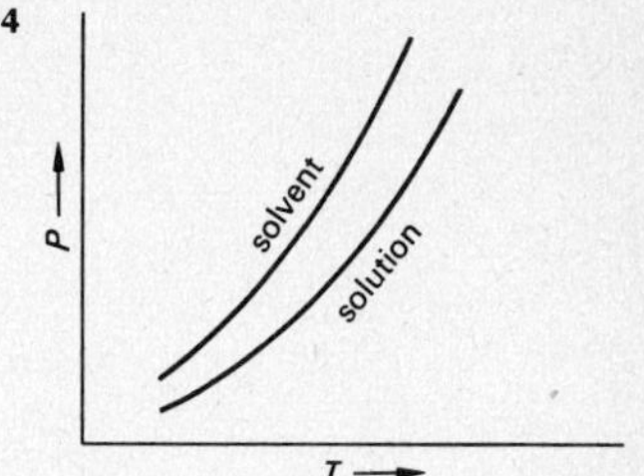

ANSWER 7 solvent

ANSWER 10 ΔP
$\Delta T_b = (\text{const})\Delta P$

ANSWER 13 solvent
solute

ANSWER 16 elevation

ANSWER 19

$$\frac{0.929 \text{ g}}{89.3 \text{ g solvent}} = 10.4 \text{ g } (1000 \text{ g})^{-1}$$

$$m = \frac{\Delta T_b}{k_b} = \frac{0.460 \text{ deg}}{5.03 \text{ deg}} = 0.0915$$

$$MW_L = \frac{10.4 \text{ g}}{0.0915 \text{ mole}} = 114 \text{ g mole}^{-1}$$

ITEM 3

As you know from Section 13–2, the addition of a nonvolatile solute to a solvent __________ the vapor pressure of the solvent.

ITEM 6

Since the vapor pressure curve for the solution lies below and to the right of the curve for the pure solvent, the boiling point of the solution is __________ than that of the pure solvent.

ITEM 9

It follows, as the figure at the right suggests, that the boiling point elevation, ΔT_b, and the vapor pressure lowering, ΔP, are __________ to one another. ($\Delta T_b = \Delta P \tan \alpha$; $\tan \alpha$ is a constant if the vapor pressure lines are parallel.)

ITEM 12

With the introduction of the proportionality constant, k_b, the relationship between the boiling point elevation and the molality of a solution can be written as the equation __________ = __________.

ITEM 15

The molality of the solution at the right is __________. The boiling point elevation allows us to calculate k_b for benzene (using the equation $\Delta T_b = k_b m$): k_b = __________.

ITEM 18

A solution of 46.8 g of solute in 1000 g of water has a ΔT_b value of 0.38 deg. The molality of the solution is __________. In 1000 g of water there are __________ g or __________ mole(s) of solute. Thus, the weight per mole, or molecular weight, can be calculated to be __________.

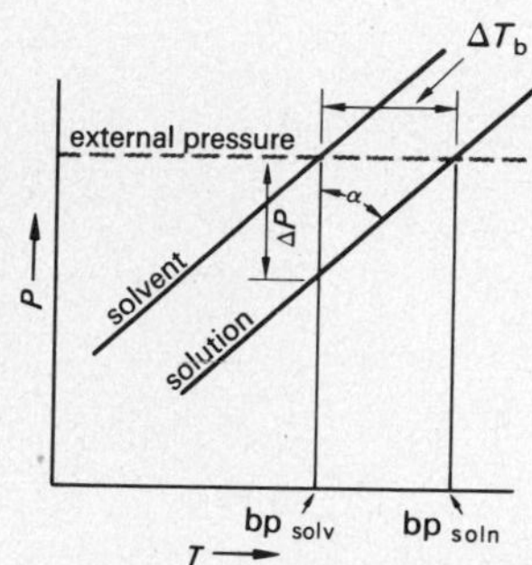

29 g naphthalene ($C_{10}H_8$)
+
1000 g benzene (C_6H_6)

boiling point 0.58°C higher than that of pure benzene

$\Delta T_b = 0.51\,m$
for aqueous solutions

ANSWER 2 vapor pressure

ANSWER 5

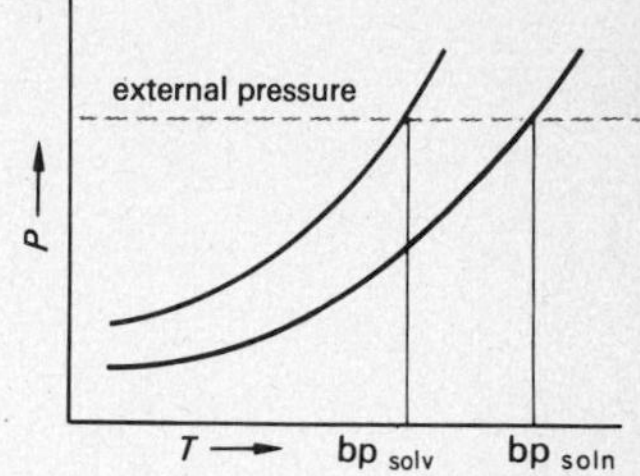

ANSWER 8

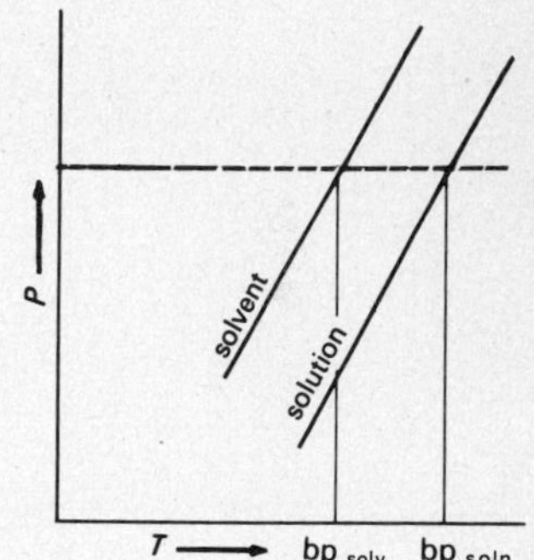

ANSWER 11 molality
m

ANSWER 14 molality

ANSWER 17

$$m = \frac{1.37 \text{ deg}}{0.51 \text{ deg per 1-molal solution}} = 2.7$$

ANSWER 20

$$\Delta T_b = k_b m = 63.4 - 61.3 = 2.1°\text{C}$$

$$m = \left[\frac{1.68 \text{ g}/(98.2 \text{ g mole}^{-1})}{30.0 \text{ g}}\right]1000 = 0.570$$

$$k_b = \frac{2.1 \text{ deg}}{0.570\,m} = 3.7 \text{ deg per 1-molal solution}$$

ANSWER 3 lowers (or decreases)

(*Item 4 is on page 690.*)

ANSWER 6 higher

(*Item 7 is on page 690.*)

ANSWER 9 directly proportional

(*Item 10 is on page 690.*)

ANSWER 12 $\Delta T_b = k_b m$

(*Item 13 is on page 690.*)

ANSWER 15

$$\text{moles } C_{10}H_8 = \frac{29 \text{ g}}{128 \text{ g mole}^{-1}} = 0.23$$

$$k_b = \frac{\Delta T_b}{m} = \frac{0.58 \text{ deg}}{0.23m}$$

$$= 2.5 \text{ deg per 1-molal solution}$$

(*Item 16 is on page 690.*)

ANSWER 18

$$\frac{0.38 \text{ deg}}{0.51 \text{ deg for 1-molal solution}} = 0.75m$$

46.8 g

0.75 mole

$$\frac{46.8 \text{ g}}{0.75 \text{ mole}} = 62 \text{ g mole}^{-1}$$

(*Item 19 is on page 690.*)

13–4 FREEZING POINT DEPRESSION

A third colligative property is the lowering, or depression, of the freezing point that generally accompanies solution formation. Freezing point depressions usually can be measured more easily than either vapor pressure lowerings or boiling point elevations and are often used to determine the molecular weight of a newly synthesized or newly isolated compound. Like the boiling point elevation, the freezing point depression is related to the molality of the solution, and this relationship can be deduced from the effect of the solute on the vapor pressure of the solvent.

Items 16 through 20 provide a test of your knowledge of this section.

ITEM 1

In the diagram at the right, curve AD represents the increase in the vapor pressure of the liquid phase with increase in temperature. Curve BA represents the increase in the vapor pressure of the ________ phase with increase in temperature.

ITEM 4

Addition of a nonvolatile solute to a liquid lowers the vapor pressure in accordance with Raoult's law, but in general, the solute has no effect on the vapor pressure of the solid phase. Label the vapor pressure curves at the right *solid solvent, liquid solvent,* and *liquid solution.* (This assumes that all the solute remains in solution.)

ITEM 7

The difference in temperature between the freezing point of the solution and the ________ is known as the *freezing point depression,* T_f, that accompanies solution formation.

ITEM 10

From trigonometry you will remember that for the right triangle

a

α

b

$a = b \tan \alpha$. In the diagram at the right, then, the line $GH = T_f \tan \alpha$ and the line $(\Delta P + GH) = \Delta T_f \tan \gamma$. Therefore, $\Delta P = (\Delta P + GH) - GH =$ ________.

ITEM 13

The constant k_f for a solvent can be calculated from the freezing point and the heat of fusion, but its value is nearly always determined experimentally. The freezing point of benzene is 5.48°C. If a solution of 0.448 g of naphthalene, $C_{10}H_8$, in 9.82 g of benzene freezes at 3.66°C, what is k_f for benzene?

ITEM 16

The freezing point of pure benzene is 5.5°C, and the value of k_f for benzene is 5.1. A 0.19-molal solution of a nonvolatile solute in benzene will have $\Delta T_f =$ ________ and will start to freeze at ________°C.

ITEM 19

A solvent often used for determining the molecular weight of organic compounds is camphor, which has a k_f of 40.0. If 0.015 g of compound Q dissolved in 1.00 g of camphor causes a freezing point depression of 5.3 deg, compound Q has a molecular weight of ________.

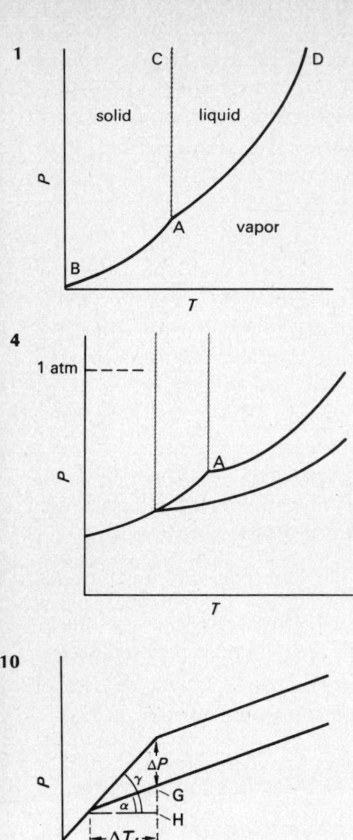
1
C
D
solid
liquid
P
A
vapor
B
T
4
1 atm
A
P
T
10
P
ΔP
γ
α
G
H
ΔT_f
T

ITEM 2

A *triple point* of a substance is defined as a temperature at which three phases are in equilibrium. In the diagram at the right, solid, liquid, and vapor phases are in equilibrium at point __________. At this temperature the vapor pressure of the liquid is equal to the vapor pressure of the __________.

ITEM 5

The diagram in Answer 4 shows that the temperature at which the solution is in equilibrium with the solid is __________ than the temperature at which the solvent is in equilibrium with the solid.

ITEM 8

Label the curves on the expanded portion from previous figures. Locate the freezing points of the solvent and the solution on the temperature axis.

ITEM 11

Since α and γ are constants, the term $(\tan \gamma - \tan \alpha)$ is a constant, and we can write $\Delta T_f \propto \Delta P$. Recalling that ΔP is proportional to m, the __________ of the solution, we write the proportionality relation $\Delta T_f \propto$ __________.

ITEM 14

The freezing point depression constant for water is 1.86 deg per 1-molal solution. A 1.00-molal solution will have a freezing point of __________°C. A 0.176-molal solution will begin to freeze at __________°C.

ITEM 17

A solution is made to contain 3.69 g of an unknown material in 28.8 g of benzene. The weight of solute per gram of benzene is __________, and the weight per 1000 g of benzene is __________.

ITEM 20

The freezing point of phenol, C_6H_5OH, is 40.5°C, and its freezing point depression constant is 7.27 deg per 1-molal solution. When 0.0820 g of a deep blue hydrocarbon is dissolved in 1.00 g of phenol, the freezing point of the solution is 35.8°C. What is the molecular weight of the hydrocarbon?

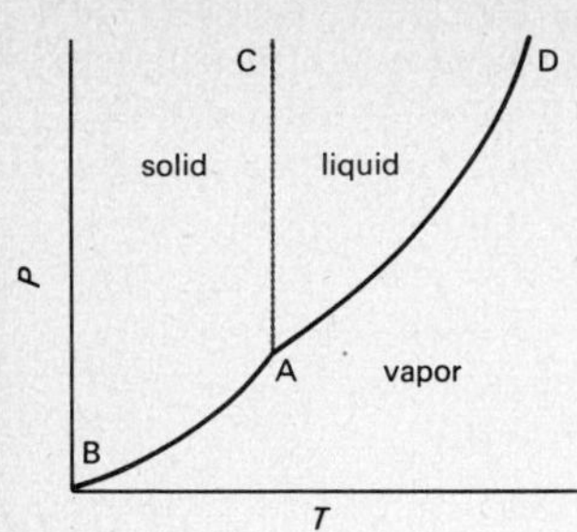

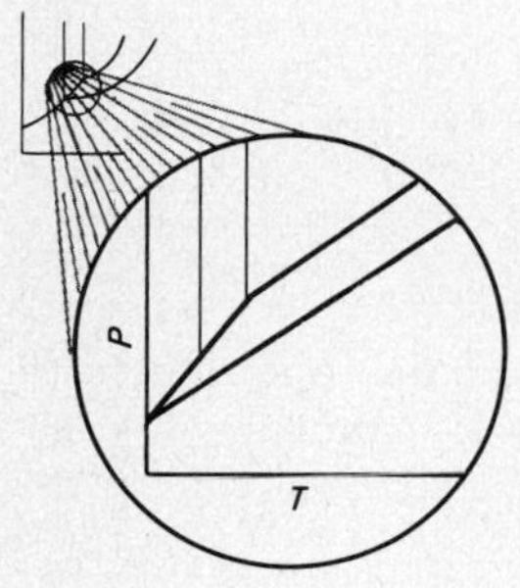

$\Delta T_f = k_f m$

ANSWER 1 solid

ANSWER 4

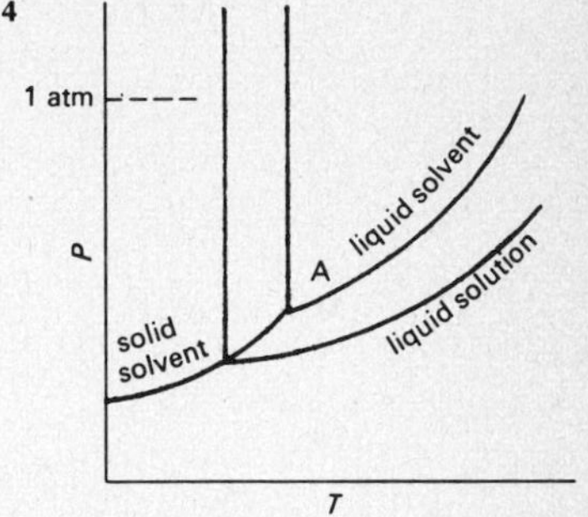

ANSWER 7 freezing point of the solvent

ANSWER 10 $\Delta T_f \tan \gamma - \Delta T_f \tan \alpha$ or $\Delta T_f(\tan \gamma - \tan \alpha)$

ANSWER 13

$$\frac{0.448 \text{ g}}{9.82 \text{ g benzene}} = 45.6 \text{ g per } 1000 \text{ g}$$

$$\frac{45.6 \text{ g}}{128 \text{ g mole}^{-1}} = 0.356 \text{ mole}$$

$$k_f = \frac{\Delta T_f}{m} = \frac{1.82 \text{ deg}}{0.356 \text{ mole}}$$

$$= 5.11 \text{ deg per 1-molal solution}$$

ANSWER 16

(5.1 deg per 1-molal solution)(0.19 molal) $= 0.97$ deg

$5.5°C - 1.0°C = 4.5°C$

ANSWER 19

$0.015 \text{ g } (1.00 \text{ g})^{-1} = 15 \text{ g } (1000 \text{ g solvent})^{-1}$

$\Delta T_f = k_f m$

$$m = \frac{5.3 \text{ deg}}{40 \text{ deg per 1-molal solution}} = 0.133$$

$$MW_Q = \frac{15 \text{ g}}{0.133 \text{ mole}} = 110 \text{ g mole}^{-1}$$

ITEM 3

At the right, the line representing the temperatures and pressures at which liquid and solid are in equilibrium is labeled __________. The *freezing point* (or *melting point*) is defined as the temperature at which liquid and solid are in equilibrium at a pressure of 1 atm. Since the liquid–solid equilibrium is only slightly pressure dependent, the freezing point is at almost exactly the same temperature as the __________.

ITEM 6

In other words, the freezing point (and the triple point) of a solution is __________ than that of the pure solvent.

ITEM 9

In a small portion of the diagram the vapor pressure curves can be approximated by straight lines and the solvent–vapor line can be assumed to be parallel to the __________ line.

ITEM 12

If the freezing point depression constant, k_f, is now introduced, an equation similar to that written for ΔT_b is obtained; that is, $\Delta T_f =$ __________.

ITEM 15

A 0.26-molal solution of sucrose in water, for which $k_f = 1.86$, will have a value of ΔT_f of __________ deg. The solution will begin to freeze at __________°C.

ITEM 18

The solution in Item 17 begins to freeze at 2.8°C. With the information at the right, we see that the molality is __________ and, since there are 128 g of solute per 1000 g of benzene, that the molecular weight of the solute is __________.

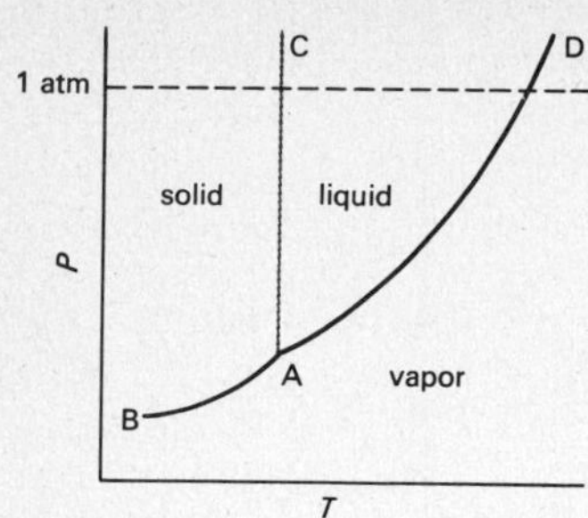

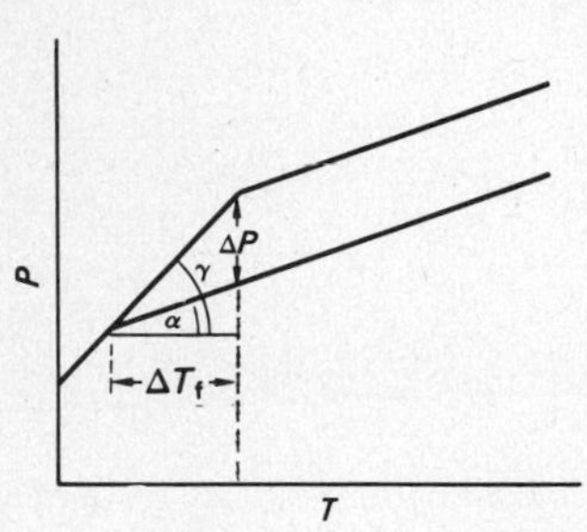

Freezing point of pure benzene = 5.5°C
k_f for benzene = 5.1

ANSWER 2 A
solid

ANSWER 5 lower

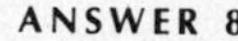
ANSWER 8

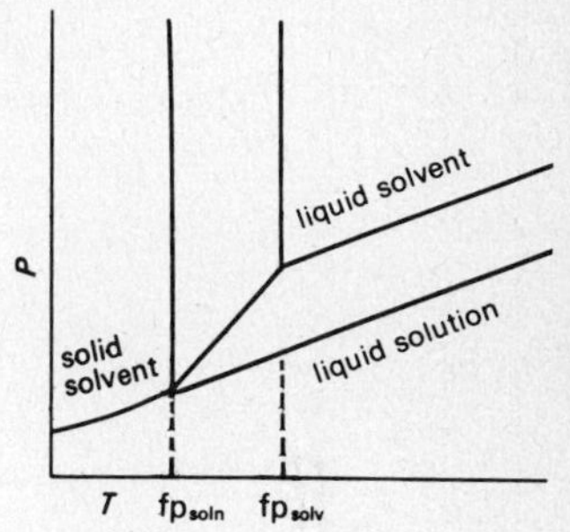

ANSWER 11 molality
m

ANSWER 14

-1.86
$\Delta T = k_f m$
$= (1.86° \text{ per 1-molal solution}) (0.176 \text{ mole})$
$= 0.327°C$
$0.00°C - 0.327°C = -0.327°C$

ANSWER 17 $\dfrac{3.69}{28.8} = 0.128 \text{ g}$

$0.128(1000) = 128 \text{ g}$

ANSWER 20

$T_f = k_f m$
$m = (40.5 - 35.8)/7.27$
$= 0.65 \text{ mole in 1000 g solvent}$
wt in 1000 g solvent
$= (0.0820/1.00)(1000) = 82.0 \text{ g}$

$$MW = \frac{82.0 \text{ g}}{0.65 \text{ mole}} = 126$$

ANSWER 3 AC
triple point

(*Item 4 is on page 700.*)

ANSWER 6 lower

(*Item 7 is on page 700.*)

ANSWER 9 solution–vapor

(*Item 10 is on page 700.*)

ANSWER 12 $k_f m$

(*Item 13 is on page 700.*)

ANSWER 15

$0.26m$(1.86 deg per 1-molal solution) = 0.48
0.00°C − 0.48°C = −0.48°C

(*Item 16 is on page 700.*)

ANSWER 18

$$\frac{5.5 \text{ deg} - 2.8 \text{ deg}}{5.1 \text{ deg per 1-molal solution}} = 0.53 \text{ molal}$$

$$\frac{128 \text{ g}}{0.53 \text{ mole}} = 240 \text{ g mole}^{-1}$$

(*Item 19 is on page 700.*)

13-5 OSMOTIC PRESSURE

The final property of a solution that depends on the number of solute molecules present, and therefore is a colligative property, is the osmotic pressure obtained when the solution is separated from the pure solvent by a semipermeable membrane. (A *semipermeable membrane* allows the solvent to pass through but prevents the passage of the solute.)

Osmotic pressure can be exhibited by arranging an apparatus as in the upper figure below and allowing the system to come to equilibrium. Solvent flows from the pure solvent compartment into the solution compartment until a certain pressure, shown by the column of liquid, builds up and prevents further flow of solvent. This pressure is called the *osmotic pressure*. The osmotic pressure can be illustrated also with an apparatus shown in the lower figure below. In this apparatus a pressure can be applied to the solution until a sensitive flow indicator shows that no solvent is flowing. Again the additional pressure that must be applied to the solution to prevent the pure solvent from flowing into the solution is the osmotic pressure.

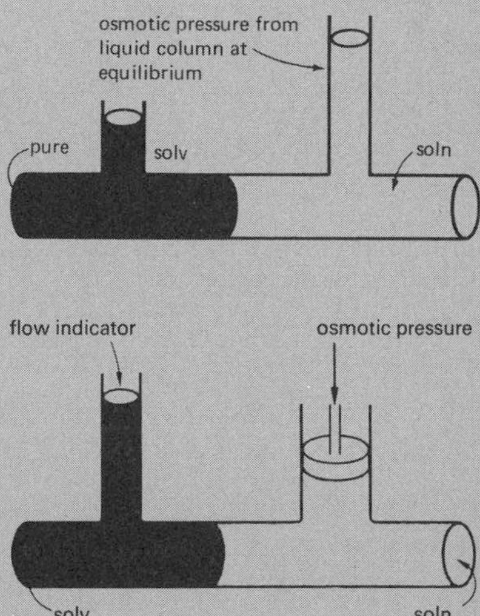

ITEM 1

Osmosis is defined as the spontaneous flow of solvent into a solution, or from a more dilute to a more concentrated solution, that occurs when the two liquids are separated from each other by a __________ membrane.

ITEM 5

The proportionality relationship π __________ can be written between π and m.

ITEM 9

The osmotic pressure of a 1.00-molal aqueous solution at 0°C is 22.4 atm. Then the general osmotic pressure equation for dilute aqueous solutions at 0°C is $\pi =$ __________.

ITEM 13

Both experimental data and theoretical calculations show that, for very dilute solutions, the proportionality constant between π and n_A/V has the value RT for any solvent. Thus, we can write the equation __________.

ITEM 17

To prevent solvent from flowing through a membrane and into the solution described in Item 16, we would have to apply pressure on the solution until the excess pressure is __________ atm.

ITEM 21

If the weight of solute per liter of solution in the preceding items is 8.32 g, the molecular weight must be __________.

ITEM 25

The sensitivity of osmotic pressure is particularly helpful when high molecular weights are determined because a large weight of solute corresponds to a relatively small number of __________ of solute.

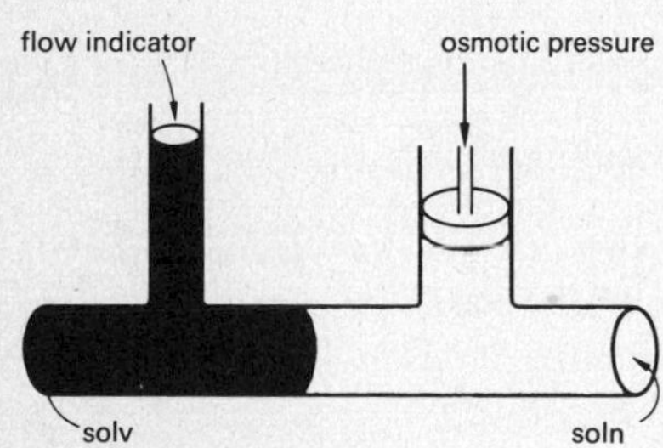

$\pi = 28.3$ torr
$n_A/V = 0.00152$ mole liter^{-1}

ITEM 2

A perfect semipermeable membrane allows __________ to pass through it but does not permit passage of a __________.

ITEM 6

If this proportionality were converted to an equality by the introduction of a constant, the value of the constant would depend only on the properties of the __________.

ITEM 10

The equation $\pi = (\text{const})m$ can be used in the same way as were the equations $\Delta P =$ __________, $\Delta T_b =$ __________, and $\Delta T_f =$ __________.

ITEM 14

The constant R is identical to the gas constant that appears in the ideal gas equation $PV = nRT$. Rearranged to this form, the osmotic pressure relationship is __________.

ITEM 18

In applying the equation $\pi V = n_A RT$ you must be careful to have the same units of pressure and volume on both sides of the equation. Therefore, the pressure units of π and the volume units of V must be the same as those implied by the value of __________ used in the equation.

ITEM 22

At 27°C, 0.00647 g of a deep blue hydrocarbon is dissolved in enough cyclohexane, C_6H_{12}, to give 10.0 ml of solution. The osmotic pressure of the solution is 94.3 torr. What is the molecular weight of the blue compound?

ITEM 26

Of the four colligative properties of solutions, __________, __________, __________, and __________, osmotic pressure measurements are used most to determine the molecular weights of high molecular weight materials.

ANSWER 1 semipermeable

ANSWER 5 $\propto m$

ANSWER 9 $(22.4)m$ atm

ANSWER 13 $\pi = RTn_A/V$

ANSWER 17 24.4

ANSWER 21 $\dfrac{8.32 \text{ g}}{0.00152 \text{ mole}} = 5470$

ANSWER 25 moles

ITEM 3

Osmotic pressure is defined as the excess pressure that must be applied to a __________ to prevent osmosis when the two liquids are separated by __________.

ITEM 7

The value of the constant in the relationship $\pi =$ (const)m can be determined by measuring the __________ for a solution of known __________.

ITEM 11

An interesting looking osmotic pressure equation is found for *very* dilute solutions if we use n_A/V (in which n_A is the moles of solute and V is the volume of solution) rather than the __________, which we have been using, to express the concentration of solute.

ITEM 15

Or, since n_A/V in units of moles per liter of solution is the same as M, the __________, the equation can be written __________.

ITEM 19

A solution of a newly synthesized material produces an osmotic pressure of 28.3 torr when separated by a membrane from pure solvent at 25°C. The value of π in atmospheres is __________, and the number of moles of solute per liter of solution is __________.

ITEM 23

Calculate the osmotic pressure of a 1.00×10^{-3}-molar aqueous solution at 25°C against pure water.

ANSWER 2 solvent
solute

ANSWER 6 solvent

ANSWER 10 (const)m or $(MW_B P_B^0/1000)m$
$k_b m$
$k_f m$

ANSWER 14 $\pi V = n_A RT$

ANSWER 18 R

ANSWER 22

$\pi V = n_A RT$

$\pi = 94.3$ torr/760 torr atm^{-1} = 0.124 atm

$$n_A = \frac{0.124\ (0.010)}{0.0820\ (300)} = 5.04 \times 10^{-5} \text{ mole in 10 ml}$$

wt in 10.0 ml of solution = 6.47×10^{-3} g

$$MW = \frac{6.47 \times 10^{-3}\text{ g}}{5.04 \times 10^{-5}\text{ mole}} = 128$$

ANSWER 26 vapor pressure lowering
boiling point elevation
freezing point depression
osmotic pressure

(*Turn to page 720.*)

ITEM 4

As with the three colligative properties already discussed, the __________, the __________, and the __________, the osmotic pressure, π, is, for dilute solutions, proportional to the __________ of the solution.

ITEM 8

When this constant is known for the solvent, it is possible to measure the __________ of a solution and then calculate its __________ to determine the molecular weight of the solute.

ITEM 12

In very dilute solutions, n_A/V is proportional to m. The relationship $\pi \propto m$ can be converted to __________.

ITEM 16

Since R has the value 0.0820 liter atm deg^{-1} $mole^{-1}$, the osmotic pressure of a solution containing 1.00 mole in 1000 ml of solution at 25°C is __________.

ITEM 20

The result of the experiment of Item 19 would permit the calculation of the molecular weight if, in addition to knowing that there is 0.00152 mole $liter^{-1}$, you knew the __________ of solute per liter.

ITEM 24

The osmotic pressure of a 1.00×10^{-3}-molal solution is approximately equal to the osmotic pressure of a 1.00×10^{-3}-molar solution and is therefore easily measureable. However, the freezing point depression would be __________, and the boiling point elevation would be __________. Since they are so small, these quantities could be measured only with considerable difficulty.

For water:
$k_f = 1.86°C$ per 1-molal solution
$k_b = 0.51°C$ per 1-molal solution

ANSWER 3 solution
a semipermeable membrane

ANSWER 7 osmotic pressure
molality

ANSWER 11 molality (m)

ANSWER 15 molarity
$\pi = MRT$ (or $\pi = M_A RT$)

ANSWER 19

$$\frac{28.3 \text{ torr}}{760 \text{ torr atm}^{-1}} = 0.0372 \text{ atm}$$

$$n_A = \pi V/RT$$

$$= \frac{0.0372 \text{ atm } (1.000 \text{ liter})}{(0.0820 \text{ liter atm deg}^{-1} \text{ mole}^{-1})(298 \text{ deg})}$$

$$= 0.00152 \text{ mole}$$

ANSWER 23

$$\pi = n_A RT/V \text{ or } \pi = MRT$$
$$= 0.0010 \text{ mole liter}^{-1}$$
$$(0.082 \text{ liter atm deg}^{-1} \text{ mole}^{-1})(298 \text{ deg})$$
$$= 0.0244 \text{ atm or } 18.5 \text{ torr}$$

ANSWER 4 vapor pressure lowering
boiling point elevation
freezing point depression
molality

(*Item 5 is on page 710.*)

ANSWER 8 osmotic pressure (π)
molality (m)

(*Item 9 is on page 710.*)

ANSWER 12 $\pi \propto n_A/V$

(*Item 13 is on page 710.*)

ANSWER 16

$\pi = n_A RT/V$ or $\pi = MRT$

$$= \frac{1.00 \text{ mole}}{1.00 \text{ liter}} (0.0820 \text{ liter atm deg}^{-1} \text{ mole}^{-1})(298 \text{ deg})$$

$$= 24.4 \text{ atm}$$

(*Item 17 is on page 710.*)

ANSWER 20 weight (number of grams)

(*Item 21 is on page 710.*)

ANSWER 24 $1.86 \times 0.00100 = 0.00186°C$
$0.51 \times 0.00100 = 0.00051°C$

(*Item 25 is on page 710.*)

Now that you have finished this chapter about solutions, you know that there are four properties of solutions that depend on the number of solute molecules present. The magnitude of each of these four colligative properties is proportional, for dilute solutions, to the molality of the solution and is independent of any other feature of the solute molecules. For this reason, the molecular weight of the solute can be determined from a knowledge of the ratio of the weight of solute to that of the solvent and of the magnitude of the vapor pressure depression, the boiling point elevation, the freezing point depression, or the osmotic pressure of the solution. It is important to remember, however, that the equations in this chapter have been developed for the ideal solution, that is, a dilute solution of a nonvolatile solute that neither associates nor dissociates in the solvent. You will find, even with these limitations, that these relationships are used widely.

A particularly interesting application is to solutions in which the solute is an acid, base, or salt and the solvent is water. Then the solute molecules dissociate, and more particles are in solution than there were molecules of added solute. Colligative properties are critical to the study of such ionizations.

REVIEW 14 RATES OF CHEMICAL REACTIONS

Many of the chemical reactions that you first do in the laboratory go from reactants to products so quickly that they appear to be instantaneous. Further study will show, however, that many reactions require an appreciable time to reach their final state. Study of the rates of such slow reactions is of obvious importance if you are depending on the reactions to yield the desired products. Furthermore, study of the rates of reactions provides remarkable insight into the way the reacting molecules, or ions, come together and rearrange to form the product molecules or ions.

The speed with which a reaction occurs depends on the concentration of the reactants and on the temperature of the system. In this review of Section 18-2 of *Chemical Principles*, only the effect of concentration will be considered.

The quantitative expression of the way in which the rate of a reaction depends on the concentrations of the reactants is the *rate equation* for the reaction. In this review you will learn how the rate equation is deduced from the observed effect of concentration on rate.

ITEM 1

When we say that a chemical reaction occurs, we mean that reactants are, to some extent, transformed into ________ in some reasonable length of time.

ITEM 11

If the rate of the reaction at the right is directly proportional to both [A] and [B], the equation defining the rate would take the form: rate $= k[A][B]$. If the rate is proportional to [A] and to $[B]^2$, the rate equation would take the form: ________.

ITEM 21

The rate equation for the reaction of Items 18 through 20 is, therefore, ________. The order of the reaction is ________.

ITEM 31

So far you have deduced the order of a reaction by finding the relation between the rate of reaction and the ________ of reactants. Rates sometimes can be determined experimentally by measuring the change in concentrations of reactants with time.

ITEM 41

The slope of the curve at three distinct times is given at the right. The rate of reaction ____creases as the reaction proceeds.

A + B → Products

$A + B \rightarrow C$

	[A]	[B]	Rate[a]
1	1.0	1.0	0.15
2	2.0	1.0	0.30
3	3.0	1.0	0.45
4	1.0	2.0	0.15
5	1.0	3.0	0.15

[a]Moles of product per liter per minute.

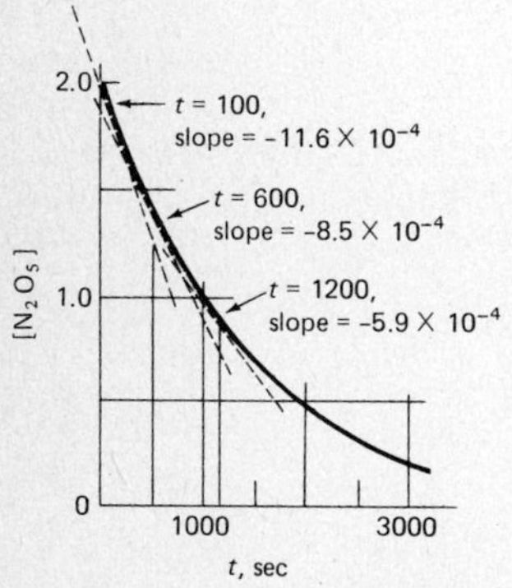

ITEM 2

When aqueous NaCl and $AgNO_3$ solutions are mixed, the AgCl is produced so rapidly that the reaction cannot be followed easily as a function of __________.

ITEM 12

The term "order of reaction" expresses the sum of the exponents to which the concentrations of the reactants are raised in the rate equation. Thus, rate = $k[A]$ represents a first-order reaction, whereas rate = $k[A]^2$ and rate = $k[A][B]$ both indicate __________ -order reactions.

ITEM 22

For this reaction, rate = $k[A]$ or k = __________. Use the data in entry 3 to determine the value and units of k.

ITEM 32

Or, consider the decomposition of N_2O_5 in carbon tetrachloride. The reaction is $N_2O_5 \rightarrow N_2O_4 + \frac{1}{2}O_2$. The N_2O_4 is soluble in CCl_4 but the O_2 is not, so it escapes from the solution as the reaction proceeds. By measuring the amount of product (O_2) formed, you can calculate how much of the __________ decomposed.

ITEM 42

The rate of the reaction at three times is listed in the table at the right. Complete the last column.

$Ag^+ + Cl^- \rightarrow AgCl$

ANSWER 1 products

ANSWER 11 rate $= k[A][B]^2$

$A + B \rightarrow C$

	[A]	[B]	Rate[a]
1	1.0	1.0	0.15
2	2.0	1.0	0.30
3	3.0	1.0	0.45
4	1.0	2.0	0.15
5	1.0	3.0	0.15

[a]Moles of product per liter per minute.

ANSWER 21 rate $= k[A]$
first

ANSWER 31 concentrations

t, sec	$[N_2O_5]$	Rate, moles liter^{-1} sec^{-1}	Rate/$[N_2O_5]$ sec
100	1.88	11.6×10^{-4}	
600	1.37	8.5×10^{-4}	6.2×10^{-4}
1200	0.955	5.9×10^{-4}	

ANSWER 41 de

ITEM 3

Other reactions, such as the one at the right, take hours to reach their final state at room temperature. Thus, the extent of reaction that has occurred during successive intervals of ________ can be measured.

ITEM 13

A reaction that proceeds according to the rate equation rate $= k[A][B]^2$ is called first order in A and second order in B. But the order of the overall reaction corresponds to the sum of the exponents of [A] and [B] and is ________.

ITEM 23

Once the value of k is known, it is a simple matter to calculate the rate at any concentration of A. What is the rate of the reaction when $[A] = 0.020$?

ITEM 33

From a knowledge of how much of the N_2O_5 has decomposed at any time, the concentration of N_2O_5 can be calculated. The table on the right shows data obtained for the reaction $N_2O_5 \rightarrow N_2O_4 + \frac{1}{2}O_2$. From such data the equation for the ________ of a reaction can be deduced.

ITEM 43

Although the rate is not constant, the quantity ________ does remain constant.

$$CH_3\overset{\overset{O}{\|}}{C}OC_2H_5 + H_2O \rightarrow CH_3\underset{\underset{O}{\|}}{C}OH + C_2H_5OH$$

$A + B \rightarrow C$

rate = 0.15[A]

t, sec	$[N_2O_5]$
0	2.00
100	1.88
200	1.77
400	1.56
800	1.21
1200	0.955
1800	0.654
2400	0.450
3000	0.310

ANSWER 2 time

ANSWER 12 second

ANSWER 22 $\frac{\text{rate}}{[A]}$

$$k = \frac{0.45 \text{ mole liter}^{-1} \text{ min}^{-1}}{3.0 \text{ moles liter}^{-1}}$$

$$= 0.15 \text{ min}^{-1}$$

ANSWER 32 reactant (N_2O_5)

ANSWER 42 6.2×10^{-4}
6.2×10^{-4}

ITEM 4

Reactions generally become more rapid as the concentrations of some of the reactants are increased. The rate of reaction is then a function of the __________ of the reactants.

ITEM 14

Write the rate equation for the reaction $A + B \rightarrow C$, which is first order in A and zeroth order in B.

ITEM 24

The data at the right refer to the reaction symbolized by $D + E \rightarrow F$. Compare entries 2 and 3. Doubling [D] (with [E] constant) ______creases the rate by a factor of __________. Thus, the rate is __________ to [D].

ITEM 34

Information about a reaction rate can be obtained by showing graphically the relationship between concentration and time. The $[N_2O_5]$ versus time curve, plotted from the data in the preceding item, is shown at the right. The tangent to this curve at $t = 600$ sec has the intercepts shown. Calculate the slope of the curve at $t = 600$ sec.

ITEM 44

Since the quantity (rate/$[N_2O_5]$) is constant, we can write rate $= k$ __________; that is, the rate is __________ to $[N_2O_5]$.

	D	+	E	→	F
	[D]		[E]		Rate[a]
1	0.37		1.26		0.036
2	0.37		0.63		0.009
3	0.74		0.63		0.018

[a]Moles of F per liter per second.

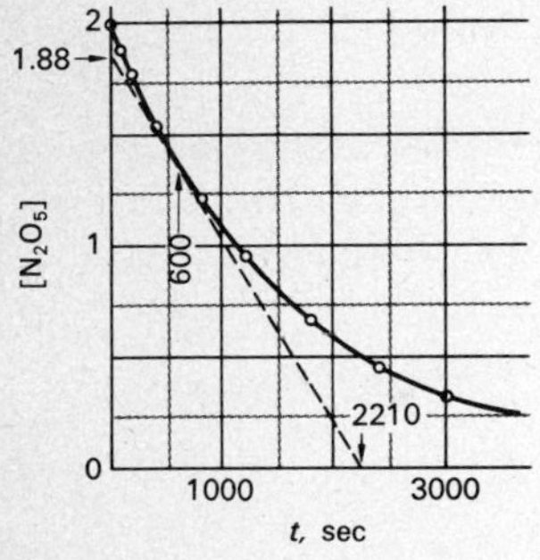

ANSWER 3 time

ANSWER 13 third

ANSWER 23

$$\text{rate} = (0.15\ \text{min}^{-1})(0.020\ \text{mole liter}^{-1})$$
$$= 0.0030\ \text{mole liter}^{-1}\ \text{min}^{-1}$$

ANSWER 33 rate

ANSWER 43 $\dfrac{\text{rate}}{[N_2O_5]}$

ITEM 5

By the *rate* of a reaction is meant the decrease in the amount of reactants, or the increase in the amount of ________, per unit interval of ________.

ITEM 15

The rate equation for the reaction $H_2 + I_2 \rightarrow 2HI$, which is first order in H_2 and first order in I_2, is ________.

ITEM 25

Compare entries 1 and 2. Halving [E] (with [D] constant) ______creases the rate by a factor of ________.

ITEM 35

As the diagram at the right suggests, the slope of the curve at any point is related to the change in the concentration of N_2O_5 with time as slope = ________.

ITEM 45

The rate of this reaction at any time is given by rate $= k[N_2O_5]$. This is a ________ -order reaction.

	D +	E →	F
	[D]	[E]	Rate[a]
1	0.37	1.26	0.036
2	0.37	0.63	0.009
3	0.74	0.63	0.018

[a]Moles of F per liter per second.

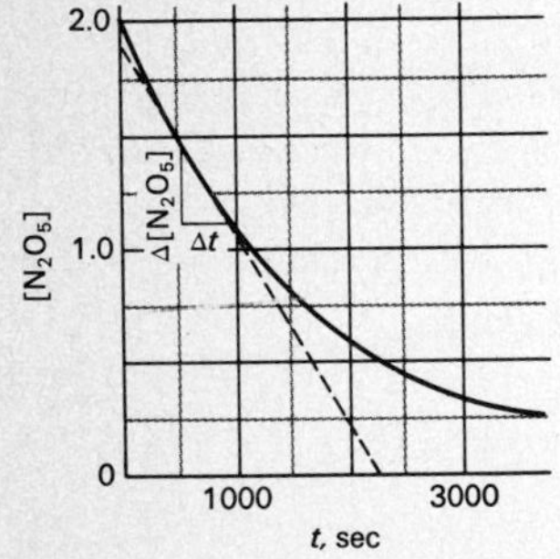

$N_2O_5 \rightarrow N_2O_4 + \frac{1}{2}O_2$

ANSWER 4 concentrations

ANSWER 14 rate $= k[A]$

ANSWER 24 in
2
proportional
(or first order with respect)

ANSWER 34

$$\text{slope} = \frac{-1.88 \text{ mole liter}^{-1}}{2210 \text{ sec}}$$

$$= -8.51 \times 10^{-4} \text{ mole liter}^{-1} \text{ sec}^{-1}$$

ANSWER 44 $[N_2O_5]$
proportional

ITEM 6

For reactions occurring in solution, the rate can be measured in terms of the change per unit time of the amount of reagent in a given volume of solution, that is, the change in the __________ of the reactants or products. Similarly, the rates of reactions of gases can be expressed as the changes per unit time in the amounts of reagents per unit volume of the system.

ITEM 16

The rate equation for the reaction $2NO + 2H_2 \rightarrow N_2 + 2H_2O$, which is second order in NO and first order in H_2, is __________.

ITEM 26

Since halving [E] changes the rate by a factor of 4, the rate is not proportional to [E] but to __________.

ITEM 36

A small change in concentration during a small change in time is expressed as $d[N_2O_5]/dt$. Thus, the slope of the curve at any point is related to small changes in the concentration of N_2O_5 during short time intervals as slope = __________.

ITEM 46

The constant, k, is known as the __________. Since the value of rate/$[N_2O_5]$ is 6.2×10^{-4}, k in the equation, rate = $k[N_2O_5]$, has the value __________.

$$D + E \rightarrow F$$

	[D]	[E]	Rate[a]
1	0.37	1.26	0.036
2	0.37	0.63	0.009
3	0.74	0.63	0.018

[a]Moles of F per liter per second.

ANSWER 5 products
time

ANSWER 15 rate $= k[H_2][I_2]$

ANSWER 25 de
4

ANSWER 35 $\dfrac{\Delta[N_2O_5]}{\Delta t}$

ANSWER 45 first

ITEM 7

For any reaction A + B → Products, the concentrations of A and B in moles per liter are represented, as usual, by [A] and [B]. If the rate of the reaction A + B → Products is doubled when the concentration of A is doubled, the rate must be directly proportional to ________.

ITEM 17

The order of a reaction usually is determined from measurements of the dependence of the ________ of the reaction on the ________ of the reactants.

ITEM 27

The rate is proportional to [D] and to $[E]^2$. The rate equation for the reaction is rate = ________. The overall order of the reaction is ________; it is ________ order in D and ________ order in E.

ITEM 37

The rate of a reaction is always expressed as a positive quantity, but it may be defined in terms of either the disappearance of a reactant or in terms of the increase in concentration of a product. Since the change in concentration with time of a reactant is a negative quantity, the rate of the N_2O_5 decomposition reaction is written as $-d[N_2O_5]/dt$, in which the minus sign indicates that $[N_2O_5]$ ________ with time.

ITEM 47

The rate of the N_2O_5 decomposition is equal to $-d[N_2O_5]/dt$. Since t was measured in seconds, the rate of the reaction has the units of ________.

	D +	E →	F
	[D]	[E]	Rate[a]
1	0.37	1.26	0.036
2	0.37	0.63	0.009
3	0.74	0.63	0.018

[a]Moles of F per liter per second.

$N_2O_5 \rightarrow N_2O_4 + \frac{1}{2}O_2$

ANSWER 6 concentration

ANSWER 16 rate $= k[NO]^2[H_2]$

ANSWER 26 $[E]^2$

ANSWER 36 $\dfrac{d[N_2O_5]}{dt}$

ANSWER 46 (specific reaction) rate constant
6.2×10^{-4}

ITEM 8

If, in the same reaction, the rate does not depend on the concentration of B, we can write the proportionality as rate $\propto$ [A]. If the proportionality constant, k, is introduced, we can write the simple equation: rate = __________.

ITEM 18

The way in which the order is determined for simple reactions is as follows: The data at the right were obtained for the reaction $A + B \rightarrow C$. Entries 1 and 2 show that doubling [A], while holding [B] constant, __________ the rate.

ITEM 28

The rate of the reaction is equal to $k[D][E]^2$. Using data from entry 1, evaluate k, the rate constant.

ITEM 38

You have seen that the rate $= -d[N_2O_5]/dt$ and the slope $= d[N_2O_5]/dt$. Therefore, the rate is related to the slope of the concentration–time curve for N_2O_5 as rate = __________.

ITEM 48

What are the units of k for the N_2O_5 decomposition reaction?

$A + B \rightarrow$ Products

$A + B \rightarrow C$

	[A]	[B]	Rate[a]
1	1.0	1.0	0.15
2	2.0	1.0	0.30
3	3.0	1.0	0.45
4	1.0	2.0	0.15
5	1.0	3.0	0.15

[a]Moles of product per liter per minute.

$D + E \rightarrow F$

	[D]	[E]	Rate[a]
1	0.37	1.26	0.036
2	0.37	0.63	0.009
3	0.74	0.63	0.018

[a]Moles of F per liter per second.

$N_2O_5 \rightarrow N_2O_4 + \frac{1}{2}O_2$

ANSWER 7 the concentration of A (or [A])

ANSWER 17 rate
concentration

ANSWER 27 $k[D][E]^2$
third
first
second

ANSWER 37 decreases

ANSWER 47 moles liter^{-1} sec^{-1}

ITEM 9

Suppose that doubling the concentration of a reactant increases the rate by a factor of 4, and that tripling the concentration of the reactant increases the rate by a factor of 9. Then the rate is proportional to the concentration of the reactant raised to the __________ power.

ITEM 19

Since doubling [A] (holding [B] constant) doubles the rate, the rate is proportional to __________.

ITEM 29

For the reaction of the preceding items, you deduced that the rate was equal to $0.061[D][E]^2$. Calculate the rate at which D would be consumed in a solution that is 1.0 molar in D and 0.020 molar in E.

ITEM 39

If, then, we plot the concentration of a reactant versus time, the rate at any time t can be determined as the negative of the __________ of the curve.

ITEM 49

At what rate does N_2O_5 decompose in a solution when the concentration of N_2O_5 is 0.600 molar?

	A + B → C		
	[A]	[B]	Rate[a]
1	1.0	1.0	0.15
2	2.0	1.0	0.30
3	3.0	1.0	0.45
4	1.0	2.0	0.15
5	1.0	3.0	0.15

[a]Moles of product per liter per minute.

D + E → F

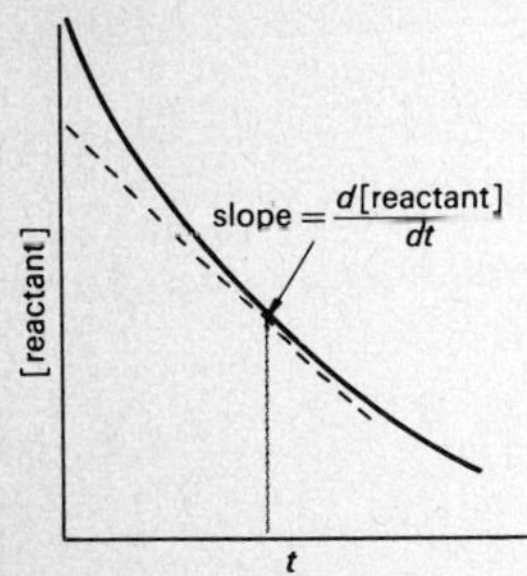

$k = 6.2 \times 10^{-4}$ sec^{-1}

ANSWER 8 k[A]

ANSWER 18 doubles

ANSWER 28

$$k = \frac{\text{rate}}{[\text{D}][\text{E}]^2}$$

$$= \frac{0.036 \text{ mole liter}^{-1} \text{ sec}^{-1}}{(0.37 \text{ mole liter}^{-1})(1.26 \text{ moles liter}^{-1})^2}$$

$$= 0.061 \text{ liter}^2 \text{ mole}^{-2} \text{ sec}^{-1}$$

ANSWER 38 −slope

ANSWER 48 $k = \frac{\text{rate}}{[N_2O_5]}$

k has units of

$$\frac{\text{moles liter}^{-1} \text{ sec}^{-1}}{\text{moles liter}^{-1}} = \text{sec}^{-1}$$

ITEM 10

If the rate of a reaction is proportional to the square of the concentration of A and independent of the concentrations of other reagents, the rate equation would have the form: rate $= k$ __________.

ITEM 20

Entries 4 and 5 in the table at the right show the dependence of the rate of the reaction on the concentration of B. Changing [B] while holding [A] constant (does, does not) affect the rate.

ITEM 30

The rate of a reaction is defined as __________.

ITEM 40

In Item 34 the slope of the curve at $t = 600$ sec was -8.51×10^{-4} mole liter^{-1} sec^{-1}, so the rate of reaction at the instant $t = 600$ sec is $-d[N_2O_5]/dt =$ __________.

ITEM 50

For any first-order reaction, where A $\rightarrow$ Products, the rate at any time is __________ to [A]. The proportionality constant is known as the __________.

Now you know that the rate equation for a reaction and the rate constant that appears in this equation can be deduced from measurements of the amount of reactant or product present at various reaction times. This will serve as an introduction to your further study of chemical reaction rates.

$$A + B \rightarrow C$$

	[A]	[B]	Rate[a]
1	1.0	1.0	0.15
2	2.0	1.0	0.30
3	3.0	1.0	0.45
4	1.0	2.0	0.15
5	1.0	3.0	0.15

[a]Moles of product per liter per minute.

ANSWER 9 second

ANSWER 19 [A]

ANSWER 29

$$\text{rate} = 0.061(1.0)(0.020)^2$$
$$= 2.4 \times 10^{-5} \text{ mole liter}^{-1} \text{ sec}^{-1}$$

(The rate of consuming D equals the rate of formation of F.)

ANSWER 39 slope

ANSWER 49

$$\text{rate} = k[N_2O_5]$$
$$= (6.2 \times 10^{-4} \text{ sec}^{-1})(0.600 \text{ mole liter}^{-1})$$
$$= 3.7 \times 10^{-4} \text{ mole liter}^{-1} \text{ sec}^{-1}$$

ANSWER 10 $[A]^2$

(*Item 11 is on page 722.*)

ANSWER 20 does not

(*Item 21 is on page 722.*)

ANSWER 30 the decrease in the amount of reactants, or the increase in the amount of products, per unit interval of time

(*Item 31 is on page 722.*)

ANSWER 40 8.51×10^{-4} mole liter^{-1} sec^{-1}

(*Item 41 is on page 722.*)

ANSWER 50 proportional
(specific reaction) rate
constant